# Bioenergy Feedstocks

# Bioenergy Feedstocks
## Breeding and Genetics

Edited by

**Malay C. Saha**
**Hem S. Bhandari**
**Joseph H. Bouton**

**WILEY-BLACKWELL**

A John Wiley & Sons, Inc., Publication

This edition first published 2013 © 2013 by John Wiley & Sons, Inc.

Wiley-Blackwell is an imprint of John Wiley & Sons, formed by the merger of Wiley's global Scientific, Technical and Medical business with Blackwell Publishing.

*Editorial offices:*   2121 State Avenue, Ames, Iowa 50014-8300, USA
 The Atrium, Southern Gate, Chichester, West Sussex, PO19 8SQ, UK
 9600 Garsington Road, Oxford, OX4 2DQ, UK

For details of our global editorial offices, for customer services and for information about how to apply for permission to reuse the copyright material in this book please see our website at www.wiley.com/wiley-blackwell.

Authorization to photocopy items for internal or personal use, or the internal or personal use of specific clients, is granted by Blackwell Publishing, provided that the base fee is paid directly to the Copyright Clearance Center, 222 Rosewood Drive, Danvers, MA 01923. For those organizations that have been granted a photocopy license by CCC, a separate system of payments has been arranged. The fee codes for users of the Transactional Reporting Service are ISBN-13: 978-0-4709-6033-2/2013.

Designations used by companies to distinguish their products are often claimed as trademarks. All brand names and product names used in this book are trade names, service marks, trademarks or registered trademarks of their respective owners. The publisher is not associated with any product or vendor mentioned in this book. This publication is designed to provide accurate and authoritative information in regard to the subject matter covered. It is sold on the understanding that the publisher is not engaged in rendering professional services. If professional advice or other expert assistance is required, the services of a competent professional should be sought.

Limit of Liability/Disclaimer of Warranty: While the publisher and author have used their best efforts in preparing this book, they make no representations or warranties with the respect to the accuracy or completeness of the contents of this book and specifically disclaim any implied warranties of merchantability or fitness for a particular purpose. It is sold on the understanding that the publisher is not engaged in rendering professional services and neither the publisher nor the author shall be liable for damages arising herefrom. If professional advice or other expert assistance is required, the services of a competent professional should be sought.

*Library of Congress Cataloging-in-Publication Data is available upon request.*

A catalogue record for this book is available from the British Library.

Wiley also publishes its books in a variety of electronic formats. Some content that appears in print may not be available in electronic books.

Cover design by Matt Kuhns

Set in 10/11.5pt Times New Roman by Aptara® Inc., New Delhi, India
Printed and bound in Malaysia by Vivar Printing Sdn Bhd

1   2013

# Contents

*The Editors*     xi
*List of Contributors*     xiii
*Preface*     xix

**1 Introduction**     1
    1.1 Historical Development     2
    1.2 Cultivar Development     2
    1.3 Breeding Approach     3
    1.4 Molecular Tools     3
    1.5 Future Outlook     4
    References     4

**2 Switchgrass Genetics and Breeding Challenges**     7
    2.1 Introduction     7
    2.2 Origin and Distribution     9
    2.3 Growth and Development, Genome Structure and Cytogenetics     9
        2.3.1 Growth and Development     10
        2.3.2 Genome Structure and Cytogenetics     12
    2.4 Genetic Diversity     12
    2.5 Phenotypic Variability and Inheritance     13
    2.6 Conventional Breeding Approaches     14
        2.6.1 Early Work     15
        2.6.2 Systematic Recurrent Selection     15
        2.6.3 Heterosis     17
    2.7 Molecular Breeding     18
        2.7.1 Molecular Markers Used for Switchgrass and Other Polyploids     18
        2.7.2 Molecular Mapping     20
        2.7.3 Association Mapping     22
        2.7.4 Transgenic Approaches     23
    2.8 Conclusions and Future Directions     23
    References     24

**3 Switchgrass Genomics**     33
    3.1 Introduction     33
    3.2 Genome Sequencing     34
        3.2.1 Other Available Sequence Resources     35
    3.3 Analysis of Expressed Sequences in Switchgrass     36

|  |  |  |  |
|---|---|---|---|
| | 3.4 | Linkage Mapping | 40 |
| | 3.5 | Cytoplasmic Genome | 42 |
| | 3.6 | Genome-enabled Improvement of Switchgrass | 42 |
| | 3.7 | Conclusions | 45 |
| | References | | 45 |
| **4** | **Germplasm Resources of *Miscanthus* and Their Application in Breeding** | | **49** |
| | 4.1 | Introduction | 49 |
| | 4.2 | Species Belonging to *Miscanthus* Genus, Their Characteristics, and Phylogenetic Relationships | 50 |
| | | 4.2.1 Section: Eumiscanthus | 50 |
| | | 4.2.2 Section: Triarrhena | 53 |
| | | 4.2.3 Section: Kariyasu | 54 |
| | 4.3 | Natural Hybrids between *Miscanthus* Species | 55 |
| | 4.4 | Karyotype Analysis | 55 |
| | 4.5 | Phylogenetic Relationships between *Miscanthus* Species | 56 |
| | 4.6 | Genetic Improvement of *Miscanthus* | 57 |
| | | 4.6.1 Germplasm Collection and Management | 57 |
| | | 4.6.2 Artificial Hybridization | 57 |
| | | 4.6.3 Polyploidization | 58 |
| | 4.7 | Variations in Several Agronomical Traits Related to Yield and Plant Performance | 58 |
| | | 4.7.1 Variation in Flowering Time | 58 |
| | | 4.7.2 Variation in Cold Tolerance | 58 |
| | | 4.7.3 Variation in Lignin, Cellulose, and Mineral Content | 59 |
| | 4.8 | Molecular Resources | 60 |
| | | 4.8.1 Development of Linkage Map for Miscanthus | 60 |
| | | 4.8.2 QTL Analysis of Traits Related to Yield and Mineral Content | 60 |
| | | 4.8.3 Molecular Markers for Hybrids Identification | 61 |
| | 4.9 | Transgenic *Miscanthus* | 61 |
| | 4.10 | Future Studies | 62 |
| | References | | 62 |
| **5** | **Breeding *Miscanthus* for Bioenergy** | | **67** |
| | 5.1 | Introduction | 67 |
| | 5.2 | *Miscanthus* as a Biomass Crop | 67 |
| | 5.3 | Breeding Strategy | 68 |
| | | 5.3.1 Collection and Characterization | 68 |
| | | 5.3.2 Hybridization | 68 |
| | | 5.3.3 Ex Situ Phenotypic Characterization | 69 |
| | | 5.3.4 Large-scale Demonstration Trials | 69 |
| | 5.4 | Genetic Diversity | 69 |
| | 5.5 | Breeding Targets | 70 |
| | | 5.5.1 Biomass Yield | 70 |
| | | 5.5.2 Morphological Traits Contributing to High Yield Potential | 75 |
| | | 5.5.3 Seed Propagation: Crop Diversification and Reducing the Cost of Establishment | 77 |

|      |       |                                                                                          |     |
|------|-------|------------------------------------------------------------------------------------------|-----|
|      | 5.6   | Incorporating Bioinformatics, Molecular Marker-Assisted Selection (MAS), and Genome-Wide Association Selection (GWAS) | 77  |
|      | 5.7   | Summary                                                                                  | 78  |
|      | Acknowledgments                                                                                  | 79  |
|      | References                                                                                       | 79  |
| **6** | **Breeding Sorghum as a Bioenergy Crop**                                                        | **83** |
|      | 6.1   | Introduction                                                                             | 83  |
|      | 6.2   | Botanical Description and Evolution                                                      | 84  |
|      |       | 6.2.1 Basic Characteristics                                                              | 84  |
|      |       | 6.2.2 Evolution and Distribution                                                         | 85  |
|      | 6.3   | Traditional Breeding and Development                                                     | 86  |
|      |       | 6.3.1 Initial Sorghum Improvement                                                        | 86  |
|      |       | 6.3.2 Development of Hybrid Sorghum and Heterosis                                        | 86  |
|      |       | 6.3.3 Current Sorghum Breeding Approaches                                                | 88  |
|      |       | 6.3.4 Germplasm Resources                                                                | 88  |
|      | 6.4   | Approaches to Breeding Sorghum as a Bioenergy Crop                                       | 90  |
|      |       | 6.4.1 Grain Sorghum                                                                      | 90  |
|      |       | 6.4.2 Sweet Sorghum                                                                      | 90  |
|      |       | 6.4.3 Biomass Sorghum                                                                    | 93  |
|      | 6.5   | Composition in Energy Sorghum Breeding                                                   | 93  |
|      | 6.6   | Genetic Variation and Inheritance                                                        | 95  |
|      |       | 6.6.1 Grain Sorghum                                                                      | 95  |
|      |       | 6.6.2 Grain Quality/Starch Composition                                                   | 96  |
|      |       | 6.6.3 Dual Purpose—Grain and Stalk                                                       | 97  |
|      |       | 6.6.4 Soluble Carbohydrates                                                              | 97  |
|      |       | 6.6.5 Breeding for Stress Tolerance                                                      | 99  |
|      | 6.7   | Wide Hybridization                                                                       | 106 |
|      |       | 6.7.1 Interspecific Hybridization                                                        | 106 |
|      |       | 6.7.2 Intergeneric Hybridization                                                         | 107 |
|      | 6.8   | Conclusions                                                                              | 107 |
|      | References                                                                                       | 107 |
| **7** | **Energy Cane**                                                                                 | **117** |
|      | 7.1   | Introduction                                                                             | 117 |
|      | 7.2   | Sugar and Energy Production Systems                                                      | 118 |
|      |       | 7.2.1 Current Global Sugarcane Production                                                | 118 |
|      |       | 7.2.2 Bioenergy Production from Sugarcane in Brazil                                      | 120 |
|      |       | 7.2.3 Overview of Main Components in Existing Sugarcane Production Systems               | 120 |
|      |       | 7.2.4 Overview and Potential Trends                                                      | 123 |
|      | 7.3   | Sugarcane Improvement                                                                    | 124 |
|      |       | 7.3.1 Taxonomy and Crop Physiology                                                       | 124 |
|      |       | 7.3.2 History of Sugarcane Breeding                                                      | 127 |
|      |       | 7.3.3 Basic Features of Sugarcane Breeding Programs                                      | 128 |
|      |       | 7.3.4 Composition of Cane for Sugar or Energy Production                                 | 130 |
|      |       | 7.3.5 Application of Molecular Genetics in Developing Energy Cane                        | 131 |

|   |   |   |   | |
|---|---|---|---|---|
| | 7.4 | Selection of Sugarcane Genotypes for Energy Production | | 134 |
| | | 7.4.1 | Overall Directions | 134 |
| | | 7.4.2 | Example of Economic Weightings for Selecting Sugarcane for Energy Products | 136 |
| | | 7.4.3 | Progress in Breeding for Energy Production | 138 |
| | 7.5 | Conclusion | | 141 |
| | Acknowledgments | | | 141 |
| | References | | | 141 |
| **8** | **Breeding Maize for Lignocellulosic Biofuel Production** | | | **151** |
| | 8.1 | Introduction | | 151 |
| | 8.2 | General Attributes of Maize as a Biofuel Crop | | 151 |
| | 8.3 | Potential Uses of Maize Stover for Bioenergy | | 153 |
| | 8.4 | Breeding Maize for Biofuels | | 154 |
| | | 8.4.1 | Selection Criteria | 154 |
| | | 8.4.2 | Stover Yield | 157 |
| | | 8.4.3 | Maximum Biomass Yield and the Effects of Time and Latitude | 159 |
| | | 8.4.4 | Stover Quality | 161 |
| | | 8.4.5 | Sustainability Parameters | 163 |
| | | 8.4.6 | Breeding Methods | 164 |
| | 8.5 | Single Genes and Transgenes | | 165 |
| | 8.6 | Future Outlook | | 167 |
| | References | | | 167 |
| **9** | **Underutilized Grasses** | | | **173** |
| | 9.1 | Introduction | | 173 |
| | 9.2 | Prairie Cordgrass | | 174 |
| | | 9.2.1 | Importance | 174 |
| | | 9.2.2 | Genetic Variation and Breeding Methods | 176 |
| | | 9.2.3 | Future Goals | 180 |
| | 9.3 | Bluestems | | 181 |
| | | 9.3.1 | Importance | 181 |
| | | 9.3.2 | Genetic Variation and Breeding Methods | 184 |
| | | 9.3.3 | Future Goals | 190 |
| | 9.4 | Eastern Gamagrass | | 191 |
| | | 9.4.1 | Importance | 191 |
| | | 9.4.2 | Genetic Variation and Breeding Methods | 192 |
| | | 9.4.3 | Future Goals | 196 |
| | References | | | 197 |
| **10** | **Alfalfa as a Bioenergy Crop** | | | **207** |
| | 10.1 | Introduction | | 207 |
| | 10.2 | Biomass for Biofuels | | 208 |
| | | 10.2.1 | Lignocellulose-based Biofuels | 208 |
| | | 10.2.2 | Plant Cell Wall Components | 209 |
| | 10.3 | Why Alfalfa? | | 211 |
| | | 10.3.1 | Background | 211 |
| | | 10.3.2 | Prospect as a Biofuel Feedstock | 212 |

|        |          |                                                      |     |
|--------|----------|------------------------------------------------------|-----|
| 10.4   | Breeding Strategies                                             | 213 |
|        | 10.4.1   | Germplasm Resources                                  | 213 |
|        | 10.4.2   | Cultivar Development                                 | 214 |
|        | 10.4.3   | Synthetic Cultivars and Heterosis                    | 214 |
|        | 10.4.4   | Molecular Breeding                                   | 215 |
|        | 10.4.5   | Trait Integration Through Biotechnology              | 216 |
| 10.5   | Breeding Targets                                                | 217 |
|        | 10.5.1   | Biomass Yield                                        | 217 |
|        | 10.5.2   | Forage Quality and Composition                       | 218 |
|        | 10.5.3   | Stress Tolerance                                     | 219 |
|        | 10.5.4   | Winter Hardiness                                     | 220 |
| 10.6   | Management and Production Inputs                                | 221 |
| 10.7   | Processing for Biofuels                                         | 222 |
| 10.8   | Additional Value from Alfalfa Production                        | 223 |
|        | 10.8.1   | Environmental Benefits                               | 223 |
|        | 10.8.2   | Alfalfa Co-products                                  | 223 |
| 10.9   | Summary                                                         | 223 |
| Acknowledgments                                                          | 224 |
| References                                                               | 224 |

**11 Transgenics for Biomass** — 233
- 11.1 Introduction — 233
  - 11.1.1 Biomass for Biofuels — 233
  - 11.1.2 Biofuels — 234
  - 11.1.3 Lignocellulosic Biomass — 234
- 11.2 Transgenic Approaches — 235
  - 11.2.1 Biolistics Transformation — 235
  - 11.2.2 Agrobacterium-mediated Transformation — 236
- 11.3 Transgenic Approaches for Biomass Improvement — 237
  - 11.3.1 Improving Biomass Yield — 237
  - 11.3.2 Modifying Biomass Composition — 240
  - 11.3.3 Regulatory Issues of Transgenic Bioenergy Crops — 242
- 11.4 Summary — 242
- Acknowledgments — 242
- References — 243

**12 Endophytes in Low-input Agriculture and Plant Biomass Production** — 249
- 12.1 Introduction — 249
- 12.2 What are Endophytes? — 249
- 12.3 Endophytes of Cool Season Grasses — 251
- 12.4 Endophytes of Warm Season Grasses — 251
- 12.5 Endophytes of Woody Angiosperms — 253
- 12.6 Other Fungal Endophytes — 253
- 12.7 Endophytes in Biomass Crop Production — 254
- 12.8 The Use of Fungal Endophytes in Bioenergy Crop Production Systems — 256
- 12.9 Endophyte Consortia — 256

| 12.10 | Source of Novel Compounds | 257 |
| 12.11 | Endophyte in Genetic Engineering of Host Plants | 258 |
| 12.12 | Conclusions | 258 |
| Acknowledgments | | 259 |
| References | | 259 |

*Index* 267

*Color plate is located between pages 172 and 173.*

# The Editors

**Malay C. Saha**  Forage Improvement Division (FID)
The Samuel Roberts Noble Foundation, Inc.
Ardmore, OK
mcsaha@noble.org

**Hem S. Bhandari**  Department of Plant Sciences
University of Tennessee
Knoxville, TN
hsbhandari@utk.edu

**Joseph H. Bouton**  Forage Improvement Division (FID)
The Samuel Roberts Noble Foundation, Inc.
Ardmore, OK
jhbouton@noble.org

# List of Contributors

**Laura Bartley**	Department of Botany and Microbiolgy
University of Oklahoma
Norman, OK

**Kishor Bhattarai**	Forage Improvement Division
The Samuel Roberts Noble Foundation
Ardmore, OK

**Arvid Boe**	South Dakota State University
Brookings, SD

**John Clifton Brown**	Institute of Biological, Environmental and Rural Sciences
(IBERS)
Aberystwyth University
Gogerddan, Aberystwyth, Ceredigion, Wales, UK

**E. Charles Brummer**	Forage Improvement Division
The Samuel Roberts Noble Foundation
Ardmore, OK

**Kelly D. Craven**	Plant Biology Division
The Samuel Roberts Noble Foundation
Ardmore, OK

**Chris Davey**	Institute of Biological, Environmental and Rural Sciences
(IBERS)
Aberystwyth University
Gogerddan, Aberystwyth, Ceredigion, Wales, UK

**Natalia de Leon**	Department of Agronomy
University of Wisconsin
Madison, WI

**Iain Donnison**	Institute of Biological, Environmental and Rural Sciences
(IBERS)
Aberystwyth University
Gogerddan, Aberystwyth, Ceredigion, Wales, UK

**Maria Stefanie Dwiyanti**  Energy Biosciences Institute
University of Illinois
Urbana, IL

**Kerrie Farrar**  Institute of Biological, Environmental and Rural Sciences (IBERS)
Aberystwyth University
Gogerddan, Aberystwyth, Ceredigion, Wales, UK

**Sita R. Ghimire**  Center for Agricultural and Environmental Biotechnology
RTI International
Research Triangle Park, NC

**J. Gonzalez-Hernandez**  South Dakota State University
Brookings, SD

**C. Frank Hardin**  Forage Improvement Division
The Samuel Roberts Noble Foundation
Ardmore, OK

**Charlotte Hayes**  Institute of Biological, Environmental and Rural Sciences (IBERS)
Aberystwyth University
Gogerddan, Aberystwyth, Ceredigion, Wales, UK

**Maurice Hinton-Jones**  Institute of Biological, Environmental and Rural Sciences (IBERS)
Aberystwyth University
Gogerddan, Aberystwyth, Ceredigion, Wales, UK

**Lin Huang**  Institute of Biological, Environmental and Rural Sciences (IBERS)
Aberystwyth University
Gogerddan, Aberystwyth, Ceredigion, Wales, UK

**Elaine Jensen**  Institute of Biological, Environmental and Rural Sciences (IBERS)
Aberystwyth University
Gogerddan, Aberystwyth, Ceredigion, Wales, UK

**Phillips Jackson**  CSIRO Plant Industry
Australian Tropical Science Innovation Precinct
Townsville, Australia

**Laurence Jones**  Institute of Biological, Environmental and Rural Sciences (IBERS)
Aberystwyth University
Gogerddan, Aberystwyth, Ceredigion, Wales, UK

| | |
|---|---|
| **Shawn M. Kaeppler** | Department of Agronomy<br>University of Wisconsin<br>Madison, WI |
| **Joe G. Lauer** | Department of Agronomy<br>University of Wisconsin<br>Madison, WI |
| **D. K. Lee** | University of Illinois<br>Urbana, IL |
| **Anne Maddison** | Institute of Biological, Environmental and Rural Sciences (IBERS)<br>Aberystwyth University<br>Gogerddan, Aberystwyth, Ceredigion, Wales, UK |
| **Heike Meyer** | Julius Kühn-Institute (JKI)<br>Federal Research Centre for Cultivated Plants<br>Bundesallee, Germany |
| **Maria J. Monteros** | Forage Improvement Division<br>The Samuel Roberts Noble Foundation<br>Ardmore, OK |
| **John Norris** | Institute of Biological, Environmental and Rural Sciences (IBERS)<br>Aberystwyth University<br>Gogerddan, Aberystwyth, Ceredigion, Wales, UK |
| **Sarah Purdy** | Institute of Biological, Environmental and Rural Sciences (IBERS)<br>Aberystwyth University<br>Gogerddan, Aberystwyth, Ceredigion, Wales, UK |
| **A. Lane Rayburn** | University of Illinois<br>Urbana, IL |
| **Paul Robson** | Institute of Biological, Environmental and Rural Sciences (IBERS)<br>Aberystwyth University<br>Gogerddan, Aberystwyth, Ceredigion, Wales, UK |
| **W. L. Rooney** | Department of Soil and Crop Sciences<br>Texas A&M University<br>College Station, Texas |
| **Aaron Saathoff** | Grain, Forage, and Bioenergy Research Unit<br>USDA-ARS and Department of Agronomy and Horticulture<br>University of Nebraska<br>Lincoln, NE |

**Cosentino Salvatore**  Department of Agriculture and Food Science
University of Catania
Via Valdisavoia, Catania, Italy

**Gautam Sarath**  Grain, Forage, and Bioenergy Research Unit
USDA-ARS and Department of Agronomy and Horticulture
University of Nebraska
Lincoln, NE

**Kai-Uwe Schwarz**  Julius Kühn-Institute (JKI)
Federal Research Centre for Cultivated Plants
Bundesallee, Braunschweig, Germany

**Gancho Slavov**  Institute of Biological, Environmental and Rural Sciences (IBERS)
Aberystwyth University
Gogerddan, Aberystwyth, Ceredigion, Wales, UK

**Tim Springer**  Southern Plains Range Research Station
Woodward, OK

**T. R. Stefaniak**  Department of Soil and Crop Sciences
Texas A&M University
College Station, TX

**J. Ryan Stewart**  Brigham Young University
Provo, UT

**Charlie Rodgers**  Ceres, Inc.
Somerville, TX

**Christian Tobias**  USDA-ARS
Western Regional Research Center
Albany, CA

**John Valentine**  Institute of Biological, Environmental and Rural Sciences (IBERS)
Aberystwyth University
Gogerddan, Aberystwyth, Ceredigion, Wales, UK

**Zeng-Yu Wang**  Forage Improvement Division
The Samuel Roberts Noble Foundation
Ardmore, OK

**Richard Webster**  Institute of Biological, Environmental and Rural Sciences (IBERS)
Aberystwyth University
Gogerddan, Aberystwyth, Ceredigion, Wales, UK

| | |
|---|---|
| **Yanqi Wu** | Department of Plant and Soil Science<br>Oklahoma State University<br>Stillwater, OK |
| **Toshihiko Yamada** | Field Science Center for Northern Biosphere<br>Hokkaido University<br>Kita-ku, Sapporo, Hokkaido, Japan |
| **Sue Youell** | Institute of Biological, Environmental and Rural Sciences (IBERS)<br>Aberystwyth University<br>Gogerddan, Aberystwyth, Ceredigion, Wales, UK |

# Preface

The world energy use grew by 39% from 1990 to 2008. It is estimated that the global demand for energy will increase by at least 50% over the next 20 years. Energy consumption growth of several developing nations remains vigorous. Hydrocarbons, petroleum, coal, and natural gas are now the chief sources of energy. All are finite resources and their natural reserves are depleting every day. In addition, during their conversion and use several greenhouse gases are emitted with a potential for climatic warming.

Bioenergy, both biofuels and biopower, produced from renewable sources are sustainable alternatives to hydrocarbons. Bioenergy use has the potential to lower greenhouse gas emissions, boost rural economy, and ensure energy security. Interest in bioenergy began in early 20th century but it was reinforced in the recent decades. Biopower includes co-firing bioenergy feedstocks with coal to reduce problem emissions. Due to government incentives during 1995–2005, commercial scale biofuels, mainly ethanol, became available in the European Union, UK, USA, Brazil, and many other countries around the world. Most of the biofuels are derived from corn grain, sugarcane, and vegetable oil feedstocks thus creating a food versus fuel controversy. Second-generation biofuels are now being made from nonfood, lignocellulosic materials such as municipal waste and wood chips, along with dedicated crops such as switchgrass and Miscanthus.

Plant breeding is critical for crop improvement. Due to intensive breeding efforts in both public and private sectors average maize grain yield has increased by 745% since 1930. Several of the dedicated feedstock crops, for example, switchgrass and Miscanthus, are only recently removed from the wild and need serious breeding efforts for improvement. Increased biomass yield and improved quality through breeding efforts can make feedstock more economical and attractive. This book on *Bioenergy Feedstocks: Breeding and Genetics* should greatly contribute to these breeding efforts.

We are grateful to John Wiley & Sons, Inc. for their prudence and supporting us in publishing this book. Contribution from many prominent scientists on bioenergy research has greatly enhanced this publication. We extend our sincere appreciation to all the chapter contributors for their invaluable contribution. We also appreciate the efforts of all who directly or indirectly supported our endeavor. We sincerely believe that this book will be a useful reference for cultivar improvement of lignocellulosic biomass feedstock crops.

*Malay C. Saha*
*Hem S. Bhandari*
*Joseph H. Bouton*

# Chapter 1
# Introduction

Joseph H. Bouton[1], Hem S. Bhandari[2], and Malay C. Saha[3]

[1] Former Director and Senior Vice President, Forage Improvement Division, The Samuel Roberts Noble Foundation, Ardmore, OK 73401 and Emeritus Professor, Crop and Soil Sciences, University of Georgia, Athens, GA 30602, USA
[2] Department of Plant Sciences, University of Tennessee, Knoxville, TN37996, USA
[3] Forage Improvement Division, The Samuel Roberts Noble Foundation, Ardmore, OK 73401, USA

By most estimates, world population growth has more than tripled during the past 100 years, going from approximately 2–7 billion persons (Anonymous, 2012). To sustain the economies needed to support this type of unprecedented population growth, readily available, cheap, scalable, and efficient energy sources were required. These sources turned out to be hydrocarbons, both oil and coal, and after the Second World War, nuclear power. Heavy hydrocarbon use resulted in their depletion and increased cost and a concurrent increase in environmental problems due to gas emissions. Although "clean" as far as gas emissions, nuclear power has its own problems associated with safety and disposal of its highly toxic waste products. Therefore, alternative energy sources such as wind, solar, and bioenergy that are capable of offsetting some of the hydrocarbons and nuclear use and mitigating their environmental problems are now being investigated and, in some cases, implemented on a commercial scale.

Lignocellulosic feedstocks derived from plant biomass emerged as a sustainable and renewable energy source that underpins the bioenergy industry (McLaughlin, 1992; Sanderson et al., 2006). Bioenergy, both biopower and biofuels, could contribute significantly to meet growing energy demand while mitigating the environmental problems. The Energy Independence and Security Act RFS2 in the United States mandates that annual biofuels' use increase to 36 billion gallons per year by 2022, of which 21 billion gallons should come from advanced biofuels (EISA, 2007). Waste products, both agricultural and forest residues, are obvious choices as base feedstocks; however, it is the use of "dedicated" energy crops where the ability to achieve the billion tons of biomass USA goals will be realized (Perlack et al., 2005). Several plant species such as switchgrass, Miscanthus, corn fodder, sorghum, energy canes, and other grass and legume species have demonstrated tremendous potential for use as dedicated bioenergy feedstocks especially for the production of advanced biofuels. Their adaptation patterns along most agro-ecological gradients also offer options for optimizing a crop species mix for any bioenergy feedstock production system.

---

*Bioenergy Feedstocks: Breeding and Genetics*, First Edition. Edited by Malay C. Saha, Hem S. Bhandari, and Joseph H. Bouton.
© 2013 John Wiley & Sons, Inc. Published 2013 by John Wiley & Sons, Inc.

## 1.1 Historical Development

The concept of bioenergy is not new. Early human civilization witnessed energy potential of plant biomass and used it in cooking and as a source of light. By 1912, Rudolf Diesel demonstrated that diesel obtained from plant biomass can be used in automobile operation (Korbitz, 1999). The shortage of crude oil during the 1970s reinforced the world's motivation toward plant biomass as alternative energy source. In Brazil, use of ethanol to power automobile dates back to the late 1920s. Brazil's National Alcohol Program under government funding was launched in 1975 to promote ethanol production from sugarcane. In 2007, Brazil produced more than 16 billion liters of ethanol (Goldemberg, 2007).

In the United States, during the past decade, billions of dollars were invested annually by the federal and state governments, venture capitalists, and major private companies for the development of new technology to convert feedstock species into renewable biofuels. Major breakthroughs have happened during the past few years and the biofuel production increased significantly. Significant improvements have also noticed on conversion technologies thus moving the biofuel from pilot scale to near-commercial scale.

At present, biofuels are produced from corn grain, sugar cane, and vegetable oil. In the United States, corn is the main feedstock used to produce ethanol. In 2010, corn-based ethanol production was about 50 billion liters (USDOE, 2011). With the increasing world's food demand there is serious economic (animal feed costs are rising) and even ethical concern with using corn grain in ethanol production. In the mid-1985s, U.S. Department of Energy (DOE) Herbaceous Energy Crops Program (HECP), coordinated by Oakridge National Laboratory (ORNL), funded research to identify potential herbaceous species as potential bioenergy feedstock. Over 30 plant herbaceous crop species including grasses and legumes were studied, and consequently switchgrass was chosen as the "model bioenergy species" (McLaughlin and Kszos, 2005). Under optimum conditions, switchgrass demonstrated annual biomass yield as high as 24 Mg ha$^{-1}$, and each ton of biomass can produce about 380 L of ethanol (Schmer et al., 2008). Carbon sequestration by 5-year-old switchgrass stand can add 2.4 Mg C ha$^{-1}$ year$^{-1}$ for 10,000 Mg ha$^{-1}$ of soil mass (Schmer et al., 2011). Other plant species with high bioenergy potential include Miscanthus, corn fodder, sorghum, sugarcane, prairie cordgrass, bluestems, eastern gamagrass, and alfalfa. Miscanthus hybrids have the potential to produce high biomass and can make a significant contribution to biofuel production and to the mitigation of climate change. Plant breeding will play an important role in improving the genetic potential of these species, as well as other potential species, and make them suitable as bioenergy feedstock.

## 1.2 Cultivar Development

Genetic improvement of plant species targeting biomass feedstock production, particularly the dedicated energy crops such as switchgrass and Miscanthus, is in a very early stage, posing both challenges and opportunities for genetic improvement. The current emphasis of most biomass feedstock cultivar development research is based on biomass yield. Due to extensive breeding efforts, maize grain yield has increased 745% from 1930 to the present (USDA-NASS, 2011). Biomass yield per unit of land is a function of many traits; thus plant breeders also have to address problems related to establishment, seed shattering, and resistance to abiotic/biotic stresses. Equally important is improvement in feedstock quality for sustainable bioeconomy.

Research is still evolving on processes to convert biomass to bioenergy/biofuel that will dictate the quality targets of dedicated bioenergy crops. One likely scenario is that both enzymatic and thermochemical conversion technologies will be required depending on the biomass feedstock availability and the targeted bioenergy end product.

## 1.3 Breeding Approach

The fundamentals of feedstock cultivar development will be the same as ones that have been successfully employed in several agricultural crops for thousands of years. Most of the potential bioenergy crops are outcrossing polyploids and great genetic diversity exists both within and among populations. This reinforces the potential for genetic improvement of these crops. Most of the named switchgrass cultivars were developed only by seed increases of desirable plants identified from the wild or selected through two or three generations under cultivation (Casler et al., 2007). The improvement of quantitative traits will require several cycles of selection (Bouton, 2008). The traits that are qualitatively inherited can be improved rapidly. Exploitation of heterosis would require identification of genes involved in heterosis and development of heterotic pools, similar to the one that was followed in hybrid breeding in maize. Different crop species would need different plant breeding methodologies depending on their mode of reproduction, ploidy systems, and germplasm availability. For example, corn has a well-developed hybrid production system using inbred lines, which may not be directly applicable to crops like switchgrass that has nearly 100% self-incompatibility. Some species of Miscanthus and sugarcane that do not produce seeds require a different approach. The hundreds of years of experience gained in the development of modern cultivars of food and other agricultural crops can directly benefit the cultivar development research of bioenergy crops.

## 1.4 Molecular Tools

Rapid development in high-throughput genotyping, genotyping based on sequencing, and computational biology continues to shape modern plant breeding into a new approach called "molecular breeding." Rapid discoveries of DNA-based markers at significantly reduced cost have impacted cultivar development methodologies in the recent years. Advances in molecular biological research have uncovered several plant biological functions and enhanced the understanding of gene function at the molecular level (Bouton, 2008). Rapidly growing genome, transcriptome, proteome, and metabolom resources of several important biofuel crops can speed the process of feedstock development which can lead to improved economics of renewable bioenergy production. Lignin polymer is found to be interfering with enzymatic digestion of lignocellulosic biomass necessitating the pretreatment of biomass feedstock, making biofuel production an economic challenge (Dien et al., 2011). However, plant biologists have been able to characterize and modify lignin pathway and produce low lignin plants by silencing genes involved in lignin pathway (Dien et al., 2011; Fu et al., 2011). Transgenic technologies have also enabled plant breeders to look beyond target species for genes conferring desirable traits, but current regulatory aspects could curtail gains from transgenics, especially for bioenergy crops, without deregulation reforms that better balance both risk and benefit (Strauss et al., 2010).

## 1.5 Future Outlook

Changing climates as seen by frequent unprecedented drought cycles have become a serious challenge in the recent decades. This will require an "adjustment philosophy" in that breeding strategies will need to continually adjust trait targets for greater stress extremes with programs concentrating on stress tolerances growing in importance (Bouton, 2010). As biomass feedstock production scales up to a commercial level, there will also be a significant shift in agricultural landscapes, leading to occurrence of new pest and diseases specific to the feedstock species. Exploration and exploitation of microbial endophytes implicated in protection of plants from a broad range of biotic and abiotic stresses are important areas for future research (Ghimire et al., 2011). Bioenergy crop breeders should therefore take proactive action to integrate all conventional and modern tools into their cultivar development research.

There are government policy issues that may assist the growth of bioenergy industry. However, these are political issues not within the scope of this book and will need to be hashed out at that level. But one thing is certain, bioenergy cultivar development research will benefit by always striving for a cost-effective product that competes in the free market with hydrocarbons and nuclear power. This should become more possible by leveraging facilities/resources established for traditional agricultural crops and implementation of regional/national/international collaborations between institutions involved in bioenergy feedstock research. Finally, sharing germplasms between participating institutes would help maintain genetic diversity of the breeding pools needed for long-term use.

## References

Anonymous. Environment, 2012. http://one-simple-idea.com/Environment1.htm. Accessed on 3 September 2012.

Bouton J. Improvement of switchgrass as a bioenergy crop. In: Vermerris W (ed.) *Genetic Improvement of Bioenergy Crops*, 2008, pp. 295–308. Springer Science, New York.

Bouton JH. Future developments and uses. In: Boller B, Posselt UK, Veronesi F (eds) Handbook of Plant Breeding, Vol. 5. Fodder Crops and Amenity Grasses, 2010, pp. 201–209. Springer, New York, Dordrecht, Heidelberg, London.

Casler MD, Stendal CA, Kapich L, Vogel KP. Genetic diversity, plant adaptation regions, and gene pools for switchgrass. Crop Sci, 2007; 47: 2261–2273.

Dien BS, Miller DJ, Hector RE, Dixon RA, Chen F, McCaslin M, Reisen P, Sarath G, Cotta MA. Enhancing alfalfa conversion efficiencies for sugar recovery and ethanol production by altering lignin composition. Bioresource Technol, 2011; 102: 6479–6486. doi:10.1016/j.biortech.2011.03.022.

The Energy Independence and Security Act [EISA], 2007. http://www.gpo.gov/fdsys/pkg/PLAW-110publ140/pdf/PLAW-110publ140.pdf. Accessed on 4 December 2012.

Fu C, Xiao X, Xi Y, Ge Y, Chen F, Bouton J, Dixon RA, Wang Z-Y. Downregulation of cinnamyl alcohol dehydrogenase (CAD) leads to improved saccharification efficiency in switchgrass. Bioenerg Res, 2011; 4: 153–164. doi:10.1007/s12155-010-9109-z.

Ghimire SR, Charlton ND, Bell JD, Krishnamurthy YL, Craven KD. Biodiversity of fungal endophyte communities inhabiting switchgrass (*Panicum virgatum* L.) growing in the native tallgrass prairie of northern Oklahoma. Fungal Divers, 2011; 47: 19–27. doi:10.1007/s13225-010-0085-6.

Goldemberg J. Ethanol for a sustainable energy future. Science, 2007; 315: 808–810.

Korbitz W. Biodiesel production in Europe and North America, and encouraging prospect. Renew Energ, 1999; 16: 1078–1083.

McLaughlin SB. New switchgrass biofuels research program for the Southeast. In: *Proceedings Ann Auto Tech Dev Contract Coord Mtng*, 1992, pp. 111–115. Dearborn, MI.

McLaughlin SB, Kszos LA. Development of switchgrass (*Panicum virgatum*) as a bioenergy feedstock in the United States. Biomass Bioenerg, 2005; 28: 515–535.

Perlack RD, Wright LL, Turnhollow AF, Graham RL, Stokes BJ, Erbach DC. Biomass as a feedstock for a bioenergy and bioproducts industry: the technical feasibility of a billion ton supply, 2005. http://feedstockreview.ornl.gov/pdf/billion_ton_vision.pdf. Accessed on 3 September 2012.

Sanderson MA, Adler PR, Boateng AA, Casler MD, Sarath G. Switchgrass as a biofuels feedstock in the USA. Can J Plant Sci, 2006; 86: 1315–1325.

Schmer MR, Liebig MA, Vogel KP, Mitchell RB. Field-scale soil property changes under switchgrass managed for bioenergy. Global Change Biol Bioenerg, 2011; 3: 439–448.

Schmer MR, Vogel KP, Mitchell RB, Perrin RK. Net energy of cellulosic ethanol from switchgrass. Proc Natl Acad Sci U S A, 2008; 105: 464–469.

Strauss SH, Kershen DL, Bouton JH, Redick TP, Tan H, Sedjo RA. Far-reaching deleterious impacts of regulations on research and environmental studies of recombinant DNA-modified perennial biofuel crops in the United States. BioScience, 2010; 60: 729–741.

USDOE. *US Billion-Ton Updates: Biomass Supply for a Bioenergy and Bioproducts Industry*. R.D. Perlack and B.J. Stokes (Leads), ORNL/TM-2011/224, 2011, p. 227. Oak Ridge National Laboratory, Oak Ridge, TN. http://www1.eere.energy.gov/biomass/pdfs/billion_ton_update.pdf. Accessed on 19 March 2013.

USDA-NASS. Quick Stats: Agricultural Statistics Data Base, 2011. http://quickstats.nass.usda.gov/results/D21A4E3A-A14A-3845-B2C7-7A62041B1648. Accessed on 4 December 2012.

# Chapter 2
# Switchgrass Genetics and Breeding Challenges

Laura Bartley[1], Yanqi Wu[2], Aaron Saathoff[3], and Gautam Sarath[3]

[1]*Department of Microbiolgy and Plant Biology, University of Oklahoma, Norman, OK 73019, USA*
[2]*Department of Plant and Soil Science, Oklahoma State University, Stillwater, OK 74078, USA*
[3]*Grain, Forage, and Bioenergy Research Unit: USDA-ARS and Department of Agronomy and Horticulture, University of Nebraska, Lincoln, NE 68583, USA*

## 2.1 Introduction

Liquid biofuel production from biomass has the potential to reduce greenhouse gas emissions from transportation and dependence on fossil fuels extracted from politically volatile regions (Somerville, 2007; Bartley and Ronald, 2009; Vega-Sanchez and Ronald, 2010). Cultivated grasses are the most abundant sustainable class of biomass that can be produced in the United States (∼57%, Perlack et al., 2005). Switchgrass (*Panicum virgatum* L.), in particular, is an attractive native species for development as a bioenergy crop given that largely unimproved varieties exhibit large biomass yield (up to 36.7 Mg ha$^{-1}$) and marked stress tolerance (Figure 2.1; Thomason et al., 2004; McLaughlin and Adams Kszos, 2005; Bouton, 2007). Even with typical, lower-yielding marginal land (5–11 Mg ha$^{-1}$), energy and emission measurements for switchgrass production give an approximately 5-fold net energy yield (output:input) and an approximately 10-fold reduction in greenhouse gas emissions compared with gasoline (Schmer et al., 2008). In order to realize greater benefits from the production of lignocellulosic fuels, there is an enormous need to apply various breeding methods and tools toward switchgrass improvement. Below, we outline switchgrass energy crop breeding goals and, in subsequent sections, provide an overview of the basic biology and genetic characteristics of switchgrass. We then discuss experiences and challenges related to switchgrass conventional and molecular breeding.

Biomass *yield* and *quality* are the two general classes of targets for genetic improvement of bioenergy crops. Selection of switchgrass for high biomass production is ongoing (Vogel et al., 2010). Recently released cultivars "BoMaster," "Cimarron," and "Colony" produce higher biomass yields than the current best commercial cultivar "Alamo" and are primarily targeted for cellulosic feedstock production (Burns et al., 2008a, 2008b, 2010; Wu and Taliaferro, 2009). Similarly, high-performing replacement proprietary cultivars for old standards such as Alamo, "Kanlow," and "Cave-in-Rock" are currently sold in commercial

---

*Bioenergy Feedstocks: Breeding and Genetics*, First Edition. Edited by Malay C. Saha, Hem S. Bhandari, and Joseph H. Bouton.
© 2013 John Wiley & Sons, Inc. Published 2013 by John Wiley & Sons, Inc.

**Figure 2.1.** A seed production field of the lowland switchgrass cultivar "Cimarron" in September, toward the end of the growing season, near Perkins, OK.

bioenergy seed trade as EG1101, EG1102, and EG2101, respectively (http://www.bladeenergy.com/SwitchProducts.aspx).

Abiotic and biotic stress tolerance and improved agronomic characteristics, such as reduced seed dormancy (Burson et al., 2009), are important for establishing and obtaining consistent biomass production. In terms of biomass quality, the goals for the two current biomass to biofuel conversion platforms, biochemical and thermochemical, are roughly opposite (Figure 2.2). For biochemical conversion methods that mostly produce alcohol fuels, the quality goal is to optimize the quantity of sugar that can be obtained from the biomass. For

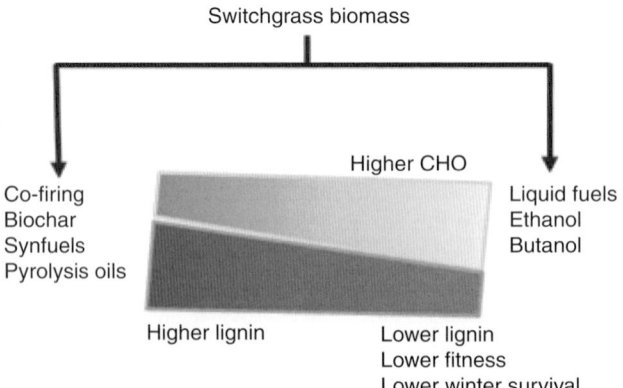

**Figure 2.2.** Potential applications and consequences of changing switchgrass biomass quality. CHO, carbohydrate.

example, transgenic switchgrass mutants with reduced lignin content in the cell wall release a larger fraction of cellulose-derived glucose when subject to *in vitro* digestion (Fu et al., 2011a, 2011b; Saathoff et al., 2011a). On the other hand, thermochemical biomass conversion methods to yield syngas or bio-oil produce higher-quality products from lignin compounds compared with carbohydrates (Bridgwater and Peacocke, 2000). With the exception of high-lignin plants for thermochemical conversion, it is worth noting that many of the characteristics desirable for switchgrass production for biofuels are similar to those desirable for forage use (Sarath et al., 2008). However, as for forage-related traits, it is anticipated that a major challenge for switchgrass breeding for many biofuel quality traits will be the occurrence of trade-offs between the genetic determinants for biomass quality and quantity.

## 2.2 Origin and Distribution

Switchgrass is a polyploid, perennial $C_4$ warm-season grass native to the tallgrass prairies of the continental United States, with a historical distribution through Central America. Consistent with the hypothesis that switchgrass survived the Pleistocene Glaciations in southern refugia (McMillan, 1959), recent analyses suggest that the southeastern United States is the center of diversity for extant switchgrass populations (Zhang et al., 2011). Much of the original native stands have disappeared over time, mostly through agricultural activities; however, pockets of native stands do exist within the continental United States (Vogel et al., 2010). Panicums belong to the subfamily Panicoideae and can be considered to be monophyletic based on molecular phylogeny (Giussani et al., 2001; Aliscioni et al., 2003). Switchgrass contains lowland and upland ecotypes, with contrasting growth habits. Lowland forms are generally found in wetter areas and tend to be robust plants with relatively high biomass yields. Upland types are found in dryer environments and tend to be smaller plants with finer leaves and lower biomass yields, but are more suitable for grazing by cattle (Vogel, 2004).

## 2.3 Growth and Development, Genome Structure and Cytogenetics

Many aspects of switchgrass biology, agronomy, and management have been discussed in detail in recent reviews (Vogel, 2004; Parrish and Fike, 2005; Bouton, 2007; Vogel et al., 2010), and the following constitutes a brief summary.

Certain key issues emerge when considering switchgrass for bioenergy; these are yield, the number of harvest per year, rates of fertilization and other chemicals required to maintain yields, and the length of time production fields will need to be maintained. For example, different growing environments could favor a two-harvest schedule for a given cultivar at a southerly latitude and limit it to one harvest at more northerly latitude. Other secondary issues to be considered are seed quality, stand establishment, and response to biotic and abiotic stresses. At present the genetic linkages between these factors are unknown. In all scenarios, the total yields and quality of the biomass will differ and have varying impact on downstream conversion to fuels (Schmer et al., 2012). Similarly, it can be anticipated that a certain maximum yield could be achieved using current breeding and management strategies; however, continued yield increases will require additional inputs. How these might influence the overall sustainability for this new crop is also unknown. However, what is clear is the need to focus breeding strategies to first improve yields and then systematically address other components required to generate high-yielding, high-quality biomass in a sustainable manner.

### 2.3.1 Growth and Development

Switchgrass plants can have long or short rhizomes, that is, horizontal stems below the soil surface, depending on the ecotype. Generally, sod-forming ecotypes, that is, uplands, have longer rhizomes, and bunch-forming ecotypes, that is, lowlands, have shorter rhizomes (Vogel, 2004). New root growth is initiated from rhizomes, and roots can be over 3 m in length. It has been estimated that switchgrass produces an equivalent amount of biomass below ground as is produced above ground over a growing season (Garten and Wullschleger, 1999; Liebig et al., 2008). Growth occurs every spring from dormant auxiliary buds present on rhizomes, crowns, and stem bases. The relative proportion of tillers produced from each of these structures is dependent on the genotype. Loss of tillering capacity, either through loss of dormant buds or damage to rhizomes, may lead to stand loss. The genetic factors controlling rhizome and root growth, meristem initiation, and tillering capacity have yet to be investigated, but promise to be exciting areas for research.

Tillers grow by extension of the internodes and formation of new nodes by an apical meristem that is protected by several layers of leaf sheath. Initial physical support and protection for the elongating stems is provided by leaf sheaths. Subsequent support is provided by extensive secondary wall deposition and lignification of various tissues within the internode (Sarath et al., 2007a; Shen et al., 2009a). Tiller extension continues during the vegetative growth of the plant and is terminated once the transition to flowering occurs. At this stage, the peduncle and flag leaf are still developing. Transition to flowering is under the control of the photoperiod (Vogel, 2004).

Heading occurs when the elongating peduncle pushes the developing inflorescence to become visible at the tops of tillers. Heading date is under genetic control and varies significantly within and among populations. Marked differences in flowering times are apparent if populations adapted to distinct latitudes are moved north or south (Casler et al., 2007b; Vogel et al., 2010). Flowering can occur over a 2-week or longer period in switchgrass, since florets at the base of the inflorescences are older than the ones near the apical portions of the panicle. In field, once a plant has transitioned to flowering, apical meristems of immature tillers will also convert from vegetative to reproductive phase. The switchgrass inflorescence is a diffused panicle 15–55 cm long with two-flowered spikelets. The upper floret is perfect and the lower floret is either staminate or empty. Florets are smooth and awnless. At anthesis, pollen grains (Figure 2.3) are shed during the day and peak times for pollen shed have been recorded in the late morning (10:00–12:00 h) or in the early afternoon (12:00–15:00 h) (Jones and Brown, 1951).

Fertilization is followed by a prolonged period (~30–40 days) of seed development prior to seed dehiscence. Mature switchgrass seeds are relatively small, 1–2 mg seed$^{-1}$ (including the hull), and seed weight is under genetic control, although it is affected by abiotic and biotic stresses during seed fill (Boe, 2003; Das and Taliaferro, 2009). Seed size has an impact on seedling growth (Green and Bransby, 1996; Smart and Moser, 1999) and could have a role in dormancy, germination, and stand establishment. Switchgrass seeds at harvest exhibit poor germination and need a period of after-ripening and cold stratification for optimal germination (Parrish and Fike, 2005, 2009). A number of internal and external factors can impact seed dormancy and germination in this species, including a strong genetic component (Shen et al.,

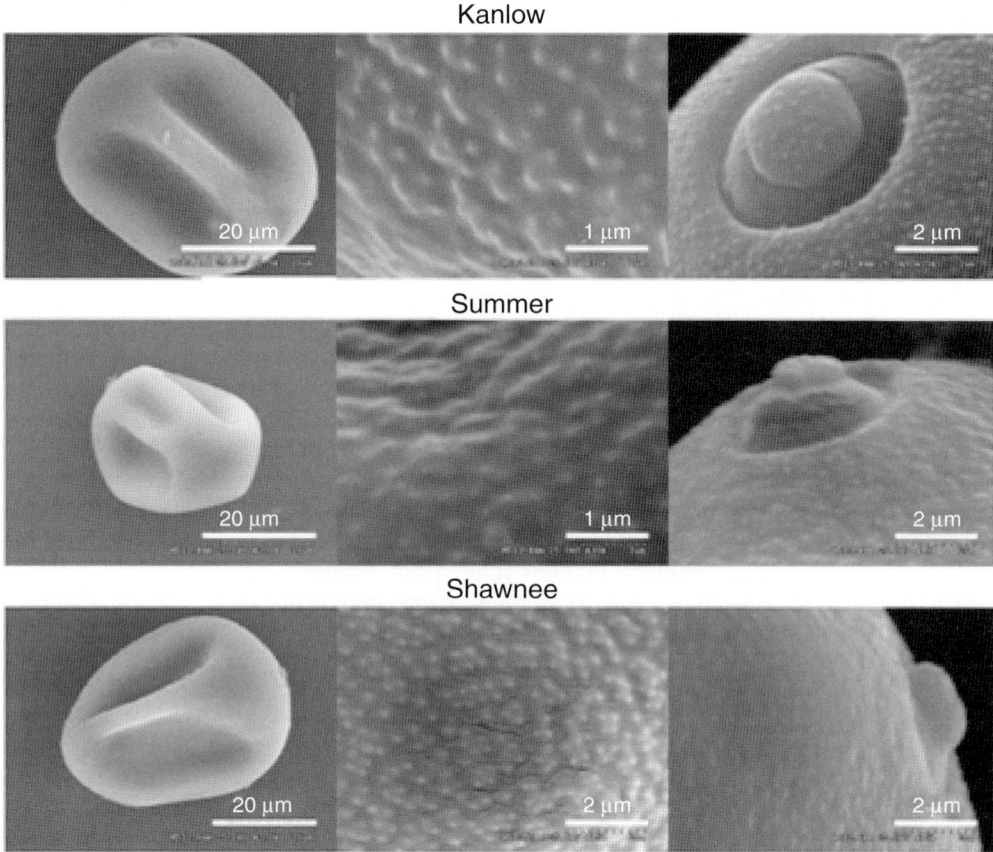

**Figure 2.3.** Switchgrass pollen from tetraploid cultivars "Kanlow" and "Summer" and from an octaploid cultivar "Shawnee." The three images for each cultivar show a low magnification image of the whole pollen grain (post-drying) and higher magnification images of the pollen wall.

2001; Sarath et al., 2006, 2007b; Sarath and Mitchell, 2008; Burson et al., 2009). Stand establishment can be problematic. Breeding for reduced dormancy and optimal germination may aid in overcoming this challenge (Burson et al., 2009), especially as failure of stand establishment in the first year can have significant impact on the overall sustainability of switchgrass production (Schmer et al., 2006; Perrin et al., 2008).

Molecular knowledge of many aspects of switchgrass biology is starting to accumulate (e.g., Tobias et al., 2008; Jakob et al., 2009; Shen et al., 2009a; Chen et al., 2010; Escamilla-Trevino et al., 2010; Matts et al., 2010; Casler et al., 2011; Fu et al., 2011a, 2011b; Saathoff et al., 2011b) and is likely to have a significant impact on genetic improvement of this important bioenergy species. The Joint Genome Institute completed a large-scale sequencing project of switchgrass expressed sequence tags (ESTs) in 2007 (Tobias et al., 2008). Other EST collections arising from next-generation sequencing have also been deposited in public databases (Srivastava et al., 2010). These sequences can be mined for many of the gene products that can affect quality and growth traits in switchgrass. For example, these collections contain almost all of the cDNAs coding for enzymes involved in lignin biosynthesis (Tobias et al., 2008; Escamilla-Trevino et al., 2010; Saathoff et al., 2011b) as well as an array of transcription factors that control a range of secondary cell wall characteristics (Shen et al., 2009b). Data mining of these sequences can be expected to intensify as researchers associate phenotype to genotype within breeding programs.

### 2.3.2 Genome Structure and Cytogenetics

The basal chromosome number for switchgrass is nine, with a $1C_x$ genome size of ~700 Mbp (Hultquist et al., 1996). The species can be distinguished into tetraploids ($2n = 4x = 36$) and octaploids ($2n = 8x = 72$), although plants with a range of ploidies have been reported (Lu et al., 1998; Costich et al., 2010). Lowland ecotypes are exclusively tetraploid whereas uplands are primarily octaploids and occasionally tetraploids, though individual plants frequently exhibit aneuploidy, especially those with higher ploidy levels (Costich et al., 2010). Plants of the same ploidies will freely intermate if flowering at the same time (Martinez-Reyna and Vogel, 2008; Vogel and Mitchell, 2008). Sequence variation among switchgrass chloroplast DNA also distinguishes two cytoplasmic types "U" (upland) and "L" (lowland) (Hultquist et al., 1996; Zalapa et al., 2011). Switchgrass is an allogamous species and its outcrossing is enforced by self-incompatibility mechanisms (Talbert et al., 1983; Martínez-Reyna and Vogel, 2002). However, in both lowland and upland populations, the magnitude of genetic variability in seed origin, that is, selfed versus crossed, is yet to be characterized.

## 2.4 Genetic Diversity

As primarily an outcrossing species, each switchgrass individual is genetically distinct. The mode of reproduction and polyploid nature tend to conserve its genetic diversity. Thus, the raw material for genetic improvement of switchgrass should be ample, despite the fact that most of the native prairie and savanna habitats persist only as remnants. With exceptions (Casler et al., 2007a; Todd et al., 2011), most recent studies of switchgrass diversity have focused on *P. virgatum* accession from the USDA's National Genetic Resources Program (NGRP). The seeds of 175 switchgrass accessions and/or cultivars are currently available through this network (USDA, 2011). Most named switchgrass cultivars represent only seed increases of desirable plants identified from the wild or selected through two or three generations under cultivation (Casler et al., 2007a). This reinforces the great potential for genetic improvement and is also indicative of the broad diversity present even among anthropogenically distributed switchgrass.

To aid breeding and assist conservation efforts, several groups have recently reported molecular diversity present within and between switchgrass populations (Missaoui et al., 2006; Casler et al., 2007a; Narasimhamoorthy et al., 2008; Cortese et al., 2010; Zalapa et al., 2011; Todd et al., 2011). Being independent of environment, molecular studies are more robust than relying solely on phenotypic observations (Collard and Mackill, 2008), though recent switchgrass phenotypic diversity analyses are also available (Casler, 2005; Casler et al., 2007a). The study by Cortese et al. (2010) is unique in simultaneously providing molecular and phenotypic data for switchgrass, which they generally found to be mutually reinforcing.

The molecular diversity studies provide evidence for the assertion made above, that switchgrass is a diverse species. Greater within than between population diversity has been reported

(Narasimhamoorthy et al., 2008; Cortese et al., 2010). Furthermore, the upland and lowland ecotypes can be readily distinguished at the molecular level (Missaoui et al., 2006; Narasimhamoorthy et al., 2008; Cortese et al., 2010; Zalapa et al., 2011; Todd et al., 2011). As expected based on observed geographic distribution of phenotypic differences (Casler et al., 2007b), recent studies have detected similarity among geographically and ecologically grouped populations (Narasimhamoorthy et al., 2008; Cortese et al., 2010; Zalapa et al., 2011). Aided by using a significant number of markers (501 alleles from 55 simple sequence repeat markers) selected for relatively high information content, Casler and colleagues were the first to distinguish ploidy level within upland ecotypes and found an abundance of octoploid-unique markers, consistent with significant genetic isolation between the ploidy levels (Zalapa et al., 2011). This study was also able to identify marker alleles diagnostic for many of the accessions examined (Zalapa et al., 2011). Still, compared with other economically useful nonbiomass grasses, that is, cereals and turf grasses, a continuing challenge for switchgrass research is to increase the number of markers and the collection sizes characterized to build a more comprehensive picture of available diversity.

## 2.5 Phenotypic Variability and Inheritance

Genetic improvement of germplasm and breeding of new cultivars are the most cost-effective means for increasing biomass yield potential in switchgrass. Biomass yield of switchgrass is a quantitative trait, which is the result of collective expression of genes, interactions among genes, effects of environmental conditions on the plant, and interactions between the plant and the environment. The environmental and genotype by environment interaction effects on biomass yield variation of switchgrass are large and can be much larger than that of the genetic effect. The results of adaptation and agronomic performance of six upland switchgrass populations grown at 12 locations in hardiness zones 3, 4, 5, and 6 indicated switchgrass populations for biomass production and other purposes should not be planted more than one hardiness zone north or south from their origin zone (Casler et al., 2007b). They also reported significant population by location interactions accounting for 10–31% of the biomass yield variation. Fuentes and Taliaferro (2002) reported variance components of biomass yield from a 7-year experiment of nine switchgrass cultivars (three lowland and six upland types) and three cultivar blends (one lowland–lowland and two lowland–upland) at Chickasha and Haskell, Oklahoma. That study indicated locations, years, cultivars, and interactions among these represent major sources of biomass yield variation, in which environment (locations and years) and its interaction with genotypes account for about 80% of the total variation. Based on these results, it is necessary to evaluate breeding products in multiple locations across target areas in multiple years.

Genetic variability and heritability for biomass yield and related traits in switchgrass breeding populations is critical for breeding programs. Breeders use such information to estimate genetic gains per selection cycle and understand the inheritance of traits. Positive responses to switchgrass breeding are contingent on the presence of adequate genetic variability within breeding populations for target traits, the heritability of the target traits, and effective selection procedures, that is, the ability of breeders to identify genetically superior plants (Taliaferro and Vogel, 1999; Vogel, 2000).

Heritability estimates may vary among breeding populations and can be different due to various statistical methods and/or experimental locations (one vs. two) used in the analysis. Narrow-sense heritability is a genetic parameter for breeders to calculate predicted genetic gains ($\Delta G$) in each selection cycle, $\Delta G = c \times k \times h_n^2 \times \sigma_p$, where $c$ is the parental control factor, $k$ the standardized selection differential, $h_n^2$ the narrow-sense heritability, and $\sigma_p$ the phenotypic standard deviation (Rose et al., 2007). Narrow-sense heritability estimates the portion of additive genetic effect among the total phenotypic variability. For switchgrass, narrow-sense heritability estimates ranging from 0.02 to 0.6 for different traits have been reported in the literature (Newell and Eberhart, 1961; Talbert et al., 1983; Hopkins et al., 1993; Boe and Lee, 2007; Rose et al., 2008). Narrow-sense heritability estimates for biomass yield in lowland populations consisting of 37 half-sib families were 0.13, based on half-sib family variation, and 0.29, from parent–offspring regression analysis (Bhandari et al., 2010). They indicated narrow-sense heritability for spring growth was 0.82 or larger; for heading days, flowering and plant spread ranged from 0.47 to 0.70; for plant height and tillering capability was from 0.26 to 0.48; and for stem thickness was 0.27 or less. In summary, reported narrow-sense heritability estimates for biomass yield vary widely, but are generally quite low. On the other hand, the estimates for biomass components, such as flowering time and plant growth habit, are moderate or high. The results may suggest that recurrent selection and quantitative trait loci (QTLs) mapping on biomass yield components for biomass yield have merits.

## 2.6 Conventional Breeding Approaches

There is considerable genetic variability in switchgrass, and most available cultivars are a synthetic pool of genotypes that are not too far removed from their progenitors. This diversity suggests that switchgrass can be improved through selective breeding for specific traits. Switchgrass breeding programs have utilized an assortment of breeding techniques to improve germplasm (Vogel and Pedersen, 1993; Missaoui et al., 2005a; Rose et al., 2008; Das and Taliaferro, 2009; Bhandari et al., 2010). However, conventional breeding cycles in switchgrass are long, and experience dictates (see Phenotypic Variability) that significant resources have to be devoted to a systematic evaluation of plants in the field (Figure 2.4). This extensive phenotyping is crucial, because the anticipated margins in producing biomass feedstock are low, and even partial failure of a new cultivar can have significant economic consequences. Thus an area of greater research emphasis has been to develop high-throughput phenotyping tools that can improve the breeding and selection process for switchgrass and other bioenergy crops (Decker et al., 2009; Fernie and Schauer, 2009; King et al., 2009; Mann et al., 2009; Santoro et al., 2010).

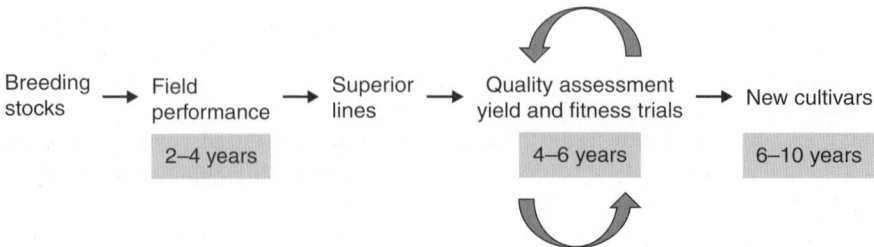

**Figure 2.4.** Conventional breeding pipeline for switchgrass.

### 2.6.1 Early Work

Most of the initial breeding work in switchgrass used between-strain genetic variability (Vogel, 2000). In this procedure, a large number of native accessions (strains) were collected from different geographic areas. The collections were then evaluated in uniform nurseries for target agronomic traits. The better accessions for targeted traits were increased and tested in replicated field experiments in additional environments. The best accessions were released directly to the growers without any additional breeding work. This procedure was used to develop early switchgrass cultivars by the state experiment stations and the Plant Materials Centers of the USDA-Soil Conservation Service (Alderson and Sharp, 1995).

### 2.6.2 Systematic Recurrent Selection

More recently, breeders have used systematic recurrent selection procedures to improve switchgrass populations (Figure 2.5). Several cultivars derived from recurrent selection using genetic gains within respective improved populations have recently been released (Burns et al., 2008a, 2008b, 2010). Recurrent selection is a powerful breeding procedure for increasing the frequency of desirable alleles conditioning complex, quantitative traits in plant populations. As discussed above, biomass yield is a complex trait conditioned by many genes (G), strongly influenced by environmental factors (E), and G by E interactions. Effectiveness of recurrent

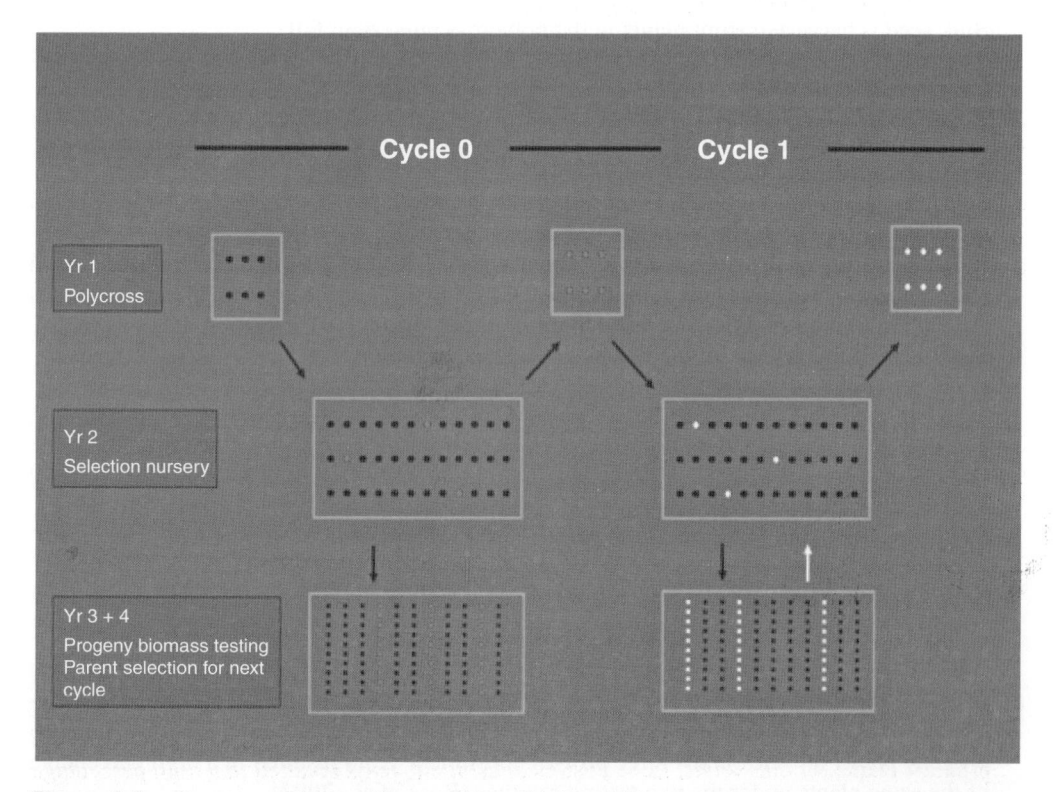

**Figure 2.5.** Recurrent selection for general combining ability procedure for switchgrass population improvement at Oklahoma State University, USA.

selection is contingent on adequate additive genetic variability in the breeding population and the ability to accurately identify superior genotypes (Taliaferro and Vogel, 1999). Application of restricted phenotypic recurrent selection for increasing biomass yields in switchgrass failed to produce substantive and steady positive results (Hopkins et al., 1993; Taliaferro and Vogel, 1999). Thus the Oklahoma State University (OSU) switchgrass breeding program has changed to recurrent selection for general combining ability (RSGCA). The RSGCA procedure employs half-sib progeny testing, in which only the maternal parent is known, as the basis for selection of (maternal) parent plants with high breeding value. Honeycomb planting design in particular has been found to be an effective phenotypic selection means to identify superior plants within and among families by removing inter-plant competition (Bouton, 2007). Though effective for increasing biomass yield, the RSGCA procedure requires 4 or 5 years to complete each cycle, and the time period required to develop and test a new cultivar can exceed 15 years (Taliaferro and Vogel, 1999; Vogel, 2000). This procedure is described in greater detail below.

In the RSGCA method employed at OSU (Figure 2.5), selection is based on a combination of phenotypic and genotypic performance with final selection based on genotype, that is, breeding value, as measured by the biomass yield performance of half-sib families. OSU RSGCA procedure details are as follows:

**Year 1**
- Parents (25–50) of top half-sib families are selected for height and profuse tillering and cloned to a field polycross nursery in spring.
- Bulk seed is harvested from plants in the polycross nursery in fall.

**Year 2**
- Seed from polycross plot is germinated in greenhouse, and early-germinating and vigorous seedlings are selected.
- Selection nursery of 1000–1200 spaced plants is established by transplanting the selected seedlings.
- Half-sib seed is hand harvested from about 100 to 200 (10–20%) plants visually judged to be superior based on height and tiller density. Field stratification, with similar plants selected in multiple subplots, is used to minimize environmental effects.
- Seed from each of the selected plants is processed separately.

**Year 3**
- Half-sib seed is germinated. About 100 half-sib families with greatest seed germination and seedling vigor are selected.
- Field evaluation nursery of the selected half-sib progeny families is established in a replicated experiment.
- Biomass yield is measured for each replicate of the half-sib families in fall.

**Year 4**
- Biomass yield of each half-sib family is measured only in the fall of the first post-establishment year, rather than for three post-establishment-year harvests. Testing of half-sib progeny plants for one versus three post-establishment years resulted in a high percentage of the same plants under the two testing regimes (Rose et al., 2007).
- Data are analysed.

- The best parents are selected on the basis of their progeny's biomass yields.
- Selected parents in the selection nursery are cloned to a field polycross nursery in spring to start a new cycle (Year 1 in new cycle).

Casler (2010) reported the gains from two cycles of within-family phenotypic selection for increased biomass yield in the WS4U upland switchgrass population. The biomass yield of all families after selection relative to the original WS4U population increased by 0.36 Mg ha$^{-1}$ cycle$^{-1}$ (Casler, 2010). The selection gains were consistent in two locations indicating the robustness of the within-family selection method in improving the biomass yield of WS4U switchgrass families. However, the two cycles of within-family selection resulted in significant reduction of genetic variation among families. Further improvement in the population would require a combination of use of within-family phenotypic selection and half-sib progeny testing of selected families to create new superior genotypes and increase favorable alleles in selected families (Casler, 2010).

### 2.6.3 Heterosis

There is strong evidence for the presence of heterotic groups among the tetraploid upland and lowland cultivars, and $F_1$ populations between such plants exhibit hybrid vigor (Figure 2.6; Taliaferro and Vogel, 1999; Martinez-Reyna and Vogel, 2008; Vogel and Mitchell, 2008). Several seminal reports have shed light on heterosis in switchgrass. Hybrids of two upland crosses and one lowland cross demonstrated 44–66% greater biomass yields than the average between the two parental (i.e., mid-parent) yields in the first-year harvest (Taliaferro

**Figure 2.6.** Hybrid vigor in lowland switchgrass. The large central plant is the progeny of a cross and is contrasted with the relatively diminutive flanking plants that are the progeny of self-pollination of one of the hybrid's parents. (*For color details, see color plate section.*)

and Vogel, 1999). More recently, Martinez-Reyna and Vogel (2008) reported that lowland-tetraploid (cv. "Kanlow") and upland-tetraploid (cv. "Summer") plants represented different heterotic groups that can potentially be used to produce $F_1$ hybrid cultivars since "Kanlow" by "Summer" hybrids exhibited mid-parent heterosis (~14–18%) for biomass yields in space-planted field trials. Using the hybrids of the same crosses ("Kanlow" × "Summer" and its reciprocal), but grown in swards, Vogel and Mitchell (2008) demonstrated significant high-parent heterosis, that is, improvement over the superior of the two parents, for biomass yields in the range from 30% to 38%. Based on this work, a protocol to use the self-incompatibility of switchgrass by growing two parents in alternating rows in seed production fields has been proposed to produce hybrid seeds (Taliaferro and Vogel, 1999; Martinez-Reyna and Vogel, 2008). However, not all reciprocal crosses between available upland and lowland cultivars yield progeny with hybrid vigor, suggesting that traits that confer heterosis are either not present in all cytoplasm or present in a repulsion phase that result in no observable yield gains (Martinez-Reyna and Vogel, 2008). More research is warranted to identify additional heterotic germplasm pools and to develop more advanced or totally new seed production methods in delivery of cultivars using heterotic effects in biomass feedstock production. The recent release of the switchgrass draft genome (PviDraft0; www.phytozome.org) and continued research into switchgrass molecular genetics will lead to discovery of genes that confer heterosis within a relatively short time (Casler et al., 2011). Mobilizing these genes into elite breeding lines should result in the continued improvement of switchgrass.

## 2.7 Molecular Breeding

Association of traits of interests, such as high biomass and low lignin content, with a DNA sequence or marker precludes the need to conduct time-consuming phenotypic evaluations in subsequent generations. Such "marker-assisted selection" has greatly accelerated the development of improved varieties in other crops (Collard and Mackill, 2008). Switchgrass researchers are in the process of developing populations with defined pedigrees for conventional trait–loci mapping, establishing marker sets that are heterozygous between the parents, and conducting genotypic and phenotypic analyses of the mapping populations. Due to their large biomass production and more straightforward genetics, it is anticipated that tetraploid lowland ecotypes will serve as the foundation for most future switchgrass cultivars for bioenergy use (Bouton, 2007; Vogel et al., 2010).

### 2.7.1 Molecular Markers Used for Switchgrass and Other Polyploids

To facilitate marker-assisted breeding in this species, researchers have begun to assemble a comprehensive set of DNA markers that can be used within breeding programs (Bouton, 2007). Early studies relied on restriction fragment length polymorphisms (RFLPs) distinguished by hybridization with low copy number sequences from other grass species, for example, cDNAs (Missaoui et al., 2005b, 2006). Other researchers have added significantly to these marker sets using a range of polymerase chain reaction (PCR)-based marker types, including simple sequence repeats (SSRs; Tobias et al., 2006, 2008; Narasimhamoorthy et al., 2008; Okada et al., 2010; Wang et al., 2011; Zalapa et al., 2011), sequence-tagged sites (STSs) designed across splice junctions (Okada et al., 2010), and amplified fragment length polymorphism (AFLP; Todd et al., 2011). These various PCR-based marker types have the advantage of often providing multiple segregating bands per genotype. AFLPs are typically dominant, while the others are codominant. Furthermore, the costs and labor associated with genotyping of these markers continue to decrease with the utilization of such high- and medium-throughput resolution methods such as fluorescence-based capillary electrophoresis and acrylamide gels (Wang et al., 2011; Zalapa et al., 2011).

Single nucleotide polymorphism (SNP) and other small DNA insertion and deletions (indels) are also being cataloged for switchgrass (Ma et al., 2008; see also http://www.maizegenetics.net/snp-discovery-in-switchgrass). The major potential advantage of using SNPs for genotyping is that they are abundant and thus have the potential to permit very high resolution mapping of traits of interest, such that the specific gene or polymorphism that causes the trait may be identified. The method already being applied to switchgrass is SNP genotyping (calling) via direct short-read sequencing, the so-called genotyping-by-sequencing (GBS, e.g., with the Illumina or SOLiD short-read platforms; M. Casler and E. Buckler, personal communication). Though still relatively costly and requiring rigorous data processing to assure accuracy (Liu et al., 2010), the steadily decreasing cost per base of sequencing will likely lead to sustained use of GBS in the future. To improve the depth of sequencing (coverage per base) and the confidence of SNP calls, researchers can take various measures to reduce the complexity of the target sequences prior to sequencing, including the following: (1) sequencing only of ESTs (Barbazuk et al., 2007); (2) avoiding abundant, repetitive genomic DNA, which also tends to be methylated, by sequencing the ends of libraries created with methylation-sensitive restriction enzymes (Elshire et al., 2011); or (3) enriching a genomic fraction of interest, such as all exons (i.e., the exome) or a previously identified QTL, via solution- or array-based sequence capture (Fu et al., 2010). Appropriate sequencing depth of reduced complexity samples can be achieved by multiplexing of samples distinguished by ligation with commercially available or home-designed short sequence "barcoding" tags (Elshire et al., 2011). Alternatively, bulk segregant analysis, in which individual genotypes are not resolved, but rather two pools of individuals – those possessing and lacking the trait of interest – are genotyped simultaneously, is also promising (Michelmore et al., 1991).

A number of less computationally demanding SNP genotyping methodologies have recently been applied to polyploids that, compared with GBS, are appropriate for genotyping of a relatively smaller number of SNPs in a large number of individuals. We refer the reader to other recent reviews for a more comprehensive treatment of SNP genotyping methodologies beyond the brief examples provided below (Bagge and Lübberstedt, 2008; Ding and Jin, 2009). One high-throughput SNP genotyping method is the selective application of the oligonucleotide-ligation-based Illumina GoldenGate assay, which has recently been applied to tetraploid and hexaploid wheats (Akhunov et al., 2009). Though amenable to very high throughput levels (Bagge and Lübberstedt, 2008), that assay is limited to only being able to detect three genotypes (e.g., CCTT, CCTC, and CCCC). In contrast, Sequenom's MassARRAY has been effectively used to genotype polyploids, including the hyper-polyploid sugarcane and tetraploid sunflower species (Bundock et al., 2009; Buggs et al., 2010). Based on single-base extension of oligonucleotides identified by mass spec, this assay can distinguish five (e.g., CCCC, CCCT, CCTT, CTTT, TTTT) and, in principle, more, genotypes (Buggs et al., 2010). High-resolution melt (HRM) curve analysis has recently been applied in hexaploid oat (Oliver et al., 2011). HRM relies on detection by a standard quantitative PCR system of the differential stability of small PCR fragments that include the SNP (or indel); although only of moderate throughput due to limited multiplexing, the simplicity and low cost of this assay make it attractive for small- to medium-scale studies (Li et al., 2010).

## 2.7.2 Molecular Mapping

A number of groups have established and commenced genotyping of conventional mapping populations. These populations consist of the progeny of controlled crosses in which the recombination of chromosomes during meiosis can be exploited to determine a genetic linkage among DNA markers in the genome and correlations between genetic markers and traits. Table 2.1 summarizes the basic details of the six mapping populations that, to the authors' knowledge, have been established through publicly funded efforts. Switchgrass is cross-pollinated with limited selfing through the existence of pre- and postfertilization gametophytic incompatibility systems, as found in other grasses (Talbert et al., 1983; Martínez-Reyna and Vogel, 2002). Generally, the amount of seeds set by self-pollination is under 1% (Talbert et al., 1983; Martínez-Reyna and Vogel, 2002). However, the reduced complexity of selfed progeny has prompted the establishment of populations of self-pollinated siblings in addition to cross-pollinated populations (Table 2.1; see the references cited therein).

Based on the analysis of the progeny of two of the mapping populations, four genetic maps have been published for switchgrass (Missaoui et al., 2005b; Okada et al., 2010). Assembly of these maps relied on the so-called "single-dose" alleles that segregate 1:1 in the progeny of a heterozygous parent. By relying on markers that are present on half of the homologous chromosomes and thus typically show simple (1:1) segregation similar to that for a diploid organism, this method overcomes many of the challenges faced using early methods for genetic map development in polyploids (Wu et al., 1992).

Missaoui et al. (2005b) assembled the first genetic maps for switchgrass based on segregation of 224 single-dose RFLP markers in 85 full-sib progeny of a cross between the following tetraploids: lowland Alamo (AP13) and upland Summer (VS16). Due to the limited size of that study, it resulted in only partial maps consisting of 11 and 16 linkage groups for the female and male maps, respectively, for 8 combined homology groups. This work also detected clear evidence of preferential pairing among homeologous chromosomes. Researchers at the Noble Foundation in Oklahoma have further expanded this population and the number of markers examined (Table 2.1; M. Saha and J. Bouton, personal communication).

More recently, Okada et al. (2010) reported complete tetraploid switchgrass maps based on the analysis of 1138 single-dose alleles from SSR and STS markers in 238 full-sib progeny of a cross between lowland Kanlow (K5) and lowland Alamo (A4). These maps each consist of 18 linkage groups arranged into 9 homeologous pairs, consistent with physical evidence of the switchgrass genome structure (Martinez-Reyna et al., 2001). Marker segregation in the progeny provided strong evidence that lowland switchgrass is an allotetraploid that shows disomic inheritance of its chromosome and has substantial subgenome differentiation. Such a phenomenon may allow for the evolution of genes with altered function between the two switchgrass genomes and is also consistent with the reduced vigor of diploid switchgrass genotypes (Young et al., 2010).

The assembled genetic maps provide resources for genetic improvement of switchgrass and, specifically, identification of QTLs. The two sets of maps described above are being integrated through genotyping of a cross of $F_1$ progeny from each of the two initial crosses (Table 2.1; C. Tobias and M. Saha, personal communication). This population would provide a highly heterozygous pool for trait selection. To meet the need of switchgrass cultivars adapted to diverse climatic conditions, other conventional mapping populations composed of upland ecotypes and a northern lowland accession are also being developed in Wisconsin and Oklahoma, respectively. A goal of these mapping studies is establishing diverse populations in which the same marker sets have been characterized to build up a set of markers that might

Table 2.1. Public switchgrass mapping populations as of May 2011.

| Cross Name | Female Parent | Male Parent | Number of Progeny | Type of Markers Used | Informative Markers Genotyped | References |
|---|---|---|---|---|---|---|
| NFGA Full-sib | AP13 Lowland cv. Alamo | VS16 Upland cv. Summer | 86; 251 180 genotyped to date | RFLP; SSR, and STS | 224; ~1000 | Missaoui et al. (2005b), M. Saha and J. Bouton, personal communication |
| ALB Full-sib | K5 Lowland cv. Kanlow | A4 Lowland cv. Alamo | 238 | SSR and STS | 637 (2093 amplicons) | Okada et al. (2010) |
| ALNF Full-sib | Pv281 (F$_1$ from ALB population) | NF472 (F$_1$ from NFGA population) | 251 | SSR and STS | 516 | C. Tobias and M. Saha, personal communication |
| WS4U[a] Full-sib | U518 Upland | U418 Upland | 75 | SNPs, genotyping by sequencing | 10K (~100K expected) | M.D. Casler and E. Buckler, personal communication |
| WS4U[a] Selfed | U518 Upland | U518 Upland | 150 | SNPs, genotyping by sequencing | 10K (~100K expected) | M.D. Casler and E. Buckler, personal communication |
| OSU NL Selfed | NL 94 LYE 16 × 13 | NL 94 LYE 16 × 13 | 132 7 genotyped to date | SSR | 423 | L. Liu, Y. Wang, T. Samuels, and Y. Wu, unpublished |

All parents are tetraploid.
[a]Casler et al. (2006).

be expected to segregate informatively in new populations, so that markers do not need to be tested in every population prior to productive mapping.

In addition to permitting marker-assisted selection of traits of interest, genetic maps based on nonrepetitive sequences, such as STSs and EST-based SSRs, establish long-range connectivity among genomic fragments that can be used to help assemble a physical genome through the course of the Switchgrass Genome Project which is focused on the AP13 genotype (see Chapter 3 for further discussion). The very high resolution physical map that the switchgrass genome will eventually represent will greatly aid switchgrass genetic improvement by facilitating molecular analysis of switchgrass gene function and the discrimination of alleles from homologs among replicated genes.

### 2.7.3 Association Mapping

Unlike conventional mapping, association mapping does not require a population with a defined structure, for example, the result of a cross between particular parents instead of a broader panel of diversity may be examined for the trait of interest. Thus, association mapping can increase the number of alleles accessible for crop improvement (Myles et al., 2009). As is desirable for association mapping, switchgrass shows high linkage disequilibrium decay (Okada et al., 2010). Though no association studies have yet been reported for switchgrass, such analyses are now underway.

Two approaches to association mapping will likely be taken. The more powerful but more costly approach, whole genome association (WGA) analysis requires large numbers of markers, on the order of tens of thousands of SNPs or hundreds of SSRs, to informatively examine the correlation between polymorphisms distributed across the genome and traits of interest. Buckler et al. (2009) have recently reported a WGA for maize flowering time. Association studies may also be conducted with a smaller number of markers that report on only portions of the genome. Genomic regions examined may be randomly chosen based on marker availability. For example, such a study has recently been reported for tetraploid alfalfa, in which significant associations were detected between biomass yield and a set of SSR markers that represent ~1% of the estimated number of SSRs required to cover the whole genome (Li et al., 2011). Another option is to focus on markers in candidate genes or a specific genomic region, for example, a QTL, that are likely to have a role in the process of interest. A recent example of this approach is the identification of associations between SNPs in lignin biosynthesis coding sequences and alterations in cell wall quality in poplar (Wegrzyn et al., 2010).

In preparation for association studies in switchgrass, Dr. E.C. Brummer and associates at the University of Georgia and the Noble Foundation (Ardmore, OK) have established a diverse panel of 36 switchgrass accessions, represented by ~500 individual plants, derived from the southern United States, between Florida and New Mexico. This collection consists of 29 accessions from the National Plant Germplasm System and 7 wild populations collected in the southeastern United States, including upland, lowland, and "intermediate" ecotypes that are primarily tetraploids. SSR markers are being used to assess the genetic structure of the collection, knowledge essential for association mapping (Myles et al., 2009). Intensive phenotypic analysis for various bioenergy-related traits is underway (E.C. Brummer and M. Saha, personal communication). Drs. M. Casler, E. Buckler, and associates have also assembled and begun phenotypic analysis of another diverse switchgrass association panel from materials collected from the northern United States (http://www.maizegenetics.net/snp-discovery-in-switchgrass; Costich et al., 2010).

## 2.7.4 Transgenic Approaches

Another approach to improving switchgrass is through transgenic means, although release of transgenic plants in the field, that is, deregulation, is yet to occur. Transgenic approaches have been advanced both for basic understanding of the species and as a means to improve its energy crop properties (Sticklen, 2008). Several groups have modified switchgrass using transgenes or through gene-silencing. It was demonstrated that switchgrass can be transformed with Agrobacterium (Somleva et al., 2002) and later the method was used to create switchgrass that produces polyhydroxybutyrate (Somleva et al., 2008). A number of papers attempting to improve transformation efficiency have recently appeared (Li et al., 2009; Li and Qu, 2009, 2011; Chen et al., 2010). Other researchers have shown that downregulation of key lignin biosynthesis genes can improve the saccharification efficiency of transformed switchgrass (Fu et al., 2011a, 2011b; Saathoff et al., 2011). Improvement of switchgrass using conventional or biotechnological means could be used to selectively change important biomass traits either for conversion into fuels or as a feedstock for green chemicals, although a thorough field evaluation will be needed to ensure survival and fitness components. Indeed, transgenic approaches may be one means of circumventing some of the anticipated trade-offs between determinants of biomass yield and quality (Figure 2.2). Readers are referred to other chapters in this volume for additional discussion of transgenic approaches as applied to switchgrass improvement and biofuel crops in general.

## 2.8 Conclusions and Future Directions

Switchgrass improvement through conventional or biotechnological means, specifically for the bioenergy sector, appears to be entering a rapid period of growth. Numerous important genomic and functional genomic resources are available or are on the cusp of becoming available, for example, a draft switchgrass genome. These resources combined with robust breeding programs distributed across universities, government, and industry can be expected to accelerate the use of marker-assisted and GWA approaches. However, several key hurdles remain. One is to improve biomass quality parameters to dovetail with specific conversion platforms. Others deal with developing molecular and biochemical insights into plant growth and development as responses to biotic and abiotic stresses. Substantial headway toward these goals will require continued funding in a resource-diminished environment and may need intensification of collaborations to solve multiple problems, using available elite germplasm. Despite these potential hurdles, the outlook is promising and the abundance of recent reports on switchgrass attests to the human and other resources being applied toward these end goals.

Biomass yield will continue to be a major focus in the foreseeable future. A constraint in this quest will be improving nutrient use by this crop that will allow for increased yields while minimizing inputs use, especially nitrogen and minerals. Research results suggest that annual application of N at 80–150 lb acre$^{-1}$ (dependent on site) will be necessary to maintain stand yields (Mitchell et al., 2008). Two recent reports on mineral concentrations and recycling in switchgrass (El-Nashaar et al., 2009; Yang et al., 2009) have shown that exploitable diversity exists within the tested germplasm. Continued basic and applied research is warranted in these areas. Switchgrass remobilizes portions of its areal N and other nutrients to the below-ground parts during its growth cycle. Maximizing remobilization of these components prior to harvest appears to be important from both a sustainability and conversion perspective (Heaton et al., 2009; Heggenstaller et al., 2009; Propheter and Staggenborg, 2010; Shahandeh et al., 2011).

This will generally entail one harvest per season, at least in more temperate regions of the United States that will be taken at a late senescent stage of the plant growth cycle.

Since biomass quality for conversion to ethanol is optimal around the boot stage of harvest (Dien et al., 2006) and quality deteriorates in a standing crop over winter (Adler et al., 2006), maintaining quality through breeding becomes imperative. Quality issues could be equally important depending on the conversion platform chosen. Much of the current research has been focused toward making switchgrass biomass more compatible with biochemical conversion processes. As long as there is pressure to utilize liquid transportation fuels from energy crops, understanding and manipulating cell wall architecture, composition, and susceptibility to different pretreatments will act as a guide in breeding programs.

Other key traits concern responses to biotic and abiotic stresses; although switchgrass has strong resistance to a number of stresses, changing the plant through breeding and selection for specific traits are likely to influence crop adaptations. These could include a loss in fitness associated with lowered lignin levels (Casler et al., 2002; Vogel et al., 2002) through changes in resistance to emerging pests, pathogen, and climate changes (Gustafson et al., 2003; Crouch et al., 2009; Grassini et al., 2009; Prasifka et al., 2009, Prasifka et al., 2010; Johnson et al., 2010; McIsaac et al., 2010; Kiniry et al., 2012).

In the short term, switchgrass improvement will continue to come from recurrent selection in collections of superior germplasms; however, this scenario is in the process of changing as molecular resources for marker-assisted selection become available. Additional game-changing technological and policy changes could have large consequences on the future genetic improvement of this key energy crop. These include the following: changes in the regulatory system for transgenic plants that would allow large-scale deployment of a cross-breeding fertile GMO feedstock (Moon et al., 2010); deployment of highly effective controlled sterility technologies to prevent gene flow among switchgrass plants (Kausch et al., 2010); and/or development of precise means for site-directed alteration of plant genomes (Osakabe et al., 2010).

## References

Adler PR, Sanderson MA, Boateng AA, Weimer PI, Jung HJG. Biomass yield and biofuel quality of switchgrass harvested in fall or spring. Agron J, 2006; 98: 1518–1525.

Akhunov E, Nicolet C, Dvorak J. Single nucleotide polymorphism genotyping in polyploid wheat with the Illumina GoldenGate assay. Theor Appl Genet, 2009; 119: 507–517.

Alderson J, Sharp WC. Grass varieties in the United States. In: USCS (ed.) *Agricultural Handbook*, Vol. 170, 1995, Washington, DC.

Aliscioni SS, Giussani LM, Zuloaga FO, Kellogg EA. A molecular phylogeny of Panicum (Poaceae: Paniceae): tests of monophyly and phylogenetic placement within the Panicoideae. Am J Bot, 2003; 90: 796–821.

Bagge M, Lübberstedt T. Functional markers in wheat: technical and economic aspects. Mol Breed, 2008; 22: 319–328.

Barbazuk WB, Emrich SJ, Chen HD, Li L, Schnable PS. SNP discovery via 454 transcriptome sequencing. Plant J, 2007; 51: 910–918.

Bartley L, Ronald PC. Plant and microbial research seeks biofuel production from lignocellulose. Calif Agric, 2009; 63: 178–184.

Bhandari HS, Saha MC, Mascia PN, Fasoula VA, Bouton JH. Variation among half-sib families and heritability for biomass yield and other traits in lowland switchgrass (Panicum virgatum L.). Crop Sci, 2010; 50: 2355–2363.

Boe A. Genetic and environmental effects on seed weight and seed yield in switchgrass. Crop Sci, 2003; 43: 63–67.

Boe A, Lee DK. Genetic variation for biomass production in prairie cordgrass and switchgrass. Crop Sci, 2007; 47: 929–934.

Bouton JH. Molecular breeding of switchgrass for use as a biofuel crop. Curr Opin Genet Dev, 2007; 17: 553–558.

Bridgwater AV, Peacocke GVC. Fast pyrolysis processes for biomass. Renew Sust Energ Rev, 2000; 4: 1–73.

Buckler ES, Holland JB, Bradbury PJ, Acharya CB, Brown PJ, Browne C, Ersoz E, Flint-Garcia S, Garcia A, Glaubitz JC, Goodman MM, Harjes C, Guill K, Kroon DE, Larsson S, Lepak NK, Li H, Mitchell SE, Pressoir G, Peiffer JA, Rosas MO, Rocheford TR, Romay MC, Romero S, Salvo S, Sanchez Villeda H, da Silva HS, Sun Q, Tian F, Upadyayula N, Ware D, Yates H, Yu J, Zhang Z, Kresovich S, McMullen MD. The genetic architecture of maize flowering time. Science, 2009; 325: 714–718.

Buggs RJ, Chamala S, Wu W, Gao L, May GD, Schnable PS, Soltis DE, Soltis PS, Barbazuk WB. Characterization of duplicate gene evolution in the recent natural allopolyploid Tragopogon miscellus by next-generation sequencing and Sequenom iPLEX MassARRAY genotyping. Mol Ecol, 2010; 19(suppl 1): 132–146.

Bundock PC, Eliott FG, Ablett G, Benson AD, Casu RE, Aitken KS, Henry RJ. Targeted single nucleotide polymorphism (SNP) discovery in a highly polyploid plant species using 454 sequencing. Plant Biotechnol J, 2009; 7: 347–354.

Burns JC, Godshalk EB, Timothy DH. Registration of 'BoMaster' switchgrass. J Plant Reg, 2008a; 2: 31–32.

Burns JC, Godshalk EB, Timothy DH. Registration of 'Performer' switchgrass. J Plant Reg, 2008b; 2: 29–30.

Burns JC, Godshalk EB, Timothy DH. Registration of 'Colony' lowland switchgrass. J Plant Reg, 2010; 4: 189–194.

Burson BL, Tischler CR, Ocumpaugh WR. Breeding for reduced post-harvest seed dormancy in switchgrass: registration of TEM-LoDorm switchgrass germplasm. J Plant Reg, 2009; 3: 99–103.

Casler MD. Ecotypic variation among switchgrass populations from the northern USA. Crop Sci, 2005; 45: 388–398.

Casler MD. Changes in mean and genetic variance during two cycles of within-family selection in switchgrass. Bioenerg Res, 2010; 3: 47–54.

Casler MD, Buxton DR, Vogel KP. Genetic modification of lignin concentration affects fitness of perennial herbaceous plants. Theor Appl Genet, 2002; 104: 127–131.

Casler MD, Stendal CA, Kapich L, Vogel KP. Genetic diversity, plant adaptation regions, and gene pools for switchgrass. Crop Sci, 2007a; 47: 2261–2273.

Casler MD, Tobias CM, Kaeppler SM, Buell CR, Wang Z-Y, Cao P, Schmutz J, Ronald P. The switchgrass genome: tools and strategies. Plant Gen, 2011; 4: 273–282.

Casler MD, Vogel KP, Beal AC. Registration of WS4U and WS8U switchgrass germplasms. Crop Sci, 2006; 46: 998–999.

Casler MD, Vogel KP, Taliaferro CM, Ehlke NJ, Berdahl JD, Brummer EC, Kallenbach RL, West CP, Mitchell RB. Latitudinal and longitudinal adaptation of switchgrass populations. Crop Sci, 2007b; 47: 2249–2260.

Chen X, Equi R, Baxter H, Berk K, Han J, Agarwal S, Zale J. A high-throughput transient gene expression system for switchgrass (Panicum virgatum L.) seedlings. Biotechnol Biofuels, 2010; 3: 9.

Collard BCY, Mackill DJ. Marker-assisted selection: an approach for precision plant breeding in the twenty-first century. Philos Trans R Soc Lond B Biol Sci, 2008; 363: 557–572.

Cortese L, Honig J, Miller C, Bonos S. Genetic diversity of twelve switchgrass populations using molecular and morphological markers. Bioenerg Res, 2010; 3: 262–271.

Costich DE, Friebe B, Sheehan MJ, Casler MD, Buckler ES. Genome-size variation in switchgrass (Panicum virgatum): flow cytometry and cytology reveal rampant aneuploidy. Plant Gen, 2010; 3: 130–141.

Crouch JA, Beirn LA, Cortese LM, Bonos SA, Clarke BB. Anthracnose disease of switchgrass caused by the novel fungal species Colletotrichum navitas. Mycol Res, 2009; 113: 1411–1421.

Das MK, Taliaferro CM. Genetic variability and interrelationships of seed yield and yield components in switchgrass. Euphytica, 2009; 167: 95–105.

Decker S, Brunecky R, Tucker M, Himmel M, Selig M. High-throughput screening techniques for biomass conversion. Bioenerg Res, 2009; 2: 179–192.

Dien BS, Jung H-JG, Vogel KP, Casler MD, Lamb JFS, Iten L, Mitchell RB, Sarath G. Chemical composition and response to dilute-acid pretreatment and enzymatic saccharification of alfalfa, reed canarygrass, and switchgrass. Biomass Bioenerg, 2006; 30: 880–891.

Ding C, Jin S. High-throughput methods for SNP genotyping. In: Komar AA (ed.) *Single Nucleotide Polymorphisms*, Vol. 578, 2009, pp. 245–254. Humana Press, New York.

El-Nashaar HM, Banowetz GM, Griffith SM, Casler MD, Vogel KP. Genotypic variability in mineral composition of switchgrass. Bioresour Technol, 2009; 100: 1809–1814.

Elshire RJ, Glaubitz JC, Sun Q, Poland JA, Kawamoto K, Buckler ES. A robust, simple genotyping-by-sequencing (GBS) approach for high diversity species. PLoS ONE, 2011; 6: e19379.

Escamilla-Trevino LL, Shen H, Uppalapati SR, Ray T, Tang YH, Hernandez T, Yin YB, Xu Y, Dixon RA. Switchgrass (Panicum virgatum) possesses a divergent family of cinnamoyl CoA reductases with distinct biochemical properties. New Phytol, 2010; 185: 143–155.

Fernie AR, Schauer N. Metabolomics-assisted breeding: a viable option for crop improvement? Trends Genet, 2009; 25: 39–48.

Fu C, Mielenz JR, Xiao X, Ge Y, Hamilton CY, Rodriguez Jr M, Chen F, Foston M, Ragauskas A, Bouton J, Dixon RA, Wang ZY. Genetic manipulation of lignin reduces recalcitrance and improves ethanol production from switchgrass. Proc Natl Acad Sci U S A, 2011a; 108: 3803–3808.

Fu Y, Springer NM, Gerhardt DJ, Ying K, Yeh CT, Wu W, Swanson-Wagner R, D'Ascenzo M, Millard T, Freeberg L, Aoyama N, Kitzman J, Burgess D, Richmond T, Albert TJ, Barbazuk WB, Jeddeloh JA, Schnable PS. Repeat subtraction-mediated sequence capture from a complex genome. Plant J, 2010; 62: 898–909.

Fu C, Xiao X, Xi Y, Ge Y, Chen F, Bouton J, Dixon RA, Wang ZY. Downregulation of cinnamyl alcohol dehydrogenase (CAD) leads to improved saccharification efficiency in switchgrass. Bioenerg Res, 2011b; 4: 153–164.

Fuentes RG, Taliaferro C. Biomass yield stability of switchgrass (Panicum virgatum) cultivars. In: Janick J, Whipkey A, (eds) *5th National Symposium on New Crops and New Uses: Strength in Diversity*, 2002. Atlanta, GA.

Garten CT, Wullschleger SD. Soil carbon inventories under a bioenergy crop (switchgrass): measurement limitations. J Environ Qual, 1999; 28: 1359–1365.

Giussani LM, Cota-Sanchez JH, Zuloaga FO, Kellogg EA. A molecular phylogeny of the grass subfamily Panicoideae (Poaceae) shows multiple origins of C-4 photosynthesis. Am J Bot, 2001; 88: 1993–2012.

Grassini P, Hunt E, Mitchell RB, Weiss A. Simulating switchgrass growth and development under potential and water-limiting conditions. Agron J, 2009; 101: 564–571.

Green JC, Bransby DI. Effects of seed size on germination and seedling growth of 'Alamo' switchgrass. Rangelands in a sustainable biosphere. In: *Proceedings of the Fifth International Rangeland Congress, Salt Lake City, UT, 23–28 July, 1995. Volume 1: Contributed Presentations*, 1996, pp. 183–184.

Gustafson DM, Boe A, Jin Y. Genetic variation for Puccinia emaculata infection in switchgrass. Crop Sci, 2003; 43: 755–759.

Heaton EA, Dohleman FG, Long SP. Seasonal nitrogen dynamics of Miscanthus x giganteus and Panicum virgatum. GCB Bioenerg, 2009; 1: 297–307.

Heggenstaller AH, Moore KJ, Liebman M, Anex RP. Nitrogen influences biomass and nutrient partitioning by perennial, warm-season grasses. Agron J, 2009; 101: 1363–1371.

Hopkins AA, Vogel KP, Moore KJ. Predicted and realized gains from selection for in vitro dry matter digestibility and forage yield in switchgrass. Crop Sci, 1993; 33: 253–258.

Hultquist SJ, Vogel KP, Lee DJ, Arumuganathan K, Kaeppler S. Chloroplast DNA and nuclear DNA content variations among cultivars of switchgrass, Panicum virgatum L. Crop Sci, 1996; 36: 1049–1052.

Jakob K, Zhou F, Paterson AH. Genetic improvement of C4 grasses as cellulosic biofuel feedstocks. In Vitro Cell Dev Biol Plant, 2009; 45: 291–305.

Johnson MVV, Kiniry JR, Sanchez H, Polley HW, Fay PA. Comparing biomass yields of low-input high-diversity communities with managed monocultures across the central United States. Bioenerg Res, 2010; 3: 353–361.

Jones MD, Brown JG. Pollination cycles of some grasses in Oklahoma. Agron J, 1951; 43: 218–222.

Kausch AP, Hague J, Oliver M, Li Y, Daniell H, Mascia P, Watrud LS, Stewart CN. Transgenic perennial biofuel feedstocks and strategies for bioconfinement. Biofuels, 2010; 1: 163–176.

King BC, Donnelly MK, Bergstrom GC, Walker LP, Gibson DM. An optimized microplate assay system for quantitative evaluation of plant cell wall-degrading enzyme activity of fungal culture extracts. Biotechnol Bioeng, 2009; 102: 1033–1044.

Kiniry JR, Johnson MVV, Bruckerhoff SB, Kaiser JU, Cordsiemon RL, Harmel RD. Clash of the titans: comparing productivity via radiation use efficiency for two grass giants of the biofuel field. Bioenerg Res, 2012; 5: 41–48.

Li JF, Park E, von Arnim AG, Nebenfuhr A. The FAST technique: a simplified Agrobacterium-based transformation method for transient gene expression analysis in seedlings of Arabidopsis and other plant species. Plant Methods, 2009; 5: 6.

Li R, Qu R. High throughput Agrobacterium-mediated switchgrass transformation. Biomass Bioenerg, 2011; 35: 1046–1054.

Li RY, Qu RD. High throughput transformation of switchgrass (Panicum virgatum) and its application for lignin reduction. In Vitro Cell Dev Biol Anim, 2009; 45: S81–S81.

Li X, Wei Y, Moore K, Michaud R, Viands D, Hansen J, Acharya A, Brummer E. Association mapping of biomass yield and stem composition in a tetraploid alfalfa breeding population. Plant Genome, 2011; 4: 24.

Li Y-D, Chu Z-Z, Liu X-G, Jing H-C, Liu Y-G, Hao D-Y. A cost-effective high-resolution melting approach using the EvaGreen dye for DNA polymorphism detection and genotyping in plants. J Integr Plant Biol, 2010; 52: 1036–1042.

Liebig M, Schmer M, Vogel K, Mitchell R. Soil carbon storage by switchgrass grown for bioenergy. Bioenerg Res, 2008; 1: 215–222.

Liu S, Chen HD, Makarevitch I, Shirmer R, Emrich SJ, Dietrich CR, Barbazuk WB, Springer NM, Schnable PS. High-throughput genetic mapping of mutants via quantitative single nucleotide polymorphism typing. Genetics, 2010; 184: 19–26.

Lu K, Kaeppler SM, Vogel KP, Arumuganathan K, Lee DJ. Nuclear DNA content and chromosome numbers in switchgrass. Great Plains Res, 1998; 8: 269–280.

Ma XF, Bouton J, Saha M, Narasimhamoorthy B, Russell T, Hernandez A, Anusauskiene L, Zapata R, Zheveleva I, Brover S, Swaller TJ. Perennial bio-energy crop, switchgrass: genetic linkage map based on high-throughput SNP and SSR genotyping. In: *Plant & Animal Genomes XVI Conference*, San Diego, CA, 2008.

Mann D, Labbé N, Sykes R, Gracom K, Kline L, Swamidoss I, Burris J, Davis M, Stewart C. Rapid assessment of lignin content and structure in switchgrass (Panicum virgatum L.) grown under different environmental conditions. Bioenerg Res, 2009; 2: 246–256.

Martínez-Reyna JM, Vogel KP. Incompatibility systems in switchgrass. Crop Sci, 2002; 42: 1800–1805.

Martinez-Reyna JM, Vogel KP. Heterosis in switchgrass: spaced plants. Crop Sci, 2008; 48: 1312–1320.

Martinez-Reyna JM, Vogel KP, Caha C, Lee DJ. Meiotic stability, chloroplast DNA polymorphisms, and morphological traits of upland × lowland switchgrass reciprocal hybrids. Crop Sci, 2001; 41: 1579–1583.

Matts J, Jagadeeswaran G, Roe BA, Sunkar R. Identification of microRNAs and their targets in switchgrass, a model biofuel plant species. J Plant Physiol, 2010; 167: 896–904.

McIsaac GF, David MB, Mitchell CA. Miscanthus and switchgrass production in central Illinois: impacts on hydrology and inorganic nitrogen leaching. J Environ Qual, 2010; 39: 1790–1799.

McLaughlin SB, Adams Kszos L. Development of switchgrass (Panicum virgatum) as a bioenergy feedstock in the United States. Biomass Bioenerg, 2005; 28: 515–535.

McMillan C. The role of ecotypic variation in the distribution of the central grassland of North America. Ecol Monogr, 1959; 29: 285–308.

Michelmore RW, Paran I, Kesseli RV. Identification of markers linked to disease-resistance genes by bulked segregant analysis: a rapid method to detect markers in specific genomic regions by using segregating populations. Proc Natl Acad Sci U S A, 1991; 88: 9828–9832.

Missaoui A, Fasoula V, Bouton J. The effect of low plant density on response to selection for biomass production in switchgrass. Euphytica, 2005a; 142: 1–12.

Missaoui AM, Paterson AH, Bouton JH. Investigation of genomic organization in switchgrass (Panicum virgatum L.) using DNA markers. Theor Appl Genet, 2005b; 110: 1372–1383.

Missaoui AM, Paterson AH, Bouton JH. Molecular markers for the classification of switchgrass (*Panicum virgatum* L.) germplasm and to assess genetic diversity in three synthetic switchgrass populations. Genet Resour Crop Evol, 2006; 53: 1291–1302.

Mitchell R, Vogel KP, Sarath G. Managing and enhancing switchgrass as a bioenergy feedstock. Biofuel Bioprod Bior, 2008; 2: 530–539.

Moon HS, Abercrombie JM, Kausch AP, Stewart Jr CN. Sustainable use of biotechnology for bioenergy feedstocks. Environ Manage, 2010; 46: 531–538.

Myles S, Peiffer J, Brown PJ, Ersoz ES, Zhang ZW, Costich DE, Buckler ES. Association mapping: critical considerations shift from genotyping to experimental design. Plant Cell, 2009; 21: 2194–2202.

Narasimhamoorthy B, Saha M, Swaller T, Bouton J. Genetic diversity in switchgrass collections assessed by EST-SSR markers. Bioenerg Res, 2008; 1: 136–146.

Newell, LC, Eberhart SA. Clone and progeny evaluation in the improvement of switchgrass, Panicum Virgatum L. Crop Sci, 1961; 1: 117–121.

Okada M, Lanzatella C, Saha MC, Bouton J, Wu R, Tobias CM. Complete switchgrass genetic maps reveal subgenome collinearity, preferential pairing, and multilocus interactions. Genetics, 2010; 185: 745–760.

Oliver RE, Lazo GR, Lutz JD, Rubenfield MJ, Tinker NA, Anderson JM, Wisniewski Morehead NH, Adhikary D, Jellen EN, Maughan PJ, Brown Guedira GL, Chao S, Beattie AD, Carson ML, Rines HW, Obert DE, Bonman JM, Jackson EW. Model SNP development for complex genomes based on hexaploid oat using high-throughput 454 sequencing technology. BMC Genomics, 2011; 12: 77.

Osakabe K, Osakabe Y, Toki S. Site-directed mutagenesis in Arabidopsis using custom-designed zinc finger nucleases. Proc Natl Acad Sci, 2010; 107: 12034–12039.

Parrish DJ, Fike JH. The biology and agronomy of switchgrass for biofuels. Crit Rev Plant Sci, 2005; 24: 423–459.

Parrish DJ, Fike JH. Selecting, establishing, and managing switchgrass (Panicum virgatum) for biofuels. Methods Mol Biol, 2009; 581: 27–40.

Perlack RD, Wright LL, Turhollow A, Graham RL, Stokes BJ, Erbach DC. Biomass as feedstock for a bioenergy and bioproducts industry: the technical feasibility of a billion-ton annual supply, 2005. Oak Ridge National Laboratory, Oak Ridge, TN.

Perrin R, Vogel K, Schmer M, Mitchell R. Farm-scale production cost of switchgrass for biomass. Bioenerg Res, 2008; 1: 91–97.

Prasifka JR, Bradshaw JD, Boe AA, Lee D, Adamski D, Gray ME. Symptoms, distribution and abundance of the stem-boring caterpillar, Blastobasis repartella (Dietz), in switchgrass. Bioenerg Res, 2010; 3: 238–242.

Prasifka JR, Bradshaw JD, Meagher RL, Nagoshi RN, Steffey KL, Gray ME. Development and feeding of fall armyworm on Miscanthus × giganteus and switchgrass. J Econ Entomol, 2009; 102: 2154–2159.

Propheter JL, Staggenborg S. Performance of annual and perennial biofuel crops: nutrient removal during the first two years. Agron J, 2010; 102: 798–805.

Rose LW, Das MK, Fuentes RG, Taliaferro CM. Effects of high- vs. low-yield environments on selection for increased biomass yield in switchgrass. Euphytica, 2007; 156: 407–415.

Rose LW, Das MK, Taliaferro CM. Estimation of genetic variability and heritability for biofuel feedstock yield in several populations of switchgrass. Ann Appl Biol, 2008; 152: 11–17.

Saathoff AJ, Sarath G, Chow EK, Dien BS, Tobias CM. Downregulation of cinnamyl-alcohol dehydrogenase in switchgrass by RNA silencing results in enhanced glucose release after cellulase treatment. PLoS ONE, 2011a; 6: e16416.

Saathoff AJ, Tobias CM, Sattler SE, Haas EJ, Twigg P, Sarath G. Switchgrass contains two cinnamyl alcohol dehydrogenases involved in lignin formation. Bioenerg Res, 2011b; 4: 120–133.

Santoro N, Cantu S, Tornqvist C-E, Falbel T, Bolivar J, Patterson S, Pauly M, Walton J. A high-throughput platform for screening milligram quantities of plant biomass for lignocellulose digestibility. Bioenerg Res, 2010; 3: 93–102.

Sarath G, Baird LM, Vogel KP, Mitchell RB. Internode structure and cell wall composition in maturing tillers of switchgrass (Panicum virgatum. L). Bioresour Technol, 2007a; 98: 2985–2992.

Sarath G, Bethke PC, Jones R, Baird LM, Hou G, Mitchell RB. Nitric oxide accelerates seed germination in warm-season grasses. Planta, 2006; 223: 1154–1164.

Sarath G, Hou G, Baird LM, Mitchell RB. Reactive oxygen species, ABA and nitric oxide interactions on the germination of warm-season C4-grasses. Planta, 2007b; 226: 697–708.

Sarath G, Mitchell R. Aged switchgrass seed lot's response to dormancy-breaking chemicals. Seed Technol, 2008; 30: 7–16.

Sarath G, Mitchell R, Sattler S, Funnell D, Pedersen J, Graybosch R, Vogel K. Opportunities and roadblocks in utilizing forages and small grains for liquid fuels. J Ind Microbiol Biotechnol, 2008; 35: 343–354.

Schmer MR, Vogel KP, Mitchell RB, Dien BS, Jung HG, Casler MD. Temporal and spatial variation in switchgrass biomass composition and theoretical ethanol yield. Agron J, 2012; 104: 54–64.

Schmer MR, Vogel KP, Mitchell RB, Moser LE, Eskridge KM, Perrin RK. Establishment stand thresholds for switchgrass grown as a bioenergy crop. Crop Sci, 2006; 46: 157–161.

Schmer MR, Vogel KP, Mitchell RB, Perrin RK. Net energy of cellulosic ethanol from switchgrass. Proc Natl Acad Sci U S A, 2008; 105: 464–469.

Shahandeh H, Chou CY, Hons FM, Hussey MA. Nutrient partitioning and carbon and nitrogen mineralization of switchgrass plant parts. Commun Soil Sci Plant Anal, 2011; 42: 599–615.

Shen H, Fu C, Xiao X, Ray T, Tang Y, Wang Z, Chen F. Developmental control of lignification in stems of lowland switchgrass variety Alamo and the effects on saccharification efficiency. Bioenerg Res, 2009a; 2: 233–245.

Shen H, Yin Y, Chen F, Xu Y, Dixon R. A bioinformatic analysis of NAC genes for plant cell wall development in relation to lignocellulosic bioenergy production. Bioenerg Res, 2009b; 2: 217–232.

Shen ZX, Parrish DJ, Wolf DD, Welbaum GE. Stratification in switchgrass seeds is reversed and hastened by drying. Crop Sci, 2001; 41: 1546–1551.

Smart AJ, Moser LE. Switchgrass seedling development as affected by seed size. Agron J, 1999; 91: 335–338.

Somerville C. Biofuels. Curr Biol, 2007; 17: R115–R119.

Somleva MN, Snell KD, Beaulieu JJ, Peoples OP, Garrison BR, Patterson NA. Production of polyhydroxybutyrate in switchgrass, a value-added co-product in an important lignocellulosic biomass crop. Plant Biotechnol J, 2008; 6: 663–678.

Somleva MN, Tomaszewski Z, Conger BV. Agrobacterium-mediated genetic transformation of switchgrass. Crop Sci, 2002; 42: 2080–2087.

Srivastava AC, Palanichelvam K, Ma J, Steele J, Blancaflor EB, Tang Y. Collection and analysis of expressed sequence tags derived from laser capture microdissected switchgrass (Panicum virgatum L. Alamo) vascular tissues. Bioenerg Res, 2010; 3: 278–294.

Sticklen MB. Plant genetic engineering for biofuel production: towards affordable cellulosic ethanol. Nat Rev Genet, 2008; 9: 433–443.

Talbert LE, Timothy DH, Burns JC, Rawlings JO, Moll RH. Estimates of genetic parameters in switchgrass1. Crop Sci, 1983; 23: 725–728.

Taliaferro C, Vogel KP. Reproductive characteristics and breeding improvement potential of switchgrass. In: *The 4th Biomass Conference of Americas*, Oakland, CA, 1999.

Thomason WE, Ruan WR, Johnson GV, Taliaferro C, Freeman KW, Wynn KJ, Mullen RW. Switchgrass response to harvest frequency and time and rate of applied nitrogen. J Plant Nutrition, 2004; 27: 1199–1226.

Tobias C. Switchgrass genomics. In: Saha MC, Bhandri HS, Bouton JH (ed.) *Bioenergy Feedstocks: Breeding and Genetics*, Wiley, USA, 2013; 33–48.

Tobias CM, Hayden DM, Twigg P, Sarath G. Genic microsatellite markers derived from EST sequences of switchgrass (*Panicum virgatum* L.). Mol Ecol Notes, 2006; 6: 185–187.

Tobias CM, Sarath G, Twigg P, Lindquist E, Pangilinan J, Penning BW, Barry K, McCann MC, Carpita NC, Lazo GR. Comparative genomics in switchgrass using 61,585 high-quality expressed sequence tags. Plant Genome, 2008; 1: 111–124.

Todd J, Wu Y, Wang Z, Samuals T. Genetic diversity in tetraploid switchgrass revealed by AFLP marker polymorphisms. Genet Mol Res, 2011; 10: 2976–2986.

USDA ARS. National Genetic Resources Program. *Germplasm Resources Information Network - (GRIN)*, 2011. National Germplasm Resources Laboratory, Beltsville, MD.

Vega-Sanchez ME, Ronald PC. Genetic and biotechnological approaches for biofuel crop improvement. Curr Opin Biotechnol, 2010; 21: 218–224.

Vogel K. Switchgrass. In: Moser LE, Sollenberger L, Burson BL (eds) *Warm-Season (C4) Grasses*, 2004. ASA-CSSA-SSSA, Madison, WI.

Vogel KP. Improving warm-season forage grasses using selection, breeding, and biotechnology. In: Moore KJ, Andersen BE (eds) *Native Warm-Season Grasses: Research Trends and Issues*, 2000, CSSA Special Publication Number 30. CSSA, ASA, Madison, WI.

Vogel KP, Hopkins AA, Moore KJ, Johnson KD, Carlson IT. Winter survival in switchgrass populations bred for high IVDMD. Crop Sci, 2002; 42: 1857–1862.

Vogel KP, Mitchell KB. Heterosis in switchgrass: biomass yield in swards. Crop Sci, 2008; 48: 2159–2164.

Vogel KP, Pedersen JF. Breeding systems for cross-pollinated perennial grasses. Plant Breed Rev, 1993; 11: 251–274.

Vogel KP, Sarath G, Saathoff AJ, Mitchell RB. Switchgrass. In: Halford NG, Karp A (eds) *Energy Crops*, 2010, pp. 341–379. RSC Publishing, Cambridge, UK.

Wang Y, Samuels T, Wu Y. Development of 1,030 genomic SSR markers in switchgrass. Theor Appl Genet, 2011; 122: 677–686.

Wegrzyn JL, Eckert AJ, Choi M, Lee JM, Stanton BJ, Sykes R, Davis MF, Tsai CJ, Neale DB. Association genetics of traits controlling lignin and cellulose biosynthesis in black cottonwood (Populus trichocarpa, Salicaceae) secondary xylem. New Phytol, 2010; 188: 515–532.

Wu KK, Burnquist W, Sorrells ME, Tew TL, Moore PH, Tanksley SD. The detection and estimation of linkage in polyploids using single-dose restriction fragments. Theor Appl Genet, 1992; 83: 294–300.

Wu Y, Taliaferro C. 'Cimarron' switchgrass: a new cultivar for bioenergy feedstock production. In: *ASA-CSSA-SSSA Annual Meeting*, Pittsburgh, PA, 2009.

Yang J, Worley E, Wang M, Lahner B, Salt D, Saha M, Udvardi M. Natural variation for nutrient use and remobilization efficiencies in switchgrass. Bioenerg Res, 2009; 2: 257–266.

Young H, Hernlem B, Anderton A, Lanzatella C, Tobias C. Dihaploid stocks of switchgrass isolated by a screening approach. Bioenerg Res, 2010; 3: 305–313.

Zalapa JE, Price DL, Kaeppler SM, Tobias CM, Okada M, Casler MD. Hierarchical classification of switchgrass genotypes using SSR and chloroplast sequences: ecotypes, ploidies, gene pools, and cultivars. Theor Appl Genet, 2011; 122: 805–817.

Zhang YW, Zalapa J, Jakubowski AR, Price DL, Acharya A, Wei YL, Brummer EC, Kaeppler SM, Casler MD. Natural hybrids and gene flow between upland and lowland switchgrass. Crop Sci, 2011; 51: 2626–2641.

# Chapter 3
# Switchgrass Genomics

Christian M. Tobias

*USDA-ARS, Western Regional Research Center, 800 Buchanan Street, Albany, CA 94710, USA*

## 3.1 Introduction

We have entered into an era of sequencing "commoditization" where reductions in the cost of acquiring sequence data and increased sequencing capacity have placed premium value on the ability to process, interpret, and draw appropriate conclusions from these large data sets. In the field of plant biology, fundamental and applied research will benefit from working efficiently within this environment. Through appropriate use of model species we will increase our fundamental understanding of biological processes through genetic, molecular, and biochemical studies. However, translating these findings into improved crop varieties is a lengthy process that few organizations have the resources to pursue. This is certainly the case with most species being viewed for purposes of bioenergy which are new crops still in the process of domestication. How such nonmodel species will benefit from inexpensive sequence data is a very current question. In many ways these species will continue to leverage research and techniques that have been successfully pioneered in human medicine or other major crops first as these receive far greater research support. However, there will always be interesting research questions that cannot be addressed well in either model systems or other major crop species, particularly with long-generation perennials.

In this chapter on switchgrass I attempt to review the state of our current understanding of its genome structure, composition, and organization. I will also reflect on how particulars of its biology impact ways in which genomic resources can be productively applied. In the very near future high-throughput sequencing projects that are already initiated will likely broaden the types of questions that can be addressed. These will require demanding new bioinformatics approaches to deal with the high polymorphism levels found in this species, the challenges posed by polyploidy (whole genome duplication), paralogy (like sequences arising through historical duplication and dispersion events), and large volumes of sequence data. There is a good reason to believe that once resources are in place to cope with these issues

---

*Bioenergy Feedstocks: Breeding and Genetics*, First Edition. Edited by Malay C. Saha, Hem S. Bhandari, and Joseph H. Bouton.
© 2013 John Wiley & Sons, Inc. Published 2013 by John Wiley & Sons, Inc.

the rate at which phenotypic data is acquired will be the most challenging remaining hurdle to predict the genotypic effects on yield, cell wall composition, interactions with pathogens, and environmental adaptation.

## 3.2 Genome Sequencing

Recently initiated projects supported by the DOE and the American Recovery and Reinvestment Act have started to produce high-quality sequence data from switchgrass using Illumina and Roche-454 sequencing technology. For draft genome assembly a tetraploid with 18 pairs of chromosomes of a lowland "Alamo" individual (AP13) has been selected that is a parent of a mapping population developed by Ali Missaoui and colleagues at the University of Georgia (Missaoui et al., 2005). De novo genome assembly in switchgrass is complicated by repetitive DNA, polyploidy, and heterozygosity and is viewed as extremely challenging. The Joint Genome Institute DOE sequencing facility in Walnut Creek, California, has been taking the lead on sequencing and assembly. Envisioning the overall genome project most groups seem to agree that an initial goal should be to attempt to resolve if possible the two subgenomes and collapse allelic variation into a single consensus sequence. This will allow separate treatment of the two subgenomes for the purposes of breeding. Optimal parameters for this are not known or whether a BAC-by-BAC, whole genome shotgun sequencing strategy, or combination of approaches ought to be used. Both of these methods have their advantages. Using a BAC-by-BAC approach would presumably simplify assembly, but the accuracy with which overlapping BACs can be identified is in question, and the process of physically ordering BAC clones is expensive. A shotgun sequencing approach using next-generation sequencing can be limited by read lengths that cannot resolve repetitive regions. However, recent technological improvements in sequencing complex genomes including paired-end technology, longer read lengths, and single-molecule sequencing approaches are making this strategy more attractive (Schatz et al., 2010). In addition to the best available sequencing technology, increased depth of sequence coverage due to heterozygosity will be required. To help resolve the individual subgenomes and haplotypes, genetic data can be incorporated for increased accuracy by determining the phase of individual DNA molecules. The population reported by Missaoui has since been expanded at the Samuel Roberts Noble foundation to over 250 $F_1$ individuals. The other parent used as the pollen donor (an upland individual: VS16) as well as the $F_1$ progeny are being utilized as a model population for genome assembly purposes. This may be accomplished by also acquiring sequence data from the $F_1$ progeny and other mapping founder in addition to AP13. However, this approach introduces additional sequences from the second founder (VS16 in this case) that must also be assembled.

An early release of the genome assembly efforts is now available (*Panicum virgatum* v0.0, DOE-JGI, http://www.phytozome.net). This assembly is based on 15x sequence coverage and represents the first attempt to build a draft genome using de novo assembly methods. Genes were annotated using the JGI annotation pipeline and 65,878 complete protein-coding sequences were found. The assembled contigs total approximately 1350 Mbp. The total number of assembled sequence contigs is 410,030, while the N50, a statistic representing the total number of scaffold sequences containing half the assembly length, was 83,229. By any standard this genome is still in the development stages and much work remains to be done to resolve repeats and to orient and order the scaffold sequences.

### 3.2.1 Other Available Sequence Resources

Availability and access to data pertaining to switchgrass other than the draft genome can best be accessed through the Biofuel Feedstock Genomics Resource (http://bfgr.plantbiology.msu.edu/index.shtml). Using publicly available sequence and comprehensive links to external resources one can find there computationally derived SNPs, transcript assemblies, links to some sequences from the National Center for Biotechnology Information (NCBI), annotations, publications, and other generally useful links. Most of the sequence data are in a relatively raw form (there is no genome browser). A Switchgrass Functional Genomics Server produced by scientists at the Samuel Roberts Noble Foundation as part of the DOE-funded Bioenergy Science Center (http://switchgrassgenomics.noble.org/index.php) provides unique transcript sequence assemblies derived from both Roche-454 and Sanger sequencing of AP13 mRNAs and an extremely valuable gene expression atlas incorporating 14 different major developmental stages of switchgrass. Genetic data for switchgrass is available for one tetraploid mapping population at Gramene (http://www.gramene.org/cmap/) using C-map. This map data can be downloaded or interacted with to perform comparisons within the Poaceae as it contains links to other species maps and genome sequences. Another site for community data is http://switchgrassgenomics.org/. This site is serving as one of the primary community-based organizational sites for switchgrass. It contains its own newsgroup, has a small but growing membership, and provides details of meetings, news, and events.

Most sequence data are also available in raw form from the NCBI and as of April 5, 2011 there were sequences comprising

- 10 Gb in total of shotgun sequences reads derived from AP13 genome acquired using the 454 GS FLX sequence platform
- 418 Mb in total derived from cDNA sequences of VS16 shoots, roots, and inflorescences acquired using the 454 GS FLX sequence platform
- 126.9 Mb in total including bases with low read quality derived from cDNA of Alamo vascular bundles and whole plant samples sequenced using 454 GS FLX
- 201 Gb in total including bases with low read quality derived from Alamo "AP13" shotgun libraries and sequenced with Illumina paired ends (2 × 114 bp)
- 6.6 Gb in total including bases with low read quality derived from cDNA of different upland populations and sequenced with Illumina
- 525, 245 EST Sanger sequences derived from the JGI community sequencing program and comprising both upland and lowland genotypes (see later)
- 195, 557 BAC end sequences derived from genotype AP13 from a BAC library produced at the Clemson University Genomics Institute

Flow cytometry indicates that the tetraploid switchgrass genome is approximately 1250–1350 Mb (Young et al., 2010). Of the AP13 sequences, if both subgenomes are considered to be well differentiated this would mean that there is likely to be over 30x genome coverage already. However, these data may need to be supplemented with additional sequence technologies that can resolve repeats more effectively.

Analysis of these sequence data is possible for those that have the interest. Much characterization can obviously be done on these data sets, but one immediate task will be cataloging the classes of repeats present in switchgrass which will likely represent over half the genome based on figures cited for sorghum (Paterson et al., 2009). This is difficult given that the length of the majority of the reads is short, but as repetitive DNA accounts for the most rapidly evolving component of plant genomes, it is important to consider for purposes of analysing the extent to which it drives genome evolution.

## 3.3 Analysis of Expressed Sequences in Switchgrass

The acquisition of gene sequences from diverse mRNA-derived libraries in the form of expressed sequence tags is often the first step in molecular analysis of an uncharacterized species, and switchgrass has not been exceptional in this pattern. In a short time the number of EST resources available has gone from virtually absent in 2005 to ranking 7th among all plants and 21st among all entries in dbEST. This was due to the sequencing capacity of the DOE-Joint Genome Institute which has been responsible for the bulk of EST and genome sequencing efforts in switchgrass and has helped produce the completed genomes of sorghum, foxtail millet, *Brachypodium distachyon*, poplar, and several other species. These dbEST rankings, however, do not reflect other mRNA-seq projects using high-throughput sequencing methods which are not present in dbEST.

In 2005, the first 12,000 switchgrass ESTs were published and used for the identification of genic microsatellites and GO annotations (Tobias et al., 2005). Overall similarity was noted with other grasses and particularly sorghum. Of the four libraries sequenced which included crown, callus, leaf, and stem-derived, some genes appeared to be differentially expressed based on library representation and composition of the contigs derived from clustering all libraries. The greatest numbers of highly represented, library-specific genes were present in the leaf library representing genes encoding components of the photosynthetic apparatus, while least number of highly represented, library-specific genes were present in the crown library. The leaf and crown libraries also showed the least overlap in terms of numbers of contigs which contained ESTs from each library (0.7%). Phenylpropanoid pathway genes were most represented in the stem library.

Di- and trinucleotide EST-SSR developed from these ESTs were characterized for amplification and evaluated in terms of diversity for a small subset of individual genotypes. These markers demonstrated an ability to reliably amplify and were also shown to be highly diverse. While the average diversity of dinucleotide repeats was greater than trinucleotide repeats as has been reported elsewhere, this difference was not significant for switchgrass and may have been attributable to random variation. Most markers (65%) produced 1–2 bands per individual (Tobias et al., 2006).

These ESTs were followed by a larger-scale project where sequencing was carried out at the DOE-Joint Genome Institute. A subset of 61,585 of these ESTs was published in the Plant Genome that was derived from callus, crown, and seedling tissues (Tobias et al., 2008). Analysis of these ESTs was presented in the form of assignment of GO terms and alignment with the published sorghum genome. In total, 73% of the consensus sequences could be aligned with the sorghum genome using a threshold $E$-value of $1 \times 10^{-20}$. Evidence for genome duplication by examining the rates of synonymous codon changes in pairs of similar coding sequences was performed. Synonymous change rates were higher than pairs of similar sorghum sequences, but comparisons produced a single peak corresponding to a synonymous substitution rate of 0.1. This single peak is taken as evidence that the two subgenomes of switchgrass are quite similar overall. The method can detect ancient whole genome duplications such as have occurred in poplar and maize. In the case of switchgrass, the genome duplication is likely to be too recent to be detected in this manner, or there may be extremely biased expression of one homeolog as a

**Pairwise synonymous substitution rate**

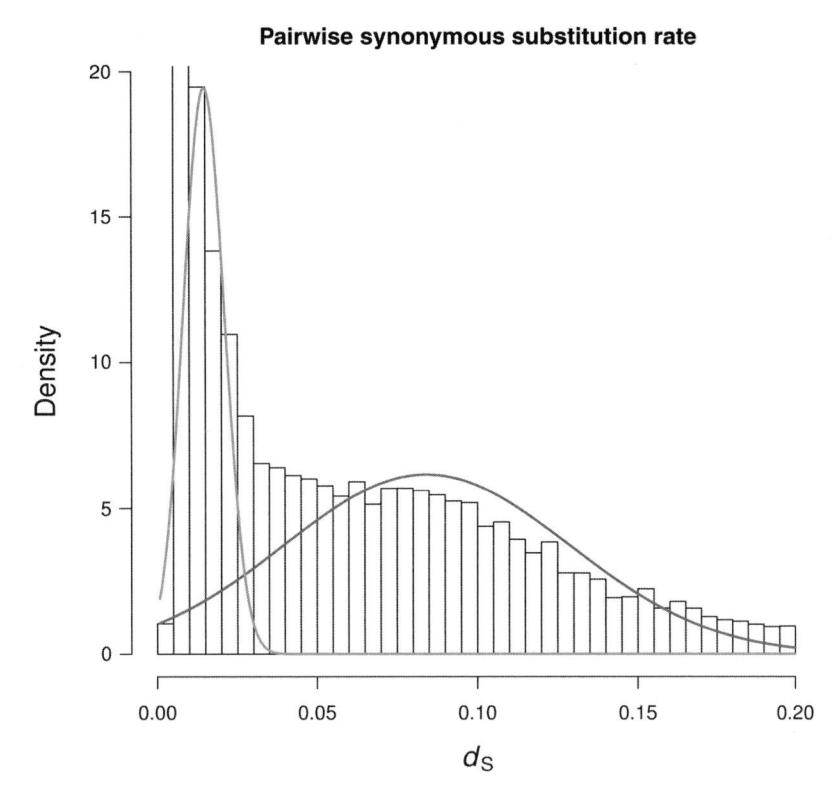

[Figure: histogram of Density vs $d_S$ ranging from 0.00 to 0.20]

**Figure 3.1.** Synonymous substitutions drawn from a single genotype EST sample. ESTs isolated from a callus tissue library CCHY derived from a single genotype were translated using prot4EST and sorghum protein sequences as a guide. EST clusters were identified using blast and then significant hits were codon aligned with ClustalW. Pairwise synonymous substitution rates were calculated by maximum likelihood using the codon substitution model. Likely allelic variation is indicated in green, whereas variation between homeologs is indicated in red.

result of genome stabilization/diploidization pressure that reduces representation of alternative sequences.

Analysis of EST pairs from a single individual's complement of ESTs represented in a library derived from callus tissue revealed a somewhat clearer picture. Here, after EST clustering, translation, and codon-based analysis of synonymous substitutions ($d_S$) a mixture of two distributions can be inferred as shown in Figure 3.1. These distribution means of 0.01 and 0.08 synonymous substitutions per synonymous site are a best guess approximation of allelic variation and subgenome variation, respectively, in switchgrass and a similar analysis in the diploid genome of sorghum did not produce two distinct peaks. The prior analysis with the clustered ESTs was likely to have missed the subtle divergence due to collapse and assembly of allelic diversity within individual EST sequences into consensus EST contigs. The same data, after clustering and aligning with related sorghum genes, were selected for tree topologies where two distinct groups of EST consensus contigs corresponded to a single sorghum gene. This approach produced 646 out of 4057 clusters which contained a switchgrass sequence as shown in Figures 3.2a–d. This was also demonstrated in the case of the switchgrass cinnamyl

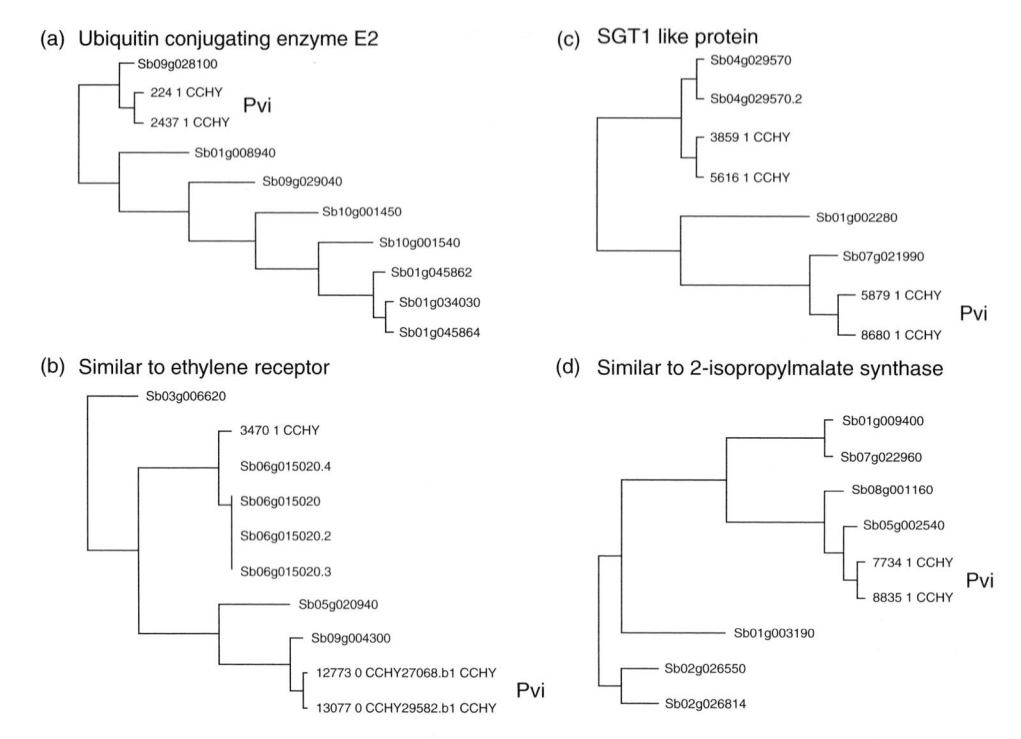

**Figure 3.2.** Dendrograms of switchgrass EST clusters similar to sorghum-coding sequences. EST sequences from a single genotype that was used to produce a callus cDNA library were grouped along with similar sorghum-coding sequences into clumps of approximately 10 genes, aligned with ClustalW and hierarchically clustered using UPGMA. Trees were identified that contained topologies where a single sorghum gene corresponded to two distinct switchgrass EST clusters or contigs. Four randomly selected trees are indicated. The relevant regions are colored red with the switchgrass sequences labeled Pvi. (a) Genes similar to ubiquitin-conjugating enzyme E2; (b) genes similar to ethylene receptor; (c) genes similar to SGT1; (d) genes similar to 2-isopropylmalate synthase.

alcohol dehydrogenase that is involved in lignification (Saathoff et al., 2010). Two sequences sharing 95% identity at the amino acid level were cloned from different tissue sources and were distinct enough to represent potential homeologs.

Detailed phylogenetic analysis of a number of gene families in switchgrass as well as transcriptional analysis of specific cell types has been performed to a limited extent. Type III peroxidase genes present in ESTs have been examined as these were well represented in libraries and produced a skewed distribution. This interesting skewed distribution was due to the overexpression of one peroxidase in callus tissue that was highly related to auxin-regulated peroxidases identified in highly embryogenic callus or upregulated by fungal infection. In total 23 full-length peroxidases were identified. Significant differential expression of other highly represented gene sets in libraries derived from different tissues has been observed. Genes involved in cell wall biosynthesis were found to be represented as a higher percentage of overall sequences in stem cDNA libraries.

A more directed sequencing approach used laser capture microdissection (LCM) to produce cDNA libraries from stem and leaf sheath vascular bundles (Srivastava et al., 2010). In all, 5734 EST sequences were generated from these cDNA libraries that represented minute quantities of tissue from vascular bundles. Five out of seven genes tested including aspartate amino transferase, copper chaperone, peroxidase 30 homolog, methionine synthase, and dirigent protein gene were preferentially expressed in the vascular bundle as verified by *in situ* hybridization. Many other sequences were identified as highly similar to ones enriched using LCM technique in other species such as maize.

MicroRNAs (miRNAs) have been recognized as important regulatory factors affecting post-transcriptional RNA processing and physiological processes such as response to environmental stress, phase transition, hormonal signaling, and polarity (Mallory and Vaucheret, 2006; Sunkar et al., 2007). In order to begin to ascertain their conserved roles in switchgrass 269 conserved miRNAs were identified through sequencing of a small RNA library in the cultivar Alamo. Of these conserved sequences the miR172 family was the most abundant appearing 90 times. These 269 sequences represented a total of 16 different conserved families of miRNAs based on secondary structure predictions and homology. Evidence for four other miRNAs was found through RNA blot analysis and analysis of primary unprocessed miRNA transcripts from EST sequence data. Expression analysis of two specific miRNAs (miR156 and miR172) in different tissues revealed opposite patterns of expression in young versus mature leaves of switchgrass that was consistent with their conserved counterparts in Arabidopsis. This supports the concept that production and accumulation of these miRNAs may be similarly regulated in both species. Other miRNAs were differentially expressed in specific tissues. Two miRNAs known to be induced under nutrient-limited conditions in Arabidopsis and *Medicago truncatula* did not appear to be upregulated in switchgrass seedlings under the stress conditions used.

Another study of switchgrass miRNAs used a comparative genome-based analysis of ESTs (Xie et al., 2010). This work identified 121 miRNAs belonging to 44 different families by sequence similarity using a Smith–Waterman alignment algorithm. Among the predicted targets of these miRNAs included 39 genes potentially involved in cellulose biosynthesis. In this study, the most abundant class of miRNA was miR444 which appeared 13 times. This family has been identified in Brachypodium and wheat and rice, but not in other plants and therefore is believed to be monocot specific. Based on target analysis miR444 was conjectured to co-participate in regulation of starch and sucrose metabolism. This study did not identify any miR172 family members perhaps due to the inherent biases of the different approaches used by the two reports.

Target genes of miRNAs can be predicted based on sequence complementarity. By analysis of existing EST sequences, some candidate target genes were identified and represent genes in the same classes of transcription factors and regulatory proteins that would be expected from work on other plants as well as some novel potential targets. These included ESTs that shared sequence similarity with transcription factor families belonging to SPL, AP2, MYB, HD-Zip, CBF, SCL, and MADS-box classes. Some other targets were identified that shared similarity to other types of regulatory proteins. A TIR1-like sequence was identified as one such target. In Arabidopsis TIR1-like proteins are thought to be involved in turnover of Aux/IAA transcriptional repressors which, in turn, coordinate auxin responsiveness. Some predicted target cleavage sites were verified in this study using 5' RACE of switchgrass mRNA (Srivastava et al., 2010). In another instance an NAC domain-containing transcription factor was shown to have complementarity to an miRNA class in its 5' UTR.

The NAC domain-containing class of transcription factors was the subject of another study that identified many of these class members from EST data in a variety of species that were likely to be involved in cell wall formation (Shen et al., 2009). This work did not produce any new sequence data but used a hidden Markov model for the conserved N terminal protein motif and existing data to identify 92 members of the family in switchgrass alone and over 1000 from 11 species which included several considered for their high biomass accumulation. NAC proteins have been demonstrated to play roles in a wide array of processes in plants and are represented among the largest families of plant transcription factors. Over 100 members have been identified in Arabidopsis, and they play roles in the development of roots and flowers, senescence (Uauy et al., 2006), tillering (Mao et al., 2007), and cell wall development and differentiation (Mitsuda et al., 2007; Zhong et al., 2008). Based on prior literature nine of these individual genes represent good candidates for involvement in cell wall modification and possibly as targets for transgenic modification.

## 3.4 Linkage Mapping

In order to perform comparative QTL studies and to integrate the switchgrass genome with the consensus genome of grasses, it is necessary to have linkage maps where the same markers are mapped across species or where the sequences of the probes are known to allow for analysis of marker collinearity with sequenced genomes. In cross-pollinated species, the approach that has been taken for linkage mapping is a pseudo-test backcross strategy (Grattapaglia and Sederoff, 1994). This approach is coupled with the mapping of markers that are present as a single dose in polyploids that would then segregate 1:1 in the progeny of a mapping population (Wu et al., 1992). Under a polysomic inheritance model, there are three alternative alleles, which make it more difficult to detect markers linked in repulsion. A chi-squared test can be used to determine if the ratio of coupling to repulsion phase markers detected deviates significantly from that expected ratio under either disomic or polysomic inheritance (Wu et al., 1992).

In 2005 one mapping study that contained the original set of 85 individual $F_1$ from the AP13 × VS16 cross used RFLP markers that had been used previously in rice, bermuda grass, and Pennisetum (Missaoui et al., 2005). A total of 57 single-dose markers were placed on the VS16 map creating 16 linkage groups and 45 markers were placed on the AP13 map creating 11 linkage groups. Most markers remained unlinked. Limited synteny between rice, sorghum, and switchgrass was noted, and a high degree of preferential pairing based on the ratio of linkages detected in coupling versus repulsion phase was seen. Thus it was concluded that inheritance was primarily disomic.

A different study on a larger mapping population was conducted in 2010 which used 284 $F_1$ progeny from a different cross of two lowland individuals derived from Alamo (A4) and Kanlow (K5) (Okada et al., 2010). This study used genomic simple-sequence repeat (SSR), EST-SSR, and STS markers that enabled sequence-based comparisons. It produced male and female linkage maps that were grouped in nine homology groups with pairs of chromosomes arbitrarily designated A and B to represent the distinct subgenomes which were aligned based on markers that detected polymorphic alleles in each homeolog. One interesting feature of this study was the sequence-based comparisons that were possible to the genomes of sorghum and rice using the most highly scoring corresponding blast hits to mapped markers. This approach led to a low-resolution view of the collinearity between rice, sorghum, and switchgrass on a whole genome level; it also allowed direct comparison of male and female maps and of the arbitrarily designated A and B subgenomes. The male and female maps were aligned using both multiallelic SSR and smaller numbers of less informative markers that were single dose in both parents and therefore segregated 3:1 present:absent in the progeny. The two subgenomes were aligned using multiallelic SSR-loci that detected polymorphisms in both genomes.

Comparison of the linkage maps with rice and sorghum genomes has indicated that large stretches of markers are collinear with particularly long stretches detected between switchgrass homology groups VII and IX and sorghum chromosomes 6 and 1. These regions extended for 35% and 91%, respectively, of the total length of the linkage group in some cases. The differences in base chromosome numbers of 10 versus 9 for sorghum and switchgrass were found to result from a whole chromosome translocation of the intervals corresponding to sorghum chromosomes 8 and 9 into switchgrass homology group III. By using rice as a common reference genome, it was possible to align the published RFLP map of foxtail millet (*Setaria italica*) (Devos et al., 1998). Figure 3.3 shows the results of this map alignment. It is clear from this alignment and the extensive collinearity with sorghum that all nine of the

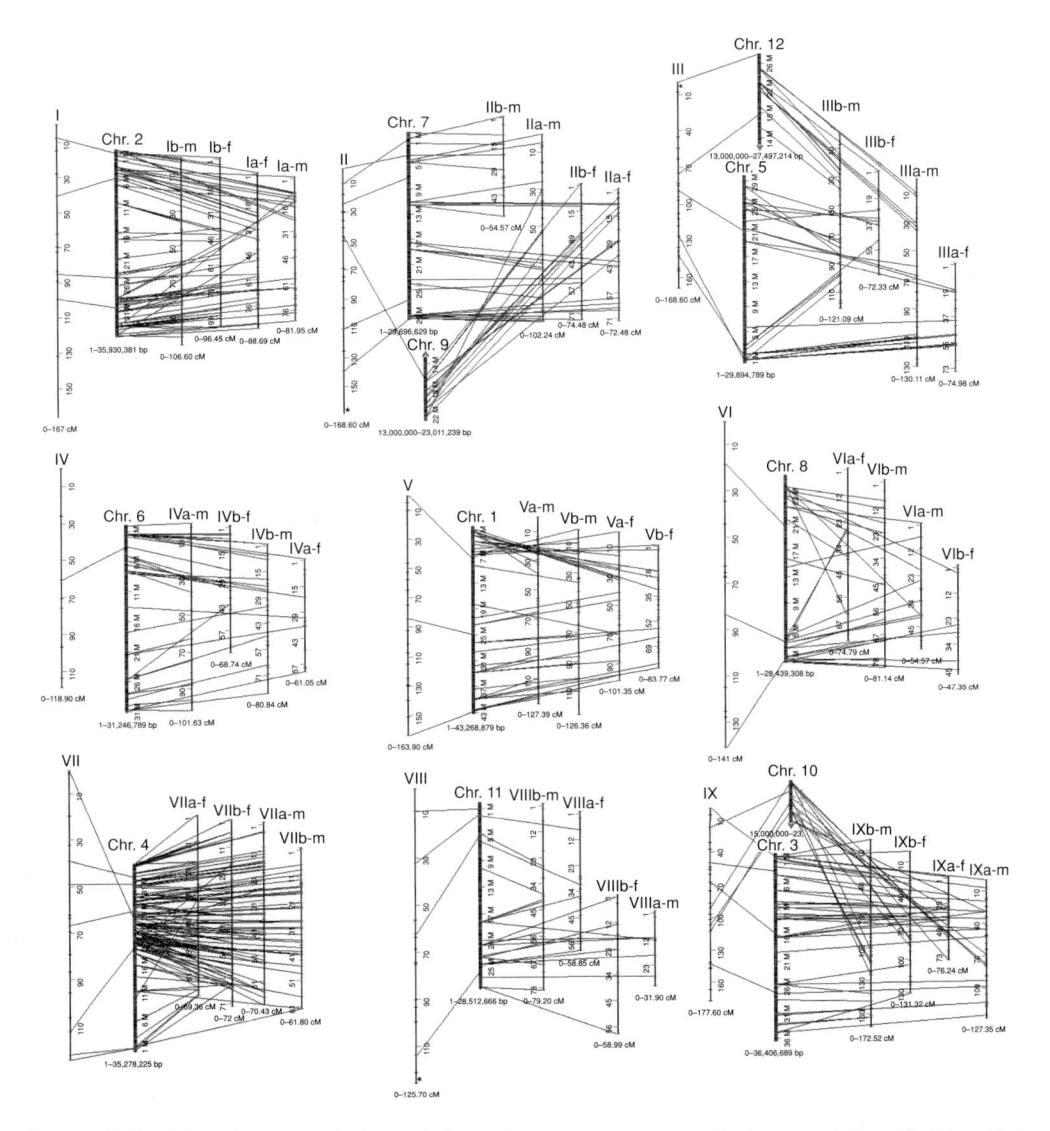

**Figure 3.3.** Map alignment of nine switchgrass homology groups with rice genome and foxtail millet RFLP map. The nine switchgrass homology groups are coupled together and aligned with the corresponding rice genomic intervals and foxtail millet RFLP map. The sequence-based correspondences between EST-SSR and rice gene indices are shown as lines between individual switchgrass linkage groups (right four vertical lines) and the rice genome sequence (middle thick vertical line). The RFLP probe-based correspondences between the foxtail millet linkage map (left vertical lines) and rice are also shown.

switchgrass homology groups can be integrated into the consensus grass genome (Devos and Gale, 2000). At the level of map resolution there appears to be complete collinearity between foxtail millet and switchgrass.

Development of marker systems for switchgrass is ongoing with the most progress being made thus far in development and application of SSRs for two main reasons: they are inexpensive to develop and their multiallelic nature can be more informative particularly in polyploids. In addition to SSR markers developed directly from EST data, over 1030 SSR-containing sequences from GA-, CA-, CAG-, and AAG-enriched genomic libraries were identified at Oklahoma State University (Wang et al., 2011). This project demonstrated that rates of successful amplification, polymorphism rate, and mean number of amplicons per individual were all greater for genomic SSR than those reported for EST-SSR. Dinucleotide SSRs were more polymorphic, indicating that for many purposes these markers are extremely useful. Their results combined with preexisting data sets indicated a total of approximately 1860 unique SSR primer pairs have been proven to amplify in switchgrass.

## 3.5 Cytoplasmic Genome

Chloroplast and mitochondrial markers are useful for direct inter-ploidal comparisons allowing the study of population genetics of genome duplication, environmental adaptation, and evolution in a maternally inherited haploid genome where sequences are highly conserved. Knowledge of intergenic sequences can also assist design of improved chloroplast transformation which may become an important technique in the future for gene containment strategies or for the high expression of valuable coproducts.

Although no reports have been made of mitochondrial sequences, Young et al. have sequenced the quadripartite upland and lowland chloroplast genomes. Perfect conservation of gene number and order with other Poaceae was observed with most of the variation between sequences associated with single nucleotide polymorphisms or small indels associated with homopolymeric SSR in intergenic regions. The few large indels that were present between upland and lowland were associated with intergenic regions and a deletion in the rpoC2 gene which is hypervariable in grasses. The evolutionary distance separating upland and lowland chloroplast genomes was estimated using rates of synonymous codon substitution in the single copy regions and published molecular clock figures for grass chloroplasts. The estimates of sequence divergence indicated that the overall difference between chloroplast genomes of lowland and upland genotypes were similar in degree to that seen between indica and japonica rice varieties or present in upland- and lowland-type populations of the wild rice *Oryza rufipogon* (Tang et al., 2004). The divergence time was estimated to be between 523,000 and 845,000 years ago.

## 3.6 Genome-enabled Improvement of Switchgrass

Switchgrass breeding has been focused on recurrent selection strategies. For further information in this area, I refer to the chapter on switchgrass breeding in this book and to past reviews of forage breeding (Vogel and Pedersen, 1993; Vogel and Jung, 2001; Bouton, 2007) and to simulation analysis of breeding strategies applied to switchgrass (Casler and Brummer, 2008). These breeding strategies will continue but will be increasingly using an integrated approach that applies genomic technology.

High levels of genetic diversity in switchgrass enforced by outcrossing and enhanced by recent whole genome duplication complicate genetic dissection of QTL. Also, many traits of interest to bioenergy crops are likely to be controlled by many QTLs with small effects. In QTL mapping strategies applied to outcrossing species, shared alleles or QTL between parents, cases where markers may be informative in only one parent, and unknown QTL dosage are problems that must be addressed. To be detected in a two-generation mapping strategy, the marker and QTL must both be heterozygous. These constraints and the need in some cases to analyse half-sib populations reduce the power to detect QTL and therefore call for larger population sizes (Weller et al., 1990; Luo, 1993). In addition to genetic complexity, the extended establishment period in switchgrass leads to the full yield potential being attained in the second and third growing seasons (Sharma et al., 2003). This makes collection of second- and third-year yield data important for breeding purposes and consequently lengthens the selection cycle and time to identify QTL for many relevant traits. Rather than exploiting QTL for breeding in this fashion, an alternative approach that efficiently utilizes dense multi-locus genotypic data for predicting performance may gain favor. This approach coined genome-wide selection (GWS) has proven successful in animals and is currently an area of intense research in plants (Meuwissen et al., 2001). GWS has never been attempted with perennial grasses or polyploid species and has only recently become feasible as genotyping costs have dropped substantially (Grattapaglia and Resende, 2010; Jannink et al., 2010). In theory, it will enable a shorter selection cycle as is shown in Figure 3.4.

Traits such as biomass yield and cell wall composition are known to be controlled by a large number of alleles typically with small, additive effects. However, statistical methods used for QTL studies and genome-wide association studies are better at identification of individual loci with large effects and do not attempt to explain all genetic variations. The approach of GWS differs in that the statistical methods used attempt to explain all the genetic variation rather than selecting a subset of markers with significant effects that explain only a portion of it. GWS relies on availability of inexpensive genotypic data to provide genomic estimates of individual breeding values (GEBVs) in the absence of phenotypic information by using a predictive model produced from phenotypes and genotypes acquired from a large training population. Recombination and selection of favorable alleles associated with desirable traits is then possible using the GEBVs, and selection of individuals with the highest GEBVs for several generations without phenotyping may significantly shorten the selection cycle in perennial crops. Although some important progress has been made, the statistical aspects of GWS, especially for outcrossing polyploid species, are still in development. The type of model to use, the number of markers required, and the size of the training population needed for best selection response are still not well understood in plants and are predicted to be heavily dependent on trait heritability, effective population size, and recombinational length of the genome. Supportive data acquired from large populations is lacking in most plants.

Several methods of acquiring genotypic information may be appropriate for switchgrass breeding purposes. Sequencing of reduced-representation libraries (RRLs) holds promise as an inexpensive genotyping method in plants and relies on bioinformatics approaches to identify SNPs in populations based on registration of short read ends to restriction sites thereby decreasing genome coverage and increasing read depth in a given region (Lijavetzky et al., 2007; Van Tassell et al., 2008; Wiedmann et al., 2008). Several mapping, QTL, and association studies have utilized these and related methods (Baird et al., 2008; Deschamps et al., 2010; Wu et al., 2010). By using sequencing rather than hybridization or other approach, both SNP discovery and genotyping are accomplished using the same platform without an intervening

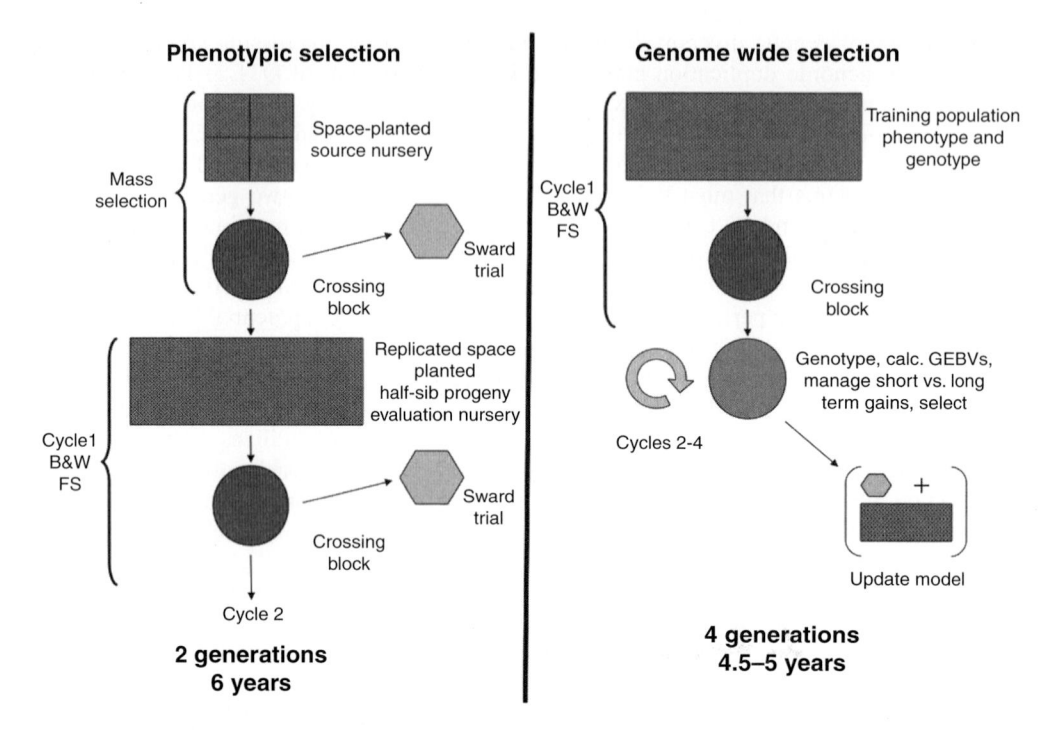

**Figure 3.4.** Phenotypic selection currently used for forage breeding efforts versus genome-wide selection using dense genotypic information. The approach indicated on the left is one breeding scheme used by forage breeders for population-based selection. The approach indicated on the right uses dense genetic data and phenotypic records from training populations to build predictive models that are used for the calculation of individual genomic estimates of breeding values (GEBVs) that are predictive of the phenotype that may be difficult, costly, or require multiple growing seasons to score directly.

assay development phase. With enough coverage depth, the SNPs can be modeled as codominant markers or as quantitative features using sequence-representation-based techniques as is currently in use for detecting genome structural variation (Yoon et al., 2009). This may be critically important in polyploids where more than two alleles may be present or one allele may be fixed in one subgenome but polymorphic in another. Other methods particularly useful for polyploid crops rely on using repetitive mobile elements and their unique variable junctions (Wanjugi et al., 2009; Witherspoon et al., 2010). In the case of hexaploid wheat these junctions are not conserved between A, B, and D subgenomes and allowed the development of D genome-specific markers through their identification in *Aegilops tauschii* (the D genome donor progenitor). Quantitative SNP genotyping with systems such as Sequenome's MassARRAY system also holds promise.

All of these methods rely on distinguishing paralogous and homeologous sequence variation from true allelic diversity to improve the accuracy of genotypic data. This is especially true in polyploid, allogamous genetic systems such as switchgrass. But to date there has been no direct measurement of subgenome differentiation at the sequence level in this species. Genetically, the two subgenomes of switchgrass are well differentiated in the sense that they exhibit a high degree of preferential pairing. Simple-sequence repeat markers have also shown significant differentiation with approximately 90% of markers mapped within a single subgenome (Okada et al., 2010). Evaluation of BACs from homeologous regions should provide an accurate assessment of nucleotide divergence. Further data will soon be available from ongoing DOE-JGI and DOE-Bioenergy Center projects from selected switchgrass BAC sequencing and draft gene-space assemblies that will produce better estimates of this subgenome differentiation and will allow estimates of the resources that are required for genotyping as well as for whole genome assembly. The issues become more complex in switchgrass cytotypes that have more than 18 pairs of chromosomes. Hexaploid and octoploid accessions of switchgrass show irregular meiotic pairing and likely exhibit polysomic inheritance as well as frequent aneuploidy (Barnett and Carver, 1967; Costich et al., 2010). This will complicate molecular breeding efforts. To what extent these issues influence any marker-based selection system will likely require extensive evaluation of populations with different chromosome numbers.

## 3.7 Conclusions

There is now sufficient sequence data from the switchgrass transcriptome to provide a preliminary "parts list" to guide further study. Work to identify the regulation of these parts, their involvement in networks, and their importance to specific biological processes is just beginning. Now, there is also a wealth of next-generation sequencing data from a few specific genotypes including ~11 million ESTs, a gene expression atlas, BAC sequences, and a preliminary genome assembly that needs to be organized into a structural framework that encompasses conserved noncoding regions, repetitive DNA classification, and larger-scale assemblies spanning gene-rich regions. There is also a need to begin to integrate this information with the related genomes of species such as foxtail millet, *B. distachyon*, maize, sorghum, and rice, that all serve as useful models for switchgrass. To what extent these functions will be performed by sequencing centers, community-based efforts, the three DOE-Bioenergy Research Centers, or other groups is not really so much the issue as how the availability of genome resources can speed the process of feedstock development and lead to improved economics of renewable bioenergy production. This area is waiting for good ideas and applications of sound science and technology and is certainly off to a good start.

## References

Baird N, Etter PD, Atwood TS, Currey MC, Shiver AL, Lewis ZA, Selker EU, Cresko WA, Johnson EA. Rapid SNP discovery and genetic mapping using sequenced RAD markers. PLoS ONE, 2008; 3(10): e3376. Available at: http://www.scopus.com/inward/record.url?eid=2-s2.0-54449098752&partnerID=40. Accessed on 19 March 2013.

Barnett F, Carver, R. Meiosis and pollen stainability in switchgrass, *Panicum virgatum* L. Crop Sci, 1967; 7: 301–304.

Bouton JH. Molecular breeding of switchgrass for use as a biofuel crop. Curr Opin Genet Dev, 2007; 17(6): 553–558.

Casler MD, Brummer EC. Theoretical expected genetic gains for among-and-within-family selection methods in perennial forage crops. Crop Sci, 2008; 48(3): 890–902.

Costich DE, Friebe B, Sheehan MJ, Casler MD, Buckler ES. Genome-size variation in switchgrass (*Panicum virgatum*): flow cytometry and cytology reveal rampant aneuploidy. Plant Genome, 2010; 3(3): 130.

Deschamps S, Rota M, Ratashak JP, Biddle P, Thureen D, Farmer A, Luck S, Beatty M, Nagasawa N, Michael L, Llaca V, Sakai H, May G, Lightner J, Campbell MA. Rapid genome-wide single nucleotide polymorphism discovery in soybean and rice via deep resequencing of reduced representation libraries with the illumina genome analyzer. Plant Genome, 2010; 3(1): 53.

Devos KM, Gale MD. Genome relationships: the grass model in current research. Plant Cell, 2000; 12(5): 637–646.

Devos K, Wang ZM, Beales J, Sasaki T, Gale MD. Comparative genetic maps of foxtail millet (*Setaria italica*) and rice (*Oryza sativa*). Theor Appl Genet, 1998; 96(1): 63–68.

Grattapaglia D, Resende M. Genomic selection in forest tree breeding. Tree Genet Genomes, 2010; 7(2): 241–255.

Grattapaglia D, Sederoff R. Genetic linkage maps of Eucalyptus grandis and Eucalyptus urophylla using a pseudo-testcross: mapping strategy and RAPD markers. Genetics, 1994; 137(4): 1121–1137.

Jannink J-L, Lorenz AJ, Iwata H. Genomic selection in plant breeding: from theory to practice. Brief Funct Genomics, 2010; 9(2): 166–177.

Lijavetzky D, Cabezas JA, Ibáñez A, Rodríguez V, Martínez-Zapater JM. High throughput SNP discovery and genotyping in grapevine (*Vitis vinifera* L.) by combining a resequencing approach and SNPlex technology. BMC Genomics, 2007; 8: 424. Available at: http://www.scopus.com/scopus/inward/record.url?eid=2-s2.0-38549148784&partnerID=40

Luo Z. The power of two experimental designs for detecting linkage between a marker locus and a locus affecting a quantitative character in a segregating population. Genet Sel Evol, 1993; 25(3): 249.

Mallory AC, Vaucheret H. Functions of microRNAs and related small RNAs in plants. Nat Genet, 2006; 38(7): 850.

Mao C, Ding W, Wu Y, Yu J, He X, Shou H, Wu P. Overexpression of a NAC-domain protein promotes shoot branching in rice. New Phytol, 2007; 176(2): 288–298.

Meuwissen T, Hayes B, Goddard M. Prediction of total genetic value using genome-wide dense marker maps. Genetics, 2001; 157(4): 1819–1829.

Missaoui AM, Paterson AH, Bouton JH. Investigation of genomic organization in switchgrass (*Panicum virgatum* L.) using DNA markers. Theor Appl Genet, 2005; 110(8): 1372–1383.

Mitsuda N, Iwase A, Yamamoto H, Yoshida M, Seki M, Shinozaki K, Ohme-Takagi M. NAC transcription factors, NST1 and NST3, are key regulators of the formation of secondary walls in woody tissues of Arabidopsis. Plant Cell, 2007; 19(1): 270–280.

Okada M, Lanzatella C, Saha M, Bouton J, Wu R, Tobias C. Complete switchgrass genetic maps reveal subgenome collinearity, preferential pairing and multilocus interactions. Genetics, 2010; 185(3): 745–760.

Saathoff AJ, Tobias CM, Sattler SE, Haas EE, Twigg P, Sarath G. Switchgrass contains two cinnamyl alcohol dehydrogenases involved in lignin formation. BioEnerg Res, 2010; 4(2): 120–133. Available at: http://www.springerlink.com/content/31775260×74831v5/fulltext.html. Accessed on 1 October 2010.

Schatz MC, Delcher AL, Salzberg SL. Assembly of large genomes using second-generation sequencing. Genome Res, 2010; 20(9): 1165–1173.

Sharma N, Piscioneri I, Pignatelli V. An evaluation of biomass yield stability of switchgrass (Panicum virgatum L.) cultivars. Energy Convers Manage, 2003; 44(18): 2953–2958.

Shen H, Yin Y, Chen F, Xu Y, Dixon RA. A bioinformatic analysis of NAC genes for plant cell wall development in relation to lignocellulosic bioenergy production. BioEnerg Res, 2009; 2(4): 217–232.

Srivastava A, Palanichelvam K, Ma J, Steele J, Blancaflor EB, Tang Y. Collection and analysis of expressed sequence tags derived from laser capture microdissected switchgrass (*Panicum virgatum* L. Alamo) vascular tissues. BioEnerg Res, 2010; 3:278–294.

Sunkar R, Chinnusamy V, Zhu J, Zhu JK. Small RNAs as big players in plant abiotic stress responses and nutrient deprivation. Trends Plant Sci, 2007; 12(7): 301–309.

Tang J, Xia H, Cao M, Zhang X, Zeng W, Hu S, Tong W, Wang J, Wang J, Yu J, Yang H, Zhu L. A comparison of rice chloroplast genomes. Plant Physiol, 2004; 135(1): 412–420.

Tobias CM, Sarath G, Twigg P, Lindquist E, Pangilinan J, Penning BW, Barry K, McCann MC, Carpita NC, Lazo GR. Comparative genomics in switchgrass using 61,585 high-quality expressed sequence tags. Plant Genome, 2008; 1(2): 111–124.

Tobias C, Hayden DM, Twigg P, Sarath G. Genic microsatellite markers derived from EST sequences of switchgrass (*Panicum virgatum* L.). Mol Ecol Notes, 2006; 6(1): 185–187.

Tobias C, Twigg P, Hayden DM, Vogel KP, Mitchell RM, Lazo GR, Chow EK, Sarath G. Analysis of expressed sequence tags and the identification of associated short tandem repeats in switchgrass. Theor Appl Genet, 2005; 111(5): 956–964.

Uauy C, Distelfeld A, Fahima T, Blechl A, Dubcovsky J. A NAC gene regulating senescence improves grain protein, zinc, and iron content in wheat. Science, 2006; 314(5803): 1298–1301.

Van Tassell C, Smith TP, Matukumalli LK, Taylor JF, Schnabel RD, Lawley CT, Haudenschild CD, Moore SS, Warren WC, Sonstegard TS. SNP discovery and allele frequency estimation by deep sequencing of reduced representation libraries. Nat Methods, 2008; 5(3): 247–252.

Vogel K, Jung H-J. Genetic modification of herbaceous plants for feed and fuel. Crit Rev Plant Sci, 2001; 20(1): 15–49.

Vogel K, Pedersen J. Breeding systems for cross-pollinated perennial grasses. Plant Breed Rev, 1993; 11: 251–274.

Wang YW, Samuels TD, Wu YQ. Development of 1,030 genomic SSR markers in switchgrass. Theor Appl Genet, 2011; 122(4): 677–686.

Wanjugi H, Coleman-Derr D, Huo N, Kianian SF, Luo MC, Wu J, Anderson O, Gu YQ. Rapid development of PCR-based genome-specific repetitive DNA junction markers in wheat. Genome, 2009; 52(6): 576–587.

Paterson AH, Bowers JE, Bruggmann R, Dubchak I, Grimwood J, Gundlach H, Haberer G, Hellsten U, Mitros T, Poliakov A, Schmutz J, Spannagl M, Tang H, Wang X, Wicker T, Bharti AK, Chapman J, Feltus FA, Gowik U, Grigoriev IV, Lyons E, Maher CA, Martis M, Narechania A, Otillar RP, Penning BW, Salamov AA, Wang Y, Zhang L, Carpita NC, Freeling M, Gingle AR, Hash CT, Keller B, Klein P, Kresovich S, McCann MC, Ming R, Peterson DG, Mehboob-ur-Rahman, Ware D, Westhoff P, Mayer KF, Messing J, Rokhsar DS. The sorghum bicolor genome and the diversification of grasses. Nature, 2009; 457(7229): 551–556.

Weller JI, Kashi Y, Soller M. Power of daughter and granddaughter designs for determining linkage between marker loci and quantitative trait loci in dairy cattle. J Dairy Sci, 1990; 73(9): 2525–2537.

Wiedmann RT, Smith TPL, Nonneman DJ. SNP discovery in swine by reduced representation and high throughput pyrosequencing. BMC Genet, 2008; 9: 81.

Witherspoon D, Xing J, Zhang Y, Watkins WS, Batzer MA, Jorde LB. Mobile element scanning (ME-Scan) by targeted high-throughput sequencing. BMC Genomics, 2010; 11(1): 410.

Wu K, Burnquist W, Sorrells ME, Tew TL, Moore PH, Tanksley SD. The detection and estimation of linkage in polyploids using single-dose restriction fragments. Theor Appl Genet, 1992; 83(3): 294–300.

Wu X, Ren C, Joshi T, Vuong T, Xu D, Nguyen HT. SNP discovery by high-throughput sequencing in soybean. BMC Genomics, 2010; 11(1): 469.

Xie F, Frazier TP, Zhang B. Identification and characterization of microRNAs and their targets in the bioenergy plant switchgrass (*Panicum virgatum*). Planta, 2010; 232(2): 417–434.

Yoon S, Xuan Z, Makarov V, Ye K, Sebat J. Sensitive and accurate detection of copy number variants using read depth of coverage. Genome Res, 2009; 19(9): 1586–1592.

Young HA, Hernlem BJ, Anderton AL, Lanzatella-Craig C, Tobias CM. Dihaploid stocks of switchgrass isolated by a screening approach. Bioenerg Res, 2010; 3(4): 305–313.

Zhong R, Lee C, Zhou J, McCarthy RL, Ye ZH. A battery of transcription factors involved in the regulation of secondary cell wall biosynthesis in Arabidopsis. Plant Cell, 2008; 20(10): 2763–2782.

Chapter 4

# Germplasm Resources of *Miscanthus* and Their Application in Breeding

Maria Stefanie Dwiyanti[1,2], J. Ryan Stewart[2,3], and Toshihiko Yamada[1]*

[1]*Field Science Center for Northern Biosphere, Hokkaido University, Kita 11 Nishi 10, Kita-ku, Sapporo, Hokkaido 060-0811, Japan*
[2]*Energy Biosciences Institute, University of Illinois, 1206 W Gregory Dr, Urbana, IL 61801, USA*
[3]*Brigham Young University, 150 East Bulldog Boulevard, Provo, UT 84602, USA*
*Corresponding author

## 4.1 Introduction

Bioenergy feedstocks are increasingly gaining attention as suitable alternatives to energy derived from petroleum oil. This worldwide trend appears to be driven by concerns to increase energy security and decrease greenhouse gas emissions (Hill et al., 2006). Soybean, palm oil, maize, and sugarcane all show potential as bioenergy crops, but concerns have been raised because crop conversion from food to biofuel might lead to widespread food scarcity and, concomitantly, higher food prices.

Out of several candidate bioenergy crops, a perennial grass hybrid originally from Japan, *Miscanthus* × *giganteus* (Figure 4.1), is widely considered to have notable potential in temperate regions of the world due to its high productivity (Beale and Long, 1995; Heaton et al., 2008; Dohleman and Long, 2009). In US field trials, *M.* × *giganteus* produced eight-fold higher yield than a mixed planting of prairie grass species (Heaton et al., 2008). Moreover, *M.* × *giganteus* requires low fertilizer and pesticide inputs relative to other bioenergy crops, such as the annual species, maize (*Zea mays*) (Lewandowski et al., 2000; Hansen et al., 2004; Clifton-Brown et al., 2007).

*M.* × *giganteus* clones now available in the market appear to be derived from a single plant introduced by a Danish plant collector, Aksel Olsen, into Europe from Yokohama, Japan, in 1935 (Linde-Laursen, 1993). *M.* × *giganteus* is a natural triploid hybrid between diploid *Miscanthus sinensis* and tetraploid *Miscanthus sacchariflorus* (Adati and Shiotani, 1962; Greef et al., 1997; Hodkinson et al., 2002a; Swaminathan et al., 2010). *M.* × *giganteus* is a sterile triploid, so it can be propagated only by rhizomes (Greef et al., 1997; Hodkinson et al., 2002a). Molecular marker analysis on several *M.* × *giganteus* clones showed that there is little variation between clones (Greef et al., 1997; Hodkinson et al., 2002a). Genetic uniformity increases *M.* × *giganteus* vulnerability to diseases, pests, and environmental stresses (Clifton-Brown et al., 2001). Furthermore, *M.* × *giganteus* sterility prevents development of new varieties of *M.* × *giganteus* (Clifton-Brown et al., 2001).

---

*Bioenergy Feedstocks: Breeding and Genetics*, First Edition. Edited by Malay C. Saha, Hem S. Bhandari, and Joseph H. Bouton.
© 2013 John Wiley & Sons, Inc. Published 2013 by John Wiley & Sons, Inc.

**Figure 4.1.** *Miscanthus × giganteus* "Illinois clone" cultivated at Sapporo, Japan. (*For color details, see color plate section.*)

Improving *M. × giganteus* has been attempted by restoring *M. × giganteus* fertility through polyploidization (Petersen et al., 2002; Glowacka et al., 2009; Yu et al., 2009) or genetic modification (Wang et al., 2011). Efforts at artificial crossing between *M. sinensis* and *M. sacchariflorus* have also been documented (Greef et al., 1997; Deuter, 2000). Nishiwaki et al. (2011) investigated sympatric populations of *M. sinensis* and *M. sacchariflorus* to locate natural hybrids between *M. sinensis* and *M. sacchariflorus*. The team was successful in discovering three putative hybrids that appeared to be identical in ploidy level to *M. × giganteus*. This chapter describes the genetics and breeding of *Miscanthus* species, their characteristics, and their phylogenetic relationship and progress of genetic improvement of *Miscanthus* species through conventional and molecular breeding.

## 4.2 Species Belonging to *Miscanthus* Genus, Their Characteristics, and Phylogenetic Relationships

Based on cytotaxonomical analysis, *Miscanthus* species in Japan can be classified into three sections, *Triarrhena* (Maxim.) Honda, *Eumiscanthus* Honda, and *Kariyasua* Ohwi. Species belonging to each section are summarized in Table 4.1 (Adati, 1958; Adati and Shiotani, 1962).

### 4.2.1 Section: Eumiscanthus

*M. sinensis*

*M. sinensis* (Figure 4.2) is a diploid form of *Miscanthus*, with a chromosome number of 38. The species has short and stout rhizomes, scabrous margins on leaves, no hairs or sparse hairs

**Table 4.1.** Classification of the *Miscanthus* species in Japan.

| Section | Species (Scientific Name) | Japanese Name |
|---|---|---|
| *Eumiscanthus* Honda | *M. sinensis* Andersson | Susuki |
| | *M. sinensis* Andersson form. *gracillimus* (Hitchcock) Ohwi | Ito-susuki |
| | *M. sinensis* Andersson form. *zebrinus* (Nicholson) Nakai | Takanoha-susuki |
| | *M. sinensis* Andersson form. *variegatus* Nakai | Shima-susuki |
| | *M. sinensis* Andersson var. *condensatus* | Hachijyo-susuki |
| | *Miscanthus floridulus* (Labill.) Warburg | Tokiwa-susuki |
| *Triarrhena* (Maxim.) Honda | *M. sacchariflorus* (Maxim.) Bentham | Ogi |
| | *M. sacchariflorus* (Maxim.) Bentham var. *brevibarbis* (Honda) Adati | Ogi-susuki |
| | *M. sacchariflorus* (Maxim.) Bentham var. *glaber* Adati | |
| *Kariyasua* Ohwi | *Miscanthus tinctorius* (Steudel) Hackel | Kariyasu |
| | *Miscanthus oligostachyus* Staff | Kariyasumodoki |
| | *Miscanthus intermedius* (Honda) Honda | Oohigenagakariyasumodoki |

on leaf sheaths, no branching of the culms, high stem density, and tufted structure (Adati and Shiotani, 1962). *M. sinensis*, however, is noted to have wide phenotypic variation. According to Adati (1958), the width of leaf blades varied between 1 and 2.7 cm, the length varied between 47 and 98 cm, and the culm length varied between 81 and 250 cm. *M. sinensis* have 5–7 mm spikelets with awns and callus hairs with the same length as the spikelets (Adati and Shiotani, 1962). Seeds are generally wind dispersed, which has been considered a factor in its potential invasiveness if fertile varieties are widely cultivated (Ohtsuka et al., 1993; Quinn et al., 2011). Although the pollen fertility rate of *M. sinensis* is more than 86%, the self-pollination rate is very low (Hirayoshi et al., 1955), indicating self-incompatibility in the species.

*M. sinensis* has the broadest distribution among *Miscanthus* species. Its native range includes eastern Russia, eastern China, Korea, Japan, Taiwan, and Southeast Asia (Numata, 1970; Koyama, 1987; Chen and Renvoize, 2006). *M. sinensis* is found in a wide range of habitats, from mountain slopes to coastal areas (Matumura and Yukimura, 1975). It is able to establish on varying soil types, with preferences for exposed, well-drained habitats (Matumura and Yukimura, 1975; Stewart et al., 2009). Conspecific plants naturally occur on soils with pH values ranging between 3.5 and 7.5, although most were growing in soils with pH between 4 and 6 (summarized in Kayama, 2001). However, An et al. (2008) reported that *M. sinensis* colonized on soils with pH values ranging from 2.7 to 5.4 in Rankoshi, Hokkaido, Japan. *M. sinensis* is also tolerant to high aluminum, chromium, and zinc (Kayama, 2001; Ezaki et al., 2008). Its tolerance to heavy metals can be utilized to develop a bioenergy crop suitable for polluted environments (Arduini et al., 2006).

*M. sinensis* genotypes from high-latitude and high-altitude areas flower earlier than those from low-altitude and low-latitude areas (Adati, 1958). The difference in flowering time reaches up to 2 months. In general, *M. sinensis* has lower lignin content compared to *M.* × *giganteus* (Hodgson et al., 2010, 2011). This trait will be useful for bioethanol production, because high lignin content will inhibit cellulase in breaking down cellulose (Hodgson et al., 2010,

**Figure 4.2.** Major *Miscanthus* species: (a) *M. sinensis*, (b) *M. sacchariflorus*, (c) *M. sinensis* var. *condensatus*, and (d) *M. floridulus* in wild population in Japan. (*For color details, see color plate section.*)

2011). In addition, sulfur, phosphorus, potassium, chlorine, and calcium content variations were observed between *M. sinensis* genotypes (Atienza et al., 2003c, 2003d). High levels of minerals can lead to unacceptable emissions of dioxins and also slagging problems during combustion process.

*M. sinensis var. condensatus*

*M. sinensis* var. *condensatus* (Figure 4.2) is generally a diploid, with a chromosome number of 38 (Adati and Shiotani, 1962). A natural triploid *M. sinensis* var. *condensatus* (chromosome number = 57) was found in Enoshima, Kanagawa Prefecture, Japan (Adati, 1958). The triploid exhibited irregular meiosis with frequent trivalents and univalents (Adati and Mitsuishi, 1956; Adati, 1958).

Compared to *M. sinensis*, *M. sinensis* var. *condensatus* has more condensed panicles, broader leaves with white spots, and more secretion of wax on the leaf sheaths (Adati and Mitsuishi, 1956; Adati, 1958; Adati and Shiotani, 1962). The leaves are also weak and have no scabrous margins (Yamashita et al., 2010). Crossings between *M. sinensis* and *M. sinensis* var.

*condensatus* resulted in fertile $F_1$ progenies with wide variations in their morphologies (Adati and Shiotani, 1962). In contrast to *M. sinensis, M. sinensis* var. *condensatus* appears to be self-compatible or at least can reproduce apomictically (Chiang et al., 2003). This self-compatibility is considered the result of strong selection from high-salinity habitat that limits diversity (Chiang et al., 2003). Understanding the genetic regulation of the self-compatibility character may provide insight into how to break self-incompatibility in the general *Miscanthus* species.

*M. sinensis* var. *condensatus* is mainly distributed in coastal areas and is tolerant to salt (Adati and Mitsuishi, 1956). *M. sinensis* var. *condensatus* is distributed from coastal areas in Japan, Korea, Taiwan, and the Philippines (Adati and Mitsuishi, 1956). In addition, *M. sinensis* var. *condensatus* does not senesce in winter. Adati and Mitsuishi (1956) reported that in the same habitat on Hachijo Island, *M. sinensis* senesced before winter while *M. sinensis* var. *condensatus* leaves stayed green.

*M. floridulus*

*M. floridulus* (Figure 4.2) is diploid, with a chromosome number of 38 (Adati and Shiotani, 1962). The width of the leaf blades of *M. floridulus* ranged between 2 and 2.6 cm and the length ranged between 75 and 90 cm. Unsurprisingly, this species is classified as having the largest leaves in the *Eumiscanthus* section. The panicles are about 50 cm long and the axis is generally elongated (Adati, 1958). In Japan, the plant height has been reported to reach 2.5 m, while in Taiwan plants with about 3 m height were reported (Adati, 1958). However, the spikelets are smaller than *M. sinensis*, with lengths ranging from 3 to 3.5 mm (Adati, 1958).

*M. floridulus* is distributed in tropical and subtropical regions, particularly the Pacific side of Japan (except for Hokkaido), Taiwan, Southeast Asia, and Polynesia (Adati, 1958; Deuter, 2000). Although most *M. floridulus* generally populates coastal regions, it has also been found in high-altitude areas. However, Chou et al. (2001) found that *M. floridulus* dominant in Taiwan lowlands could not grow well at 2600 m. *M. floridulus* flowers around July to August, earlier than other *Miscanthus* species (Adati, 1958; Yamashita et al., 2010). Similar to *M. sinensis* var. *condensatus*, *M. floridulus* does not senesce in winter (Adati, 1958; Yamashita et al., 2010). Because it is distributed mainly in tropical areas, Deuter (2000) suggested that *M. floridulus* could be used as a parent stock for biomass crop breeding in the tropics or areas with warm, moist climates.

### 4.2.2 Section: Triarrhena

*M. sacchariflorus*

Lafferty and Lelley (1994) reported that there are two types of *M. sacchariflorus*, one with 38 chromosomes and another with 76 chromosomes. Hirayoshi et al. (1957) found that *M. sacchariflorus* in Japan were tetraploids with a chromosome number of 76 (Figure 4.2). However, *M. sacchariflorus* in China is diploid ($2n = 38$) (Deuter, 2000). The tetraploid *M. sacchariflorus* has large and hardy stems with high lignin content, tall and branching culms, which can reach more than 3 m, low culm number, and creeping, stout rhizomes. Its leaf sheaths are densely covered with bristles when young (Adati, 1958; Adati and Shiotani, 1962). Also, it has culm nodes from which aerial branches and roots develop (Adati and Shiotani, 1962). *M. sacchariflorus* develops hollow stems to adapt to soils with high moisture (Sacks et al., 2013). It has awnless spikelets with callus hairs that are about two to four times longer than the spikelets (Adati and Shiotani, 1962).

*M. sacchariflorus* is distributed from southern Siberia to China and Korea (Clifton-Brown et al., 2001). It prefers exposed, fertile, and moist places such as flood plains, riverbanks, and lakes (Matumura and Yukimura, 1975). It is also more sensitive to frost than *M. sinensis* (Farrell et al., 2006). Jensen et al. (2011) reported that *M. sacchariflorus* accessions from different regions of Japan, China, and Korea started flowering from mid-July to late November. Although relationship between geographical origins with flowering time of *M. sacchariflorus* is yet to be determined, it is considered a short-day species (Deuter, 2000). In contrast to *M. sinensis* that forms new tillers during vegetation period, *M. sacchariflorus* forms about 80% of its tillers in spring. This may be the reason why *M. sacchariflorus* has a relatively short and more concentrated flowering period, whereas *M. sinensis* has a longer flowering time (Deuter, 2000). *M. sacchariflorus* from more northern locations has been reported to go dormant in autumn even when grown under greenhouse conditions (Deuter, 2000).

*M. sacchariflorus* has a high lignin-to-cellulose ratio similar to that of *M. × giganteus* (Hodgson et al., 2010). *M. sacchariflorus* loses its leaf sheaths early (Kaack and Schwartz, 2001) relative to *M. × giganteus*, which retains its leaf sheaths during the winter. Leaf sheaths attached to the culms improve plant resistance to lodging (Kaack and Schwartz, 2001). On the other hand, given that leaves generally have the highest mineral content in a plant (Lewandowski and Kicherer, 1997), selecting for accessions that readily senesce their leaves may be needed to improve combustion quality of the crop.

### 4.2.3 Section: Kariyasu

*M. tinctorius*

*M. tinctorius* is a diploid with 38 chromosomes (Adati, 1958). The name "Kariyasu" means "easy to cut" in Japanese, and it reflects the fact that *M. tinctorius* has long been utilized as fodder. The leaves are broader and thinner than *M. sinensis* or *M. sacchariflorus*, which makes it less likely to inflict damage to the skin when harvesting. Traditionally, it was also used for yellow dye ("tinctorius" came from "tinct" = color) (Sacks et al., 2013). *M. tinctorius* has small stature (1–1.8 m), sparse pubescence on the outer surface of leaves, smooth inner leaf surfaces, short rhizomes, awnless spikelets, and short callus hairs (Adati and Shiotani, 1962). The spikelets of *M. tinctorius* have short callus hairs, usually only half of the spikelet's length (Adati and Shiotani, 1962; Clayton et al., 2010). Short callus hairs will prevent wind dispersal of seeds, which could be crossed into *M. sinensis* to limit seed dispersal (Deuter, 2000). *M. tinctorius* is mainly distributed in the mountainous region of central Honshu, Japan (Adati, 1958). The flowering time of *M. tinctorius* is between August and October (Adati, 1958).

*M. oligostachyus*

*M. oligostachyus* is a diploid species (2n = 38). *M. oligostachyus* has short and slender rhizomes and, on average, reaches between 0.6 and 0.8 m in height. The outer leaf surface is smooth, but there is pubescence on the inner leaf surface. The spikelets have awns, and the callus hairs are 2–5 mm long (Adati and Shiotani, 1962). *M. oligostachyus* is distributed in the mountainous region of Kyushu to southeastern Tohoku in Japan (Adati, 1958; Adati and Shiotani, 1962). The flowering time of *M. oligostachyus* is from August to October (Adati, 1958).

*M. intermedius*

*M. intermedius* is a hexaploid with 114 chromosomes. *M. intermedius* has thick rhizomes, 1–2 m culm height, and its leaves are smooth on the outer surface, but hairy on the back surface. Its spikelets are 6–8 mm long, with 2–4 mm awns (Adati and Shiotani, 1962). The spikelets also have 5–7 mm long callus hairs, which are longer than those of *M. tinctorius* or *M. oligostachyus* (Adati and Shiotani, 1962). The distribution of the species is restricted to the northwestern part of Tohoku, Japan (Adati and Shiotani, 1962). *M. intermedius* flowers from August to October (Adati, 1958). Based on the morphological characteristics and cytological analysis, it is considered an amphipolyploid that originated as a cross between *M. tinctorius* and *M. oligostachyus* (Hirayoshi et al., 1956).

## 4.3 Natural Hybrids between *Miscanthus* Species

Being the center of origin of *Miscanthus* species, there are many overlapping populations of different *Miscanthus* species in Japan. In those overlapping populations, natural hybrids between *Miscanthus* species can be found. The occurrence of natural hybrids in Japan has been documented (Honda, 1939; Hirayoshi et al., 1957; Adati, 1958). Honda (1939) reported *Miscanthus* natural hybrid plants as tall as *M. sacchariflorus* but with awned spikelets similar to that of *M. sinensis*. Honda (1939) named the plant *Miscanthus ogiformis* Honda (ogi-susuki). Hirayoshi et al. (1957) found two hybrids grown from caryopses collected from *M. sinensis* in Gifu, Japan, but with morphological characteristics similar to that of *M. sacchariflorus*. He confirmed through cytological analysis that the hybrids were triploid with 57 chromosomes. Adati (1958) identified two triploid hybrids in Hyogo, Japan. Recently, three triploid hybrids were found in seeds collected from *M. sacchariflorus* that grew in Kushima, southern Japan (Nishiwaki et al., 2011). The plants had 57 chromosomes as confirmed through microscopic analysis. Through ITS and chloroplast DNA analysis, the plants were confirmed as hybrids between *M. sinensis* and *M. sacchariflorus* (Dwiyanti et al., unpublished data). Interestingly, based on ITS region analysis, the triploid hybrids were genetically different than the widely cultivated *M.* × *giganteus*.

## 4.4 Karyotype Analysis

Initially, *Miscanthus* was thought as a primitive genus in Andropogoneae subtribe of Poaceae. However, the basic chromosome number of *Miscanthus* is 19, whereas other genera within Andropogoneae have basic chromosome numbers of 9 or 10. Adati and Shiotani (1962) noted that *Miscanthus* genus has an amphidiploid type of origin from a cross between species with chromosome number 9 and species with chromosome number 10. Adati (1958) studied the karyotype of *M. sinensis* and *M. sacchariflorus*. *M. sinensis* has 38 chromosomes consisting of 22 chromosomes with submedian centromeres, 12 with median centromeres, and 4 with acrocentric centromeres. Two largest chromosomes have intercalary trabants (satellite). *M. sacchariflorus* has 76 chromosomes consisting of 42 chromosomes with submedian centromeres, 26 with median centromeres, and 8 with acrocentric centromeres. Contrary to expectations, *M. sacchariflorus* has only 2 SAT chromosomes, which is not different than diploid *Miscanthus*. It is suggested that *M. sacchariflorus* has two types of genomes, one of which is homologous to that of *M. sinensis* and another that is partially homologous to that of *M. sinensis* (Adati, 1958; Adati and Shiotani, 1962).

Linde-Laursen (1993) analysed the karyotype of *M.* × *giganteus* and discovered that *M.* × *giganteus* has only one SAT chromosome. According to Adati (1958), *M. ogiformis* also has only one SAT chromosome. Meanwhile, a triploid hybrid between *M. sinensis* and *M. sacchariflorus* from Shinoyama, Hyogo, had 57 chromosomes with 2 SAT chromosomes. Another triploid (Akaishi, Hyogo) had 57 chromosomes with 2 SAT chromosomes. It is interesting that the triploids had different number of SAT chromosomes, indicating differences in genome origin.

Besides microscopic evaluation, chromosomes of *M. sinensis* and *M. sacchariflorus* were evaluated using fluorescence *in situ* hybridization (FISH) and genomic *in situ* hybridization (GISH) (Takahashi and Shibata, 2002). FISH analysis used 18S–5.8S–26S and 5S rDNA probes to localize rRNA genes. 18S–5.8S–26S rDNA signals were detected at secondary constrictions of two SAT chromosomes of *M. sinensis*. The signals were also detected at secondary constrictions of SAT chromosomes of *M. sacchariflorus*, but Takahashi and Shibata (2002) found four SAT chromosomes in *M. sacchariflorus* instead of two as reported by Adati (1958). Although GISH probes used in the analysis were set such that hybridization occurred between sequences with more than 80% similarity, genomes constituting tetraploid *M. sacchariflorus* and diploid *M. sinensis* could not be differentiated. Since GISH probes detect repetitive DNA in the genome, the result indicated that the repetitive DNA in genomes of tetraploid *M. sacchariflorus* and diploid *M. sinensis* had undergone little change during *Miscanthus* species divergence.

## 4.5 Phylogenetic Relationships between *Miscanthus* Species

Hodkinson et al. (2002b) used amplified fragment length polymorphism (AFLP) markers to analyse genetic variation between *M. sinensis*, *M. sacchariflorus*, *M.* × *giganteus*, *M. oligostachyus*, *Miscanthus nepalensis*, *M. floridulus*, *M. sinensis* var. *condensatus*, and *Miscanthus transmorrisonensis* accessions. In accordance with Greef et al. (1997), Hodkinson et al. (2002b) found through AFLP analysis that *M.* × *giganteus* clones were clustered to a group between *M. sinensis* and *M. sacchariflorus* and detected wide variation within *M. sinensis*. *M. floridulus* and *M. sinensis* var. *condensatus* were assigned to the same cluster with *M. sinensis*, indicating close phylogenetic relationship between the three *Miscanthus* species. *Miscanthus* species endemic to Taiwan, *M. transmorrisonensis*, also had close phylogenetic relationship to *M. sinensis*. *M. oligostachyus* was separated from the rest of *Miscanthus* species. The grouping matched classification by Adati and Shiotani (1962), placing *M. sinensis* and *M. floridulus* under *Eumiscanthus*, *M. sacchariflorus* under *Triarrhena*, and *M. oligostachyus* under *Kariyasua* section.

Chou and Ueng (1992) used isozyme analysis method to analyse the phylogenetic relationship between *Miscanthus* species in Taiwan, *M. floridulus*, *M. transmorrisonensis*, *M. flavidus*, *M. sinensis*, and *M. sinensis* var. *formosanus*. Based on the isozyme analysis results, *M. sinensis* is the most primitive taxon, from which led the subsequent divergence of *M. sinensis* var. *formosanus*, *M. floridulus*, *M. flavidus*, and *M. transmorrisonensis*. Based on plastid and nrDNA, Hodkinson et al. (2002c) found that *M. sinensis*, *M. sacchariflorus*, *M. oligostachyus*, *M. transmorrisonensis*, and *M. floridulus* were monophyletic (developed from the same ancestor). Sugarcane (*Saccharum officinarum*) and its wild species *S. robustum* were closer to *M. sinensis* group than African *Miscanthus* (*M. junceus* and *M. ecklonii*) or Himalayan

*M. fucus* to *M. sinensis* group. In fact, African *Miscanthus* has basic chromosome number (x) of 15 instead of 19 as in *Miscanthus* spp. *M. junceus* and *M. ecklonii*. *M. junceus* and *M. ecklonii* are classified as *Miscanthidium*.

## 4.6 Genetic Improvement of *Miscanthus*

### 4.6.1 Germplasm Collection and Management

Most *Miscanthus* varieties commercially available were developed as ornamental plants (North America and Europe), fodder, dye, and roof material (Japan). Thus, it is important to discover *Miscanthus* varieties with characteristics suitable for bioethanol production (high yield, low lignin, and low mineral content) from wild germplasm.

Germplasms can be collected in the form of seeds or rhizomes, but seeds may be preferable over rhizomes because they can be stored for longer periods of time. Generally, seeds are stored at 5°–10°C for short-term storage and −20° or −196°C for longer storage (Gonzales-Benito et al., 1995). However, germinability may decrease after long-time storage. Matumura and Yukimura (1975) investigated germination rate of *M. sinensis*, *M. sacchariflorus*, and *M. tinctorius* seeds that were stored for 195 days at room temperature or at about 3°C for 40–195 days before germination. Seeds from the three *Miscanthus* species stored in cold temperatures were able to germinate at high rates after exposure to alternating temperature (30/20°C). Even seeds stored for 195 days at room temperature were able to germinate at relatively high percentages. However, seeds stored at room temperature for a year or more had significantly reduced germination (Deuter, 2000).

### 4.6.2 Artificial Hybridization

In parallel to finding natural triploids, artificial hybridizations have been attempted. Flowering time synchronization, parent compatibility, pollen amount, and morphological structure are important factors in determining seed-setting rate in hybridization (Hirayoshi et al., 1955; Deuter, 2000). Generally, seed-setting rate is highest in open pollination, followed by paired-cross and self-pollination (Hirayoshi et al., 1955; Deuter, 2000). Seed-setting rate by self-pollination is lower than 1%, due to self-incompatibility.

Besides *M. sinensis* and *M. sacchariflorus*, hybrids have been created between other *Miscanthus* species. These hybrids may be useful for cultivation in specific locations, such as sodic soils and tropical regions. Hybridization between *M. sinensis* and *M. sinensis* var. *condensatus* and *M. sinensis* and *M. tinctorius* resulted in $F_1$ hybrids that were self-incompatible, but could produce fertile pollen (Hirayoshi et al., 1959). A triploid and unexpected tetraploid were obtained from crossing between *M. sacchariflorus* and *M. sinensis* var. *condensatus* (Hirayoshi et al., 1960). Culm length, leaf length, leaf width, and ear size of both polyploids exceeded those of the parents. However, the triploid was sterile and hair length of spikelets, silky luster, width of lemma, length of awn, grass type, and evergreeness resembled *M. sacchariflorus*. Meanwhile, the tetraploid resembled more *M. sinensis* var. *condensatus* in those characteristics and the pollens were fertile. Matumura et al. (1986) investigated the rhizome structure of triploid, tetraploid, and their parents, and found that the triploid resembled *M. sacchariflorus* more while the tetraploid resembled *M. sinensis* var. *condensatus*. Adati and Shiotani (1962) also reported hybridization between *M. floridulus* and *M. sinensis*, producing $F_1$ hybrids with regular meiosis division with 19 bivalents at first metaphase.

### 4.6.3 Polyploidization

Colchicine and oryzalin were used to double the chromosomes of *Miscanthus* (Petersen et al., 2002; Glowacka et al., 2009). Glowacka et al. (2009) applied colchicine in various concentrations and exposure times to induce polyploids from *M. sinensis* and *M.* × *giganteus*. Higher colchicine concentrations reduced the survival rate and tillering rate of the plants. Longer exposure time did not have any effect on plant survival rate, but it significantly reduced polyploidization rate. Plant genotypes also influenced the polyploidization rate in *M. sinensis* genotypes, ranging from 0.77 to 3.82%.

Generating hexaploid *M.* × *giganteus* to restore fertility may be a way to improve *M.* × *giganteus* through conventional breeding. Yu et al. (2009) generated hexaploid plants from *M.* × *giganteus*. The team induced calli growth from immature inflorescence tissue and treated the calli with colchicine or oryzalin in various concentrations and exposure times. The rate of calli survival was generally higher in calli treated with colchicine, but more hexaploids generated from calli treated with oryzalin at most of the concentrations tested. The hexaploid plants had slightly broader stems and larger stomata size compared to the triploid plants.

## 4.7 Variations in Several Agronomical Traits Related to Yield and Plant Performance

### 4.7.1 Variation in Flowering Time

Flowering time is one of several factors to be considered in selecting high-yielding *Miscanthus* (Clifton-Brown et al., 2001). When plants start flowering, they will stop growing and relocate their nutrients into seeds or rhizomes (Wingler et al., 2010). *Miscanthus* plants that flower late or never flower tend to yield higher than those that flower early (Clifton-Brown et al., 2001). On the other hand, choosing plants with similar flowering times is important to ensure success in artificial pollination.

Jensen et al. (2011) observed the flowering time of various *M. sinensis* for 3 years at Aberystwyth, UK. The plants were collected from Japan, China, Korea, and Russia (latitude range 32.2–43.6°N). The onset of flowering time varied from mid-June to late November. Accessions from more northern areas and higher altitudes tended to flower earlier than accessions from southern areas or lower altitudes. In its original habitat, *M. sinensis* plants in northern Japan tended to start heading 2 months earlier than those in southern Japan (Adati, 1958). Similarly, *M. sinensis* in high-altitude regions tended to flower earlier than those in low-altitude regions (Adati, 1958). In addition, time range of ear emergence of *M. sinensis* plants from higher altitudes was longer in duration than *M. sinensis* plants from lower altitudes (Adati, 1958). Moreover, *M. sinensis* at higher altitudes tended to develop faster than plants at lower altitudes (Adati, 1958). Further analyses about relationships between early flowering and faster growth rates are needed.

### 4.7.2 Variation in Cold Tolerance

*M.* × *giganteus* is a promising bioenergy crop for plantation in temperate region, since the yield is still high at low temperature compared to other $C_3$ or $C_4$ plants such as maize (Dohleman and Long, 2009). However, at some north European areas, many *M.* × *giganteus* rhizomes cannot survive winter conditions, particularly in the first season after planting (Greef et al., 1997). Clifton-Brown and Lewandowski (2000) compared cold tolerance in one *M. sacchariflorus*, two *M.* × *giganteus*, and two *M. sinensis* (Sin-H6 and Sin-H9) populations. Based on artificial freezing tests, the lethal temperature at which 50% of the rhizomes killed (LT$_{50}$) for *M.* × *giganteus* and *M. sacchariflorus* genotypes was −3.4°C, while LT$_{50}$ for *M. sinensis* genotype Sin-H6 was −6.3°C, and Sin-H9 was about −5°C. It is suggested that low moisture content contributes to frost resistance (Clifton-Brown and Lewandowski, 2000).

Cold tolerance can also be associated with early shoot emergence in spring and shoot tolerance to frost. Low temperatures during spring can delay the timing for shoot emergence, subsequently reduce growth time, and reduce *Miscanthus* yields in autumn (Farrell et al., 2006). In addition, frost that comes after shoot emergence can kill shoots and reduces shoot number. *M. sinensis* is able to grow shoots as low as 7°C, lower than *M. sacchariflorus* or *M.* × *giganteus* (Farrell et al., 2006). Although leaves that are tolerant to frost and cold during spring have lower moisture content as in cold-tolerant rhizomes, *M. sinensis* genotype with the highest rhizome frost tolerance was less tolerant in terms of shoot tolerance, indicating that shoot and rhizome frost tolerance are not correlated.

### 4.7.3 Variation in Lignin, Cellulose, and Mineral Content

For bioethanol production, *Miscanthus* with low lignin content is preferred because high lignin content decreases the efficiency of fermentation process that is an important component in bioethanol production and increases the cost of pretreatment of bioethanol production (Mosier et al., 2005; Hodgson et al., 2010). Moreover, high lignin content is associated with the production of unstable and viscous bio-oils (Fahmi et al., 2008). However, high lignin concentration gives higher heating value that is preferable for combustion (Demirbas, 2001). From the analysis of cell wall composition of *M. sinensis*, *M. sacchariflorus*, and *M.* × *giganteus* accessions grown in five locations in Europe, either in winter or in autumn harvest, *M. sacchariflorus* and *M.* × *giganteus* constantly showed higher lignin content and lower hemicellulose content than *M. sinensis* (Hodgson et al., 2010). There is lignin content variation between *M. sinensis* genotypes, with EMI09 showing the lowest lignin content in autumn and winter harvest (6.9% and 8.8%, respectively) and EMI11 showing the highest (8.2% and 9.1%, respectively). These genotypes can be used as parents for genetic mapping populations to elucidate genetic regulation of lignin content in *M. sinensis*.

Potassium, chlorine, sulfur, phosphorus, and calcium are important minerals for plants. However, during biomass combustion process, these minerals can react with each other and with silica, forming slag (sticky glass-like silicates and oxides) or corrosive alkali sulfates. These compounds can cause serious problems to combustion power plants (Obenberger et al., 1997). Chlorine forms chloric acid and dioxin in combustion process. Therefore, *Miscanthus* plants with low mineral content are preferred for biofuel production. Lewandowski and Kicherer (1997) showed that mineral contents were lower in *M.* × *giganteus* that were grown in warm locations than in those from humid and cool locations. Fertilizer application also affected mineral content in the plants. Ash and potassium concentrations increased with increasing potassium fertilizer application. Although further studies are needed, potassium fertilizer application could increase chlorine concentration. Chlorine, potassium, and ash concentrations were lower in February than in December; probably the minerals had been leached from plant materials by rainfall during winter. Thus, plants harvested in February had better combustion quality than those harvested in December. Compared to *M. × giganteus*, *M. sinensis* has lower potassium and chlorine (Jorgensen, 1997). There were variations in mineral content among *M. sinensis* genotypes; later-ripening genotypes showed higher yield and N and K concentrations compared to early-ripening genotypes.

Silicon is important for grass species to increase the plant resistance to lodging and drought; improve disease, insect, and nematode resistance; improve soil nutrient availability; and improve reproductive fertility (Woli et al., 2011). However, silicon will react with aluminum, chlorine, potassium, and other alkalis to form slag during the combustion process. Woli et al. (2011) found that Si concentration in *M. × giganteus* plants from several locations in the United States ranged from 0.72 to 1.62%, indicating that soil type influences Si uptake. Si concentration in *M. sinensis* ranged between 0.81% and 3.56%.

## 4.8 Molecular Resources

### 4.8.1 Development of Linkage Map for Miscanthus

Genetic map is useful for identification and characterization of genes regulating traits related to biomass production. The first genetic map for *Miscanthus* was developed based on random amplified polymorphic DNA (RAPD) markers because of limited information about *Miscanthus* genome (Atienza et al., 2002). A genetic linkage map was constructed based on genotypic information of 89 individuals derived from a cross between parents $F_{1.1}$ and $F_{1.7}$, two progenies of a cross between two *M. sinensis* that showed different phenotypes. The genetic map consists of 257 markers that are mapped into 28 linkage groups with total map length = 1074.5 cM (Atienza et al., 2002). The linkage groups outnumbered the haploid chromosome number of *M. sinensis* ($x = 19$) because several linkage groups contained only a few markers. More markers need to be developed and mapped to complete the whole *Miscanthus* genome.

Besides RAPD markers, simple sequence repeats (SSR) and restriction fragment length polymorphism (RFLP) markers were also used in genetic map construction (Atienza et al., 2002). Because SSR and RFLP markers were developed based on genome information of other species, genetic map based on SSR or RFLPs can be compared to genetic maps of other grass species, which will be useful in comparative studies (Hernandez et al., 2001). Thirty-nine RFLP probes from oat, barley, wheat, rice, and maize were tested for cross-hybridization with *M. sinensis*. Among the probes tested, maize has the highest hybridization rate (100%), followed by rice (80%), oat (40%), and barley (42.8%). SSR markers developed from maize were also reproducible and showed polymorphisms at high rate in *M. sinensis*, *M. sacchariflorus*, *M. × giganteus*, and *M. sinensis* var. *condensatus* (Hernandez et al., 2001). Therefore, maize could be a useful source for RFLP and SSR markers for *Miscanthus* (Hernandez et al., 2001). Besides maize, genetic markers may be developed based on genomic information from sorghum and grass model plant *Brachypodium distachyon* (Swaminathan et al., 2010; Zhao et al., 2011). SSR markers can also be developed from *M. sinensis* RAPD markers using PCR isolation microsatellite array method (Hung et al., 2009).

### 4.8.2 QTL Analysis of Traits Related to Yield and Mineral Content

Atienza et al. (2003a, 2003b) analysed quantitative trait loci (QTLs) related to yield (based on stem diameter, total height, flag-leaf height, total yield, stem yield, tops yield, and leaves yield). Three QTLs for stem diameter, three QTLs for total height, and one QTL for flag-leaf height were observed in 2000 (Atienza et al., 2003a). In 2001, one QTL for stem diameter, three QTLs for total height, and two QTLs for flag-leaf height were detected. Multiple QTLs were detected for total yield, stem yield, tops yield, and leaf yield, but only two QTLs for total yield and one QTL for tops yield were consistently observed in both years (Atienza et al., 2003b). Since *Miscanthus* takes 3 years to reach full establishment (Lewandowski et al., 2000; Clifton-Brown and Lewandowski, 2000), QTLs detected in the first year (field establishment period) may correlate to genes that regulate early plant growth, while QTLs detected in the third year (full establishment) may correlate to genes that regulate plant growth after establishment (Atienza et al., 2003a, 2003b).

QTLs related to chlorine (Cl), potassium (K), calcium (Ca), sulfur (S), and phosphorus (P) have been studied by Atienza et al. (2003c, 2003d). Mineral content changes depending on plant age (Clifton-Brown and Lewandowski, 2002). This was reflected in the QTL result in which different QTLs were observed in the first year and the second year of analysis. Therefore, in performing QTL analysis in *Miscanthus*, geneticists must consider both QTLs that are stable across years and environments and QTLs that only appear at specific development stage.

### 4.8.3 *Molecular Markers for Hybrids Identification*

DNA and molecular markers are effective in determining the genetic background of plants of a given species at an early development stage. Thus, these markers may be applied to detect hybrids at seed or seedling stages. Until now, AFLP marker and variations in ITS region of 18S–5.8S–25S ribosomal DNA have been utilized to confirm that *M.* × *giganteus* is a hybrid between *M. sinensis* and *M. sacchariflorus* (Hodkinson et al., 2002a). Besides AFLP and ITS region sequence, inter-simple sequence repeats (ISSR) markers may be used to detect hybrids because they can distinguish different *Miscanthus* species (Hodkinson et al., 2002b). Although three types of markers are available for hybrids detection, geneticists may need to develop SSR and SNP markers because these are the most convenient to analyse large populations.

## 4.9 Transgenic *Miscanthus*

Since *M.* × *giganteus* is sterile, and genomic information of *Miscanthus* is still very limited, another approach to improve *M* × *giganteus* quality is genetic modification. The genetic transformation method for *Miscanthus* is still in its infancy, because at present, there is only one detailed report of genetic transformation in *Miscanthus* (Wang et al., 2011). Particle bombardment-mediated transformation systems were used to transform *M. sinensis*. Wang et al. (2011) tested the compact callus production potential in 18 accessions of *M. sinensis*, which were collected from various sites in Japan. Compact callus production rate is important in transgenic production because in sugarcane, compact callus is embryogenic while the other type of callus, that is, "soft callus" is not (Ho and Vasil, 1983). Among 18 accessions, an accession from Tanegashima Island, Japan, showed the highest rate of compact callus production (25%). *Agrobacterium tumefaciens*-mediated transformation of *Miscanthus* was reported by Engler and Chen (2009) and Kim (2010). In Engler and Chen (2009), calli were generated from germinating seeds of *M. sinensis*. Kim (2010) used the transformation method to downregulate the lignin biosynthesis pathway of *M.* × *giganteus*.

## 4.10 Future Studies

High yield of *Miscanthus* makes it promising as a biomass crop; however, *Miscanthus* is a new crop that needs development. Currently, information about *Miscanthus* characteristics as bioenergy crop is limited to *M.* × *giganteus* and several *M. sinensis* accessions. Collecting *Miscanthus* germplasms from various latitudes, soil types, and climate regions and evaluating traits related to yield, flowering time, cold tolerance, lignin, cellulose, and mineral content are needed to broaden genetic resources for *Miscanthus* breeding. Besides *M. sinensis* and *M. sacchariflorus*, other *Miscanthus* species can also be utilized as breeding materials for their unique characteristics such as salt tolerance of *M. sinensis* var. *condensatus* or adaptability to tropical climate of *M. floridulus*. Hybridization between different *Miscanthus* species is possible because of high outcrossing rate and self-incompatibility of *Miscanthus*. Further studies are needed to analyse the hybridization compatibility between different *Miscanthus* species.

Perennial habit and complexity of the genome are limiting factors in elucidating genetic basis of *Miscanthus* agronomical traits and quality traits. Recent advances of "-omics" technologies will accelerate the progress of whole genome sequencing, genetic marker development, and elucidation of physiological process in *Miscanthus*.

## References

Adati S. Studies on genus *Miscanthus* with special reference to the Japanese species for breeding purpose as fodder crops. Bull Fac Agron Mie Univ, 1958; 17: 1–112 (in Japanese).

Adati S, Mitsuishi S. Wild growing forage plants of the Far East, especially Japan, suitable for breeding purposes. Part III. Cultivation of Hatizyo-Susuki (*Miscanthus sinensis* var. *condensatus* Makino) in Hatizyo-Island. Bull Fac Agric Mie Univ, 1956; 12: 7–12 (in Japanese).

Adati S, Shiotani I. The cytotaxonomy of the genus *Miscanthus* and its phylogenic status. Bull Fac Agric Mie Univ, 1962; 25: 1–14.

An GH, Miyakawa S, Kawahara A, Osaki M, Ezawa T. Community structures of arbuscular mycorrhizal fungi associated with pioneer grass species *Miscanthus sinensis* in acid sulfate soils: habitat segregation along pH gradients. Soil Sci Plant Nutr, 2008; 54: 517–528.

Arduini I, Ercoli L, Mariotti M, Masoni A. Response of *Miscanthus* to toxic cadmium applications during the period of maximum growth. Environ Exp Bot, 2006; 55: 29–40.

Atienza SG, Satovic Z, Petersen KK, Dolstra O, Martin A. Preliminary genetic linkage map of *Miscanthus sinensis* with RAPD markers. Theor Appl Genet, 2002; 105: 946–952.

Atienza SG, Satovic Z, Petersen KK, Dolstra O, Martin A. Identification of QTLs associated with yield and its components in *Miscanthus sinensis* Anderss. Euphytica, 2003a; 132: 353–361.

Atienza SG, Satovic Z, Petersen KK, Dolstra O, Martin A. Identification of QTLs influencing agronomic traits in *Miscanthus sinensis* Anderss. I. Total height, flag-leaf height and stem diameter. Theor Appl Genet, 2003b; 107: 123–129.

Atienza SG, Satovic Z, Petersen KK, Dolstra O, Martin A. Identification of QTLs influencing combustion quality in *Miscanthus sinensis* Anderss. II. Chlorine and potassium content. Theor Appl Genet, 2003c; 107: 857–863.

Atienza SG, Satovic Z, Petersen KK, Dolstra O, Martin A. Influencing combustion quality in *Miscanthus sinensis* Anderss.: identification of QTLs for calcium, phosphorus and sulphur content. Plant Breed, 2003d; 122: 141–145.

Beale CV, Long SP. Can perennial C$_4$ grasses attain high efficiencies of radiant energy conversion in cool climates? Plant Cell Environ, 1995; 18: 641–650.

Chen SL, Renvoize SA. *Miscanthus*. In: Wu ZY, Raven PH, Hong DY (eds) *Flora of China*, Vol. 22, 2006, pp. 581–583. Science Press, Missouri Botanical Garden Press, St Louis.

Chiang YC, Schaal BA, Chou CH, Huang S, Chiang TY. Contrasting selection modes at the *Adh1* locus in outcrossing *Miscanthus sinensis* vs. inbreeding *Miscanthus condensatus* (Poaceae). Am J Bot, 2003; 90: 561–570.

Chou CH, Chiang TY, Chiang YC. Towards an integrative biology research: a case study on adaptive and evolutionary trends of *Miscanthus* populations in Taiwan. Weed Biol Manage, 2001; 1: 81–88.

Chou CH, Ueng JJ. Phylogenetic relationship among species of *Miscanthus* populations in Taiwan. Bot Bull Acad Sin, 1992; 33: 63–73.

Clayton WD, Harman KT, Williamson H. GrassBase—the online world grass flora, 2010. http://www.kew.org/data/grasses-db.html. Last accessed 11 December 2012.

Clifton-Brown JC, Breuer J, Jones MB. Carbon mitigation by the energy crop, Miscanthus. Global Change Biol, 2007; 13: 2296–2307.

Clifton-Brown JC, Lewandowski I. Overwintering problems of newly established *Miscanthus* plantations can be overcome by identifying genotypes with improved rhizome cold tolerance. New Phytol, 2000; 148: 287–294.

Clifton-Brown JC, Lewandowski I. Screening Miscanthus genotypes in field trials to optimise biomass yield and quality in Southern Germany. Eur J Agron, 2002; 16: 97–110.

Clifton-Brown JC, Lewandowski I, Andersson B, Basch G, Christian DG, Kjeldsen JB, Jorgensen U, Mortensen JV, Riche AB, Schwarz KU, Tayebi K, Teixeira F. Performance of 15 Miscanthus genotypes at five sites in Europe. Agron J, 2001; 93: 1013–1019.

Demirbas A. Relationships between lignin contents and heating values of biomass. Energ Convers Manage, 2001; 42: 183–188.

Deuter M. Breeding approaches to improvement of yield and quality in *Miscanthus* grown in Europe. In: Lewandowski I, Clifton-Brwon J (eds) *European Miscanthus Improvement (FAIR3 CT-96-1392)*, Final Report, 2000, pp. 28–52. University of Hohenheim, Germany.

Dohleman FG, Long SP. More productive than maize in the Midwest: how does *Miscanthus* do it? Plant Physiol, 2009; 150: 2104–2115.

Engler D, Chen J. Transformation and engineered trait modification in *Miscanthus* species. US Patent US 2011/0047651 A1, 2009.

Ezaki B, Nagao E, Yamamoto Y, Nakashima S, Enomoto T. Wild plants, *Andropogon virginicus* L. and *Miscanthus sinensis* Anders., are tolerant to multiple stresses including aluminum, heavy metals and oxidative stresses. Biotic Abiotic Stress, 2008; 27: 951–961.

Fahmi R, Bridgwater AV, Donnison I, Yates N, Jones JM. The effect of lignin and inorganic species in biomass on pyrolysis oil yields, quality and stability. Fuel, 2008; 87: 1230–1240.

Farrell AD, Clifton-Brown JC, Lewandowski I, Jones MB. Genotypic variation in cold tolerance influences the yield of *Miscanthus*. Ann Appl Biol, 2006; 149: 337–345.

Glowacka K, Jezowaski S, Kaczmarek Z. Polyploidization of *Miscanthus sinensis* and *Miscanthus x giganteus* by plant colchicine treatment. Ind Crops Products, 2009; 30: 444–446.

Gonzales-Benito ME, Iriondo JM, Pita JM, Perez-Garcia F. Effects of seed cryopreservation and priming on germination in several cultivars of *Apium graveolens*. Ann Bot, 1995; 75: 1–4.

Greef JM, Deuter M, Jung C, Schondelmaier J. Genetic diversity of European *Miscanthus* species revealed by AFLP fingerprinting. Genet Resources Crop Evol, 1997; 44: 185–195.

Hansen EM, Christensen BT, Jensen LS, Kristensen K. Carbon sequestration in soil beneath long-term *Miscanthus* plantation as determined by $^{13}$C abundance. Biomass Bioenerg, 2004; 26: 97–105.

Heaton EA, Dohleman FG, Long SP. Meeting US biofuel goals with less land: the potential of *Miscanthus*. Global Change Biol, 2008; 14: 2000–2014.

Hernandez P, Dorado G, Laurie DA, Martin A. Microsatellites and RFLP probes from maize are sufficient sources of molecular markers for the biomass energy crop *Miscanthus*. Theor Appl Genet, 2001; 102: 616–622.

Hill J, Nelson E, Tilman D, Polasky S, Tiffany D. Environmental, economic, and energetic costs and benefits of biodiesel and ethanol biofuels. Proc Natl Acad Sci U S A, 2006; 103: 11206–11210.

Hirayoshi I, Nishikawa K, Hakura A. Cyto-genetical studies on forage plants (VIII) 3x- and 4x-hybrid arisen from the cross *Miscanthus sinensis* var. *condensatus* x *Miscanthus sacchariflorus*. Res Bull Fac Agric Gifu Univ, 1960; 12: 82–88 (in Japanese).

Hirayoshi I, Nishikawa K, Kato R. Cytogenetic studies on forage plants (IV) self-incompatibility in *Miscanthus*. JPN J Breed, 1955; 5: 167–170 (in Japanese).

Hirayoshi I, Nishikawa K, Kubono M. Cyto-genetical studies on forage plants (V) polyploidy and distribution in *Miscanthus* sect. *Kariyasua* Ohwi. Res Bull Fac Agric Gifu Univ, 1956; 7: 9–14 (in Japanese).

Hirayoshi I, Nishikawa K, Kubono M, Murase T. Cyto-genetical studies on forage plants (VI) on the chromosome number of Ogi (*Miscanthus sacchariflorus*). Res Bull Fac Agric Gifu Univ, 1957; 8: 8–13 (in Japanese).

Hirayoshi I, Nishikawa K, Kubono M, Sakaida T. Cyto-genetical studies on forage plants (VII) chromosome conjugation and fertility of *Miscanthus* hybrids including *M. sinensis*, *M. sinensis* var. *condensatus* and *M. tinctorius*. Res Bull Fac Agric Gifu Univ, 1959; 11: 86–91 (in Japanese).

Ho W, Vasil IK. Somatic embryogenesis in sugarcane (*Saccharum officinarum* L.): growth and regeneration from embryogenic cell suspension cultures. Ann Bot, 1983; 51: 719–726.

Hodgson EM, Lister SJ, Bridgwater AV, Clifton-Brown J, Donnison IS. Genotypic and environmentally derived variation in the cell wall composition of *Miscanthus* in relation to its use as a biomass feedstock. Biomass Bioenerg, 2010; 34: 652–660.

Hodgson EM, Nowakowski DJ, Shield I, Riche A, Bridgwater AV, Clifton-Brown JC. Variation in *Miscanthus* chemical composition and implications for conversion by pyrolysis and thermo-chemical bio-refining for fuels and chemicals. Bioresource Technol, 2011; 102: 3411–3418.

Hodkinson TR, Chase MW, Takahashi C, Leitch IJ, Bennet MD, Renvoize SA. The use of DNA sequencing (ITS and *trnl-F*), AFLP, and fluorescent in situ hybridization to study allopolyploid *Miscanthus* (Poaceae). Am J Bot, 2002a; 89: 279–286.

Hodkinson TR, Chase MW, Takahashi C, Leitch IJ, Bennet MD, Renvoize SA. Characterization of a genetic resources collection for *Miscanthus* (Saccharineae, Andropogoneae, Poaceae) using AFLP and ISSR PCR. Ann Bot, 2002b; 89: 627–636.

Hodkinson TR, Chase MW, Takahashi C, Leitch IJ, Bennet MD, Renvoize SA. Phylogenetics of *Miscanthus*, *Saccharum* and related genera (Saccharineae, Andropogoneae, Poaceae) based on DNA sequencing from ITS nuclear ribosomal GNA and plastics *trnL* intron and *tml-F* intergenic spacers. J Plant Res, 2002c; 115: 381–392.

Honda M. Nuntia ad floram Japonie 38 (New report of plants in Japan XXXVIII). Bot Mag, 1939; 53: 144 (in Japanese).

Hung KH, Chiang TY, Chiu CT, Hsu TW, Ho CW. Isolation and characterization of microsatellite loci from a potential biofuel plant *Miscanthus sinensis* (Poaceae). Conserv Genet, 2009; 10: 1377–1380.

Jensen E, Farrar K, Thomas-Jones S, Hastings A, Donnison I, Clifton-Brown J. Characterization of flowering time diversity in *Miscanthus* species. Global Change Biol Bioenerg, 2011; 3: 387–400. doi:10.1111/j.1757-1707.2011.01097.x.

Jorgensen U. Genotypic variation in dry matter accumulation and content of N, K and Cl in *Miscanthus* in Denmark. Biomass Bioenerg, 1997; 12: 155–169.

Kaack K, Schwartz K-U. Morphological and mechanical properties of *Miscanthus* in relation to harvesting, lodging, and growth conditions. Ind Crops Products, 2001; 14: 145–154.

Kayama M. Comparison of the aluminum tolerance of *Miscanthus sinensis* Anderss. and *Miscanthus sacchariflorus* Bentham in hydroculture. Int J Plant Sci, 2001; 162: 1025–1031.

Kim HS. Functional studies of lignin biosynthesis genes and putative flowering gene in *Miscanthus* x *giganteus* and studies on indolyl glucosinolate biosynthesis and translocation in *Brassica oleracea*. Dissertation, 2010. University of Illinois at Urbana-Champaign.

Koyama T. *Grasses of Japan and Its Neighboring Regions: An Identification Manual*, 1987, pp. 1–582. Kondansha Ltd, Tokyo.

Lafferty J, Lelley T. Cytogenetic studies of different *Miscanthus* species with potential for agricultural use. Plant Breed, 1994; 113: 246–249.

Leandowski I, Clifton-Brown JC, Andersson B, Basch G, Christian DG, Jorgensen U, Jones MB, Rich AB, Scurlock JMO, Huisman W. *Miscanthus*: European experience with a novel energy crop. Biomass Bioenerg, 2000; 19: 209–227.

Lewandowski I, Kicherer A. Combustion quality of biomass: practical relevance and experiments to modify the biomass quality of *Miscanthus* x *giganteus*. Eur J Agron, 1997; 6: 163–177.

Linde-Laursen I. Cytogenetic analysis of *Miscanthus* 'Giganteus', an interspecific hybrid. Hereditas, 1993; 119: 297–300.

Matumura M, Hakumura Y, Saijoh Y. Ecological aspects of *Miscanthus sinensis* var. *condensatus* × *M. sacchariflorus* and their 3x- 4x-hybrids (2) growth behaviour of the current year's rhizomes. Res Bull Fac Agric Gifu Univ, 1986; 51: 347–362 (in Japanese).

Matumura M, Yukimura T. Fundamental studies on artificial propagation by seeding useful wild grasses in Japan. VI. Germination behaviors of three native species of genus *Miscanthus*; *M. sacchariflorus*, *M. sinensis*, and *M. tinctorius*. Res Bull Fac Agric Gifu Univ, 1975; 38: 339–349 (in Japanese).

Mosier N, Wyman C, Dale B, Elander R, Lee YY, Holtzapple M, Ladisch M. Features of promising technologies for pretreatment of lignocellulosic biomass. Bioresource Technol, 2005; 96: 673–686.

Nishiwaki A, Mizuguti A, Kuwabara S, Toma Y, Ishigaki G, Miyashita T, Yamada T, Matuura H, Yamaguchi S, Rayburn AL, Akashi R, Stewart JR. Discovery of natural *Miscanthus* (Poaceae) triploid plants in sympatric populations of *Miscanthus sacchariflorus* and *Miscanthus sinensis* in southern Japan. Am J Bot, 2011; 98: 154–159.

Numata N. Geographical distribution and ecology of *Miscanthus sinensis*. New Pesticide, 1970; 24: 8–16 (in Japanese).

Obenberger I, Bidermann F, Widmann W, Riedl E. Concentrations of inorganic elements in biomass fuels and recovery in the different ash fractions. Biomass Bioenerg, 1997; 12: 211–224.

Ohtsuka T, Sakura T, Ohsawa M. Early herbaceous succession along a topographical gradient on forest clear-felling sites in mountainous terrain, central Japan. Ecol Res, 1993; 8: 329–340.

Petersen KK, Hagberg P, Kristiansen K, Forkmann G. *In vitro* chromosome doubling of *Miscanthus sinensis*. Plant Breed, 2002; 121: 445–450.

Quinn LD, Matlaga DP, Stewart JR, Davis AS. Empirical evidence of long-distance dispersal in *Miscanthus sinensis* and *Miscanthus* x *giganteus*. Invasive Plant Sci Manage, 2011; 4: 142–150.

Sacks EJ, Juvik JA, Lin Q, Stewart JR, Yamada T. The gene pool of *Miscanthus* species and its improvement. In: Paterson, AH (ed.) *Genomics of the Saccharinae*, 2013, pp. 73–101. Springer, New York.

Stewart RJ, Toma Y, Fernandez FG, Nishiwaki A, Yamada T, Bollero G. The ecology and agronomy of *Miscanthus sinensis*, a species important to bioenergy crop development in its native range in Japan: a review. Global Change Biol Bioenerg, 2009; 1: 126–153.

Swaminathan K, Alabady MS, Varala K, De Paoli E, Ho I, Rokhsar DS, Arumuganathan AK, Ming R, Green PJ, Meyers BC, Moose SP, Hudson ME. Genomic and small RNA sequencing of *Miscanthus* × *giganteus* shows the utility of sorghum as a reference genome sequence for Andropogoneae grasses. Genome Biol, 2010; 11: R12.

Takahashi C, Shibata F. Analysis of *Miscanthus sacchariflorus* and *M. sinensis* chromosomes by fluorescence in situ hybridization using rDNA and total genomic DNA probes. Chromosome Sci, 2002; 6: 7–11.

Wang X, Yamada T, Kong F-J, Abe Y, Hoshino Y, Sato H, Akamizo T, Kanazawa A, Yamada T. Establishment of an efficient *in vitro* culture and particle bombardment-mediated transformation systems in *Miscanthus sinensis* Anderss., a potential bioenergy crop. Global Change Biol Bioenerg, 2011; 3: 322–332.

Wingler A, Purdy SJ, Edwards SA, Chardon F, Masclaux-Daubresse C. QTL analysis for sugar-regulated leaf senescence supports flowering-dependent and – independent senescence pathways. New Phytol, 2010; 185: 420–433.

Woli KP, David MB, Tsai J, Voigt TB, Darmody RG, Mitchell CA. Evaluation silicon concentrations in biofuel feedstock crops *Miscanthus* and switchgrass. Biomass Bioenerg, 2011; 35: 2807–2813.

Yamashita H, Gau M, Eguchi K, Takai T. Exploration and collection of *Miscanthus* species in Kumamoto Prefecture, Japan. Annu Rep Explor Introd Plant Genet Resources, 2010; 26: 58–64 (in Japanese).

Yu CY, Kim HS, Rayburn L, Widholm JM, Juvik JA. Chromosome doubling of the bioenergy crop, *Miscanthus* x *giganteus*. Global Change Biol Bioenerg, 2009; 1: 404–412.

Zhao H, Yu J, You FM, Luo M, Peng J. Transferability of microsatellite markers from *Brachypodium distachyon* to *Miscanthus sinensis*, a potential biomass crop. J Integr Plant Biol, 2011; 53: 232–245.

# Chapter 5
# Breeding *Miscanthus* for Bioenergy

John Clifton Brown[1], Paul Robson[1], Chris Davey[1], Kerrie Farrar[1], Charlotte Hayes[1], Lin Huang[1], Elaine Jensen[1], Laurence Jones[1], Maurice Hinton-Jones[1], Anne Maddison[1], Heike Meyer[2], John Norris[1], Sarah Purdy[1], Charlie Rodgers[3], Kai-Uwe Schwarz[2], Cosentino Salvatore,[4] Gancho Slavov[1], John Valentine[1], Richard Webster[1], Sue Youell[1], and Iain Donnison[1]

[1]*Institute of Biological, Environmental and Rural Sciences (IBERS), Aberystwyth University, Gogerddan, Aberystwyth, Ceredigion, Wales, SY23 3EB, UK*
[2]*Julius Kühn-Institute (JKI) Federal Research Centre for Cultivated Plants, Bundesallee 50, D-38116 Braunschweig, Germany*
[3]*Ceres, Inc., 3199 Co. Rd. 269 E., Somerville, TX 77879, USA*
[4]*Department of Agriculture and Food Science, University of Catania, Via Valdisavoia 5, 95123 Catania, Italy*

## 5.1 Introduction

Highly adapted *Miscanthus* hybrids can make a significant global contribution to the mitigation of anthropogenic climate change via sustainable substitution of fossil fuels within the next decade. In this chapter we review the biological potential of *Miscanthus* resulting from a plethora of naturally occurring variations in traits relevant to yield and we outline the breeding approaches currently being applied to *Miscanthus*. The chapter concludes with a perspective on the socioeconomic value of bioenergy from *Miscanthus*.

## 5.2 *Miscanthus* as a Biomass Crop

*Miscanthus* is native to Eastern Asia, its distribution stretching from the equator to approximately 50°N. It is commonly found growing on roadsides, at field boundaries in the plains, and on the mountainsides up to 3000 m. *Miscanthus* is a member of the $C_4$ grass tribe Andropogoneae, which includes maize, sorghum, and sugarcane (Daniels and Roach, 1987). $C_4$ photosynthesis is well known to confer advantages, in comparison with $C_3$ photosynthesis, through increased water and nitrogen use efficiency per gram of biomass produced (Brutnell et al., 2010). *Miscanthus* is unusual within the $C_4$ grasses in including accessions that retain high yields in cool climates. For example, physiologists have measured exceptionally high rates of photosynthesis in *M.* × *giganteus* leaves grown at temperatures as low as 14°C (Wang et al., 2008). *Miscanthus* is therefore of interest as a sustainable biomass crop across a wide range of diverse geographic zones and is not restricted to the tropics as are other high-yielding crops such as sugarcane. Furthermore, *Miscanthus* is perennial, harvested in early spring when the majority of nutrients have been withdrawn from the stems and translocated to the rhizome. This nutrient translocation is an important trait as the net energy of a biomass crop is calculated by subtracting the inputs from the outputs, and thus a crop with no annual planting cycle and low fertilizer requirements is highly desirable. Moreover, the soil carbon sequestration potential of a perennial crop is far higher than that of an annual crop where annual ploughing depletes soil organic carbon.

European interest in *Miscanthus* began in the early twentieth century with collections from Asia, primarily for taxonomic purposes. *Miscanthus* breeding originally began in Germany in the 1960s for the production of horticultural varieties. This resulted in many well-known ornamentals, such as Silberfeder, Zebrinus, and Variegatus, which can be bought in garden centers. A few of these ornamentals are of interest for biomass, especially for growth on poorer soils. The potential of *Miscanthus* as a dedicated biomass crop was first recognized in Denmark during the 1980s and focused on a Japanese accession collected in the 1930s widely known as *M. × giganteus* (Greef and Deuter, 1993). *M. × giganteus* has been the subject of intensive research for more than 20 years and is still the leading *Miscanthus* type grown as an energy crop today. Establishment of breeding programs for improved bioenergy potential began in the late 1980s with several germplasm collection missions in Asia. Dr. Martin Deuter, a breeder based in the private company Tinplant (Klein Wanzleben, Germany), made selections and crosses for more than a decade. Deuter's early hybrids were tested against *M. × giganteus* in the European *Miscanthus* Improvement programme (EMI, 1997–2000) and several showed strong potential with improved drought and frost tolerance (Clifton-Brown et al., 2001; Lewandowski et al., 2003).

Today, there are at least five major *Miscanthus* breeding programs outside Asia: in Europe, USA, and Canada. We (the authors of this chapter) are all contributors from one of these breeding programs, based in IBERS, Aberystwyth, UK.

## 5.3 Breeding Strategy

*Miscanthus* breeding programs necessarily have different stages, underlined below, which culminate in the release of commercial hybrids with tailored agronomy.

### 5.3.1 Collection and Characterization

Collection and characterization of diverse germplasm with traits conferring advantages in novel hybrids are very important. A breeding program requires a broad range of germplasms collected across the range of latitudes and geographies where the target species occurs to maximize the opportunity to capture a full range of trait diversity. Trait characterization in *Miscanthus* of different geographical origin is discussed in subsequent sections.

### 5.3.2 Hybridization

*Miscanthus* is predominantly outbreeding, due to genetic self-incompatibility mechanisms and very low seed numbers are produced via self-pollination. For small quantities of seed, paired crosses are made by bagging together panicles from selected parents. For larger quantities of seed, crosses are performed in isolation chambers or in field plots. In either case, a paired cross often results in seed set on both parents. The quantities of seed produced from a cross depend on many factors including sexual compatibility, flowering synchronicity, humidity, temperature, and plant health.

Synthetic varieties are used as the main approach to produce varieties which preserve heterozygosity and minimize inbreeding. A synthetic variety was defined by Allard (1960) as "a variety that is maintained from open-pollinated seed following its synthesis by hybridisation in all combinations among a number of selected genotypes."

### 5.3.3 Ex Situ Phenotypic Characterization

Ex situ phenotypic characterization of wild germplasm and new hybrids in a range of climates is important to understand genotype × environment interactions. Field evaluations of diverse germplasm in both spaced plots and multilocation trials are used to characterize novel accessions for yield potential and chemical composition. As *Miscanthus* is a perennial, selections of outstanding crosses can only start to be made reliably after the second growing season. Phenotyping depends on the coordination of researchers at different sites and the implementation of standard protocols to ensure inter-site comparisons.

### 5.3.4 Large-scale Demonstration Trials

Large-scale demonstration trials are used to develop the agronomic practices which are needed to successfully establish, manage, and harvest the crop. Since *Miscanthus* biomass at harvest is low density, pelletization and high-density baling are being developed to improve storage and transport before the crop is used. Traits such as stem diameter are expected to define the most economic method for biomass densification and will consequently feed back into the selection of parents for hybridization.

## 5.4 Genetic Diversity

*Miscanthus* is characterized by the tough raceme rachis and bisexual paired two-flowered spikelets. Molecular and morphological systematics places *Miscanthus* closest to sugarcane, followed by sorghum (Hodkinson et al., 2002). A revised taxonomic scheme that groups species into five sections was presented recently (Clifton-Brown et al., 2011a). The two sections with species relevant to biomass production are known as section *Triarrhena* and section *Miscanthus*. Within section *Triarrhena* there is enormous phenotypic variation, with *M. sacchariflorus* plants ranging from short (<1.5 m) to very tall (up to 5 m). Section *Miscanthus* contains both the species *M. sinensis* and *M. floridulus*.

*M. sinensis* accessions show wide ranges in morphological traits such as height, stem number, and stem thickness, all of which influence yield potential. *M. floridulus* has high biomass potential but remains green throughout the year. Germplasm of *Miscanthus* species have been collected and are currently under evaluation at IBERS. The basic chromosome number of section *Miscanthus* is 19 (Hirayoshi et al., 1955; Hirayoshi and Mitsuishi, 1956; Adati, 1958; Adati and Shiotani, 1962). Diploid and tetraploid forms of *Miscanthus* are the most common. Hybrids between 2x and 4x forms have occurred naturally, producing vigorous but sterile triploids such as *M.* × *giganteus* (Figure 5.1).

**Figure 5.1.** The four main types of *Miscanthus* used in breeding for biomass are (a) *M. sacchariflorus* (shown is Japanese type, tetraploid), (b) *M. sinensis* (typical for all Asia, diploid), (c) *M. sacchariflorus* (shown is Chinese type, diploid), (d) *M. floridulus* (shown is Taiwanese type, diploid) growing in the field in Aberystwyth, Wales, UK. Dr. Lin Huang is 1.6 m tall. (*For color details, see color plate section.*)

## 5.5 Breeding Targets

### 5.5.1 Biomass Yield

Biomass yield is a complex trait and can be considered as either the composite of several morphological traits or the result of various developmental traits. Biomass yield may be increased by optimizing resource capture (photosynthesis) both temporally and spatially and by manipulating resource allocation (of photosynthate) within the plant.

## Increasing Resource Capture

Extending the length of the growing season equates to increased radiation capture and hence theoretically to higher yields. The potential impact of variation in key developmental traits determining the effective growing season is shown in Figure 5.2.

This approach was used to create parameters for an early version of the *Miscanthus* growth model (MISCANMOD) to assess the potential impacts of observed differences in early leaf emergence time in spring and leaf frost tolerance on yield. Using this model, the observed variation in a very small number of diverse accessions could potentially increase yields by 30% using climate data from Stuttgart (Farrell et al., 2006). In order to assess the actual potential under field conditions, a replicated spaced plant diversity trial with 244 genotypes was established at Aberystwyth in 2005 to characterize variation in a range of traits which are discussed further in this chapter (Figure 5.3).

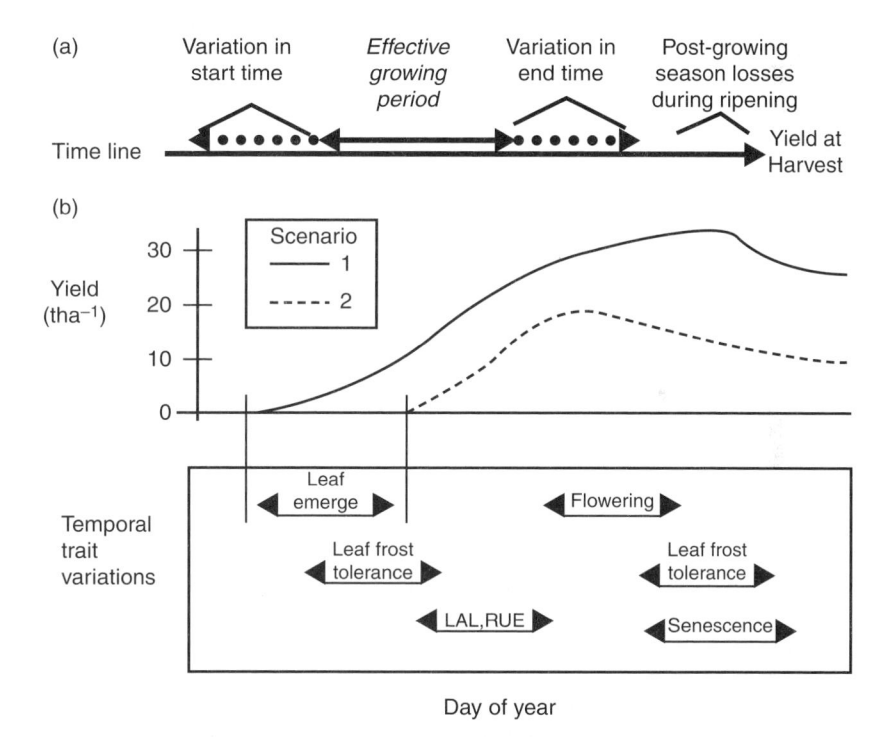

**Figure 5.2.** A conceptual time line model to assess the impact of trait variations which contribute to harvested yield (a). In scenario 1, yield trajectory during the growing season (b) shows an optimum genotype perfectly matched to the local climate. This theoretical genotype combines (1) early canopy closure, resulting from low thermal requirements for leaf emergence from the overwintering rhizome, and sufficient tolerance to spring frosts; (2) canopy architecture that absorbs all incident radiation efficiently through the canopy (leaf area index, LAI) and photosynthetic conversion (here expressed as radiation use efficiency, RUE); (3) late flowering; (4) tolerance to cold nights in autumn; and (5) effective post-growing season senescence to ensure sufficient nutrient remobilization and dry down. In scenario 2, only a fraction of the available radiation is captured and high losses occur at the end of the growing season, due to non-optimized flowering time and senescence, resulting in low harvestable yields.

**Figure 5.3.** The diversity trial at Aberystwyth. Each of 244 *M. sinensis*, *M. sacchariflorus*, and hybrid genotypes is replicated in four randomized blocks. *(For color details, see color plate section.)*

## Emergence Rate, Frost Tolerance, and Canopy Development

Emergence of shoots earlier in the year effectively lengthens the beginning of the growing season, increasing the amount of time the canopy is available to intercept solar radiation. At higher latitudes early emergence will only be advantageous if the shoots produced can tolerate any periods of frost that occur during spring. Effective canopy closure not only enhances photosynthetic light capture but also shades out competing weeds, thereby providing a double bonus to the crop. In the diversity trial in Aberystwyth we have found large variation in spring emergence, chilling tolerance, and canopy closure rates (Figure 5.4).

## Time of Flowering

Flowering affects yield quality and quantity indirectly because during this physiological transition photosynthates are directed toward the manufacture of reproductive organs, rather than leaves and stems. Early flowering will therefore result in reduced biomass accumulation and thus yield (Figure 5.2). Conversely, plants that do not flower at all may not initiate senescence before the first frosts of winter, resulting in reduced crop quality at harvest (senescence, and its impact on crop quality, is discussed in the next section). A large range in flowering times has been observed in the diversity trial of 244 genotypes at Aberystwyth with the onset of flowering extending over a 5-month period in *M. sinensis* genotypes. Genotypes that flowered earliest produced flag leaves (flowering stage 1) in late June, while the latest flowering genotypes did not reach flowering stage 1 until mid-November (Figure 5.5) (Jensen et al., 2011a). Flowering time variation in *Miscanthus* is therefore extensive and provides enormous diversity for the production of hybrids tailored to different climatic environments. Interestingly, although the relationship between flowering time and yield was statistically significant ($P > 0.001$) in this diversity trial it was not clearly defined ($R^2 = 0.121$). This was most likely due to the high morphological diversity of both *M. sinensis* and *M. sacchariflorus* genotypes from both wild and horticultural collections, confounding the relationship between yield and flowering time.

**Figure 5.4.** Variation in spring emergence from the overwintering rhizomes in four selected genotypes at Aberystwyth (photos taken on 18 April, 2009). These exemplar plants are growing in the same phenotype diversity trial at Aberystwyth as shown in Figure 5.3. Japanese genotypes (a) and (b) are variants of *M. sacchariflorus* and are later emerging than (c) and (d), which are variants of *M. sinensis*. In (c) there is evidence of low-temperature chlorosis (yellowing), while (d), a selection made in Sweden, has both early canopy closure and dark green leaves. *(For color details, see color plate section.)*

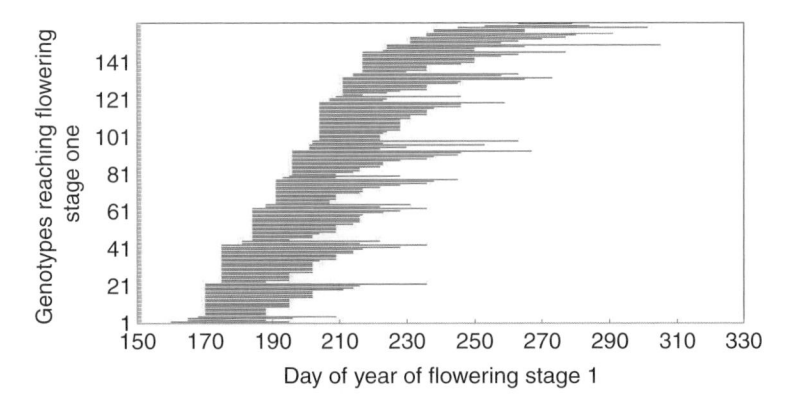

**Figure 5.5.** Flowering onset variation in *M. sinensis* genotypes within the diversity trial at Aberystwyth, Wales, UK (2007–2009). Left-hand side of bar indicates day of year on which the first flag leaf emerged (flowering stage 1) from a single genotypic replicate in any year. Right-hand side of bar indicates the latest day of the year that flag leaf emergence occurred in the same genotype in any year. Only genotypes that flowered in every replicate and in each year were included.

Close examination of the data showed early-flowering genotypes were never high yielding compared to mid- or late-flowering genotypes. Controlled environment experiments using *M. sacchariflorus* genotypes also indicate that delayed flowering leads to increased yield. Data from multilocation trials will enable further dissection of the complex environmental signals (photoperiod, temperature, and soil moisture deficit) regulating flowering induction from which a robust model to predict flowering time can be generated (Jensen et al., 2011b).

*Senescence*

Senescence impacts *Miscanthus* yield quality and quantity. For example, senescence was found to correlate well with moisture content in the final harvested *Miscanthus* crop (Robson et al., 2012). Senescence also affects nitrogen recycling in the developing canopy, thereby affecting crop sustainability. Remobilization of nitrogen and other minerals to the rhizome prior to harvest in early spring improves the energy balance of the crop and will therefore improve the C-mitigating benefits of growing *Miscanthus*. Variation observed in the diversity trial for senescence in three successive years is shown in Figure 5.6. The clustering of *M. sacchariflorus* (gray bar) and *M.* × *giganteus* (black bar) data with higher senescence scores shows the early senescence of these groups in this trial. Some delayed senescence types can stay green almost to the point of spring harvest. In such cases moisture content is high, and nutrient remobilization is incomplete, so that undesirable elements such as Cl and K are also likely to be present during combustion (Lewandowski et al., 2003). When yield was compared across common genotypes grown in several European countries, the delayed senescence phenotype was associated with low yield in northern latitudes and high yield in southern latitudes (Clifton-Brown et al., 2001). This suggests that variation in senescence may provide a route to the optimization of *Miscanthus* crop yield in different environments. The schematic in Figure 5.2 indicates that senescence may be a major determinant of canopy duration, an important determinant of yield (Dohleman and Long, 2009).

The correlation between whole plant senescence and yield in the diversity trial was statistically significant ($P > 0.001$) but was not clearly defined ($R^2 = 0.139$) and was similar to the

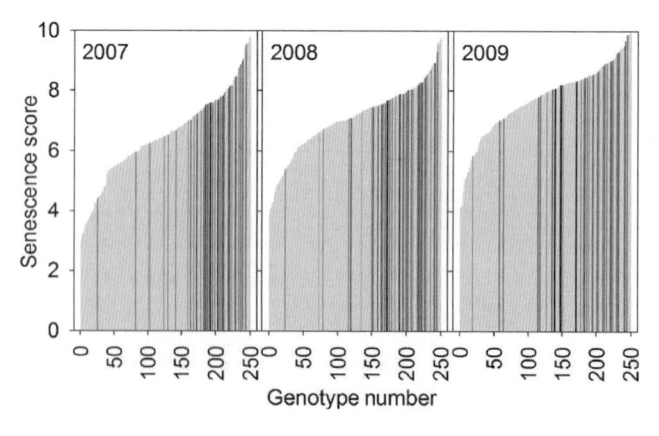

**Figure 5.6.** Variation in senescence scores in diversity trial, Aberystwyth, Wales, UK. Scores were taken approximately fortnightly and averaged over the 4 months prior to harvest for 3 years (2007–2009) (dark gray: *M. sacchariflorus*, black: *M.* × *giganteus*, light gray: *M. sinensis*).

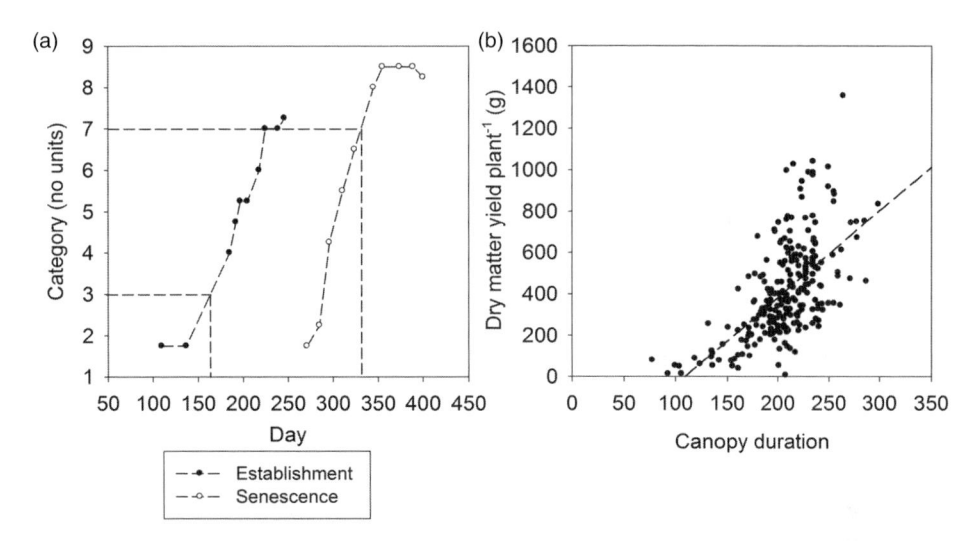

**Figure 5.7.** (a) Canopy duration determination from a single accession as the number of days between establishment stage 3 and senescence stage 7. (b) Averaged canopy duration for 244 accessions correlated with dry matter yield for 2008 growing season in diversity trait trial at Aberystwyth, Wales, UK.

correlation between flowering and yield. Senescence optimizes canopy duration both temporally and spatially, the spatial aspect being particularly significant in plot experiments. High levels of senescence may therefore promote nitrogen use efficiency by remobilizing nitrogen from leaves in lower light irradiances to those higher up the plant. The expected association between senescence and yield may consequently be confounded.

*Canopy Duration*

Composite traits may better explain variation in yield. For example, canopy duration may be calculated using the senescence score and a measure of early season establishment (Figures 5.2 and 5.7). Canopy duration correlates more strongly with yield ($R^2 = 0.37$, $P < 0.001$) than senescence or flowering alone (senescence and yield: $R^2 = 0.139$, $P < 0.001$; flowering and yield: $R^2 = 0.121$, $P < 0.001$). However, there are a large number of plants with similar canopy durations that span an almost tenfold range of dry matter yield values. This confirms the complex nature of yield determination in *Miscanthus* and the need to optimize several different traits simultaneously in order to maximize yield potential.

### 5.5.2 Morphological Traits Contributing to High Yield Potential

*Miscanthus* biomass comprises all harvested above-ground parts of the plant and is predominantly stem material. Yield may therefore be augmented by increasing stem height, stem diameter, tiller numbers per plant (estimated by transect tiller count), or multiples of these traits. Furthermore, *Miscanthus* species display either clump (*M. sinensis*) or spreading (*M. sacchariflorus*) habits, with interspecific hybrids demonstrating intermediate phenotypes. The basal diameter of the plant has implications for planting density, plant development, light interception, and, in the case of the extreme spreaders, potential invasiveness. *Miscanthus*

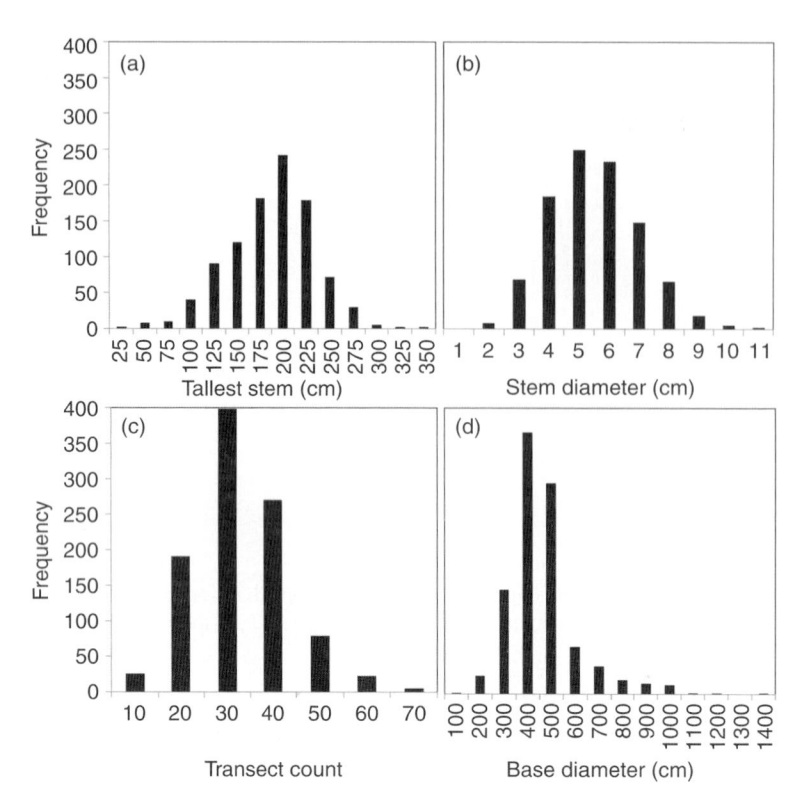

**Figure 5.8.** Frequency distributions for (a) tallest stem, (b) stem diameter, (c) transect count, (d) base diameter for four replicates of 244 genotypes on a diversity trial in Aberystwyth, Wales.

morphology is highly diverse, as shown in Figure 5.8a–d, and the same yield may therefore be achieved through different combinations of yield components, either by numerous fine stems or by fewer thicker stems. For example, two genotypes with approximately the same dry matter yield in 2007 had 19 and 188 stems, respectively.

$M. \times giganteus$ is high yielding as it has an optimized morphology consisting of numerous fairly tall, thick stems and an intermediate base. No one trait is maximized: there are taller plants and plants with thicker or more numerous stems, but few with a higher yielding combination. In practice, there are limitations to the potential for increasing each trait. There is a physical relationship between stem height and stem diameter, as thin stems above a certain height would be highly prone to lodging. Furthermore, it may be necessary to restrict stem diameter to enable the use of existing farm machinery, as a requirement for new equipment would reduce the economic return for the farmer. Independently, no individual trait predicts yield ($R^2$ values: tallest stem 0.26, stem diameter 0.26, transect count 0.01, base diameter 0.06); however, combining these traits enables a greater predictive power. Of all single measurements, canopy height has the highest correlation ($R^2 = 0.4$); however, canopy height is a complex trait consisting of both stem and leaf traits.

Numerous genes regulating stem height, diameter, and number have been reported in other species. Although no single trait independently predicts yield, increasing stem number (tillering) is the sole simple trait most likely to increase yield. Examples of genes influencing tillering include *teosinte branched*, which was discovered in maize (Doebley et al., 1995), and the *MAX* genes (reviewed by Leyser et al., (2008). A number of these genes are under study in *Miscanthus* in parallel with quantitative trait loci and association studies, which are discussed subsequently.

### 5.5.3 Seed Propagation: Crop Diversification and Reducing the Cost of Establishment

A key limitation to the current deployment of *Miscanthus* is the cost of vegetatively propagating a sterile triploid such as *M.* × *giganteus*. While there is some scope for lowering the costs of vegetative propagules such as rhizomes, conversion of the crop to one propagated by seed could significantly lower the cost of crop establishment (Christian et al., 2005; Atkinson, 2009). Furthermore, a single clone is vulnerable to pests and disease. This risk would be greatly reduced in a more genetically diverse variety based on $F_1$ hybrids or a synthetic population. Late-flowering population varieties are preferable because these maximize growing season length and consequent yield while minimizing the production of fertile seeds leading to unwanted volunteer seedlings. As in other crops, one option is to produce $F_1$ hybrid seed which would give rise to a sterile crop.

Minimum temperatures required for germination in *M. sinensis* ranged from 9.7 to 11.6°C (Clifton-Brown et al., 2011b). These are approximately 6°C higher than base temperatures for germination of *Lolium perenne* grasses bred in Wales for the United Kingdom (Clifton-Brown et al., 2011b). We concluded that in cool temperate climates such as at Aberystwyth, natural field temperatures are too low for reliable and timely establishment. Recently, trials with temporary degradable films developed for maize establishment were found to raise seedbed temperatures over +4°C. We believe by combining these novel agronomic practices and selection for lower thermal requirements for germination that seeds will replace vegetative propagation in the future.

## 5.6 Incorporating Bioinformatics, Molecular Marker-Assisted Selection (MAS), and Genome-Wide Association Studies (GWAS)

Modern-day crop improvement is an information-intensive process. For effective and efficient improvement, a range of activities from molecular biology and genetics to bioinformatics must also be integrated into a breeding program. The rate of progress made by any breeding program depends as much on the efficient handling of information generated from these activities as it does on the activities themselves. To accelerate targeted breeding, it will be necessary to handle tens of thousands of plants and assess their characteristics and input data, using mobile web-enabled devices, into a breeding database. Such an approach will streamline the breeding program at all stages as well as provide a mechanism to incorporate molecular and genomic methodologies.

*Miscanthus* breeding will benefit from the use of genetic resources, such as biparental mapping and association populations, to determine the genetic basis of desirable traits. Many of the traits already described above show significant differences across species groups. A combination of different mapping populations may therefore be required to identify major and minor QTL for traits of interest, encompassing both intra- and interspecific populations. For

**Table 5.1.** Trait heritability averaged over 3 years ($H^2$) and genetic ($V_G$) and error ($V_\varepsilon$) variances.

| Trait | $V_G$ | $V_\varepsilon$ | $H^2$ |
|---|---|---|---|
| Stem transect count | 29.39 | 17.61 | 0.63 |
| Maximum canopy height | 385.82 | 114.69 | 0.77 |
| Dry matter | 92,959.00 | 70,444.70 | 0.57 |
| Day of year for flowering stage 1 | 788.49 | 37.95 | 0.95 |
| Average senescence | 1.24 | 0.11 | 0.92 |
| Lignin | 0.45 | 0.18 | 0.72 |

example, intra- and interspecific mapping populations are already being utilized at Aberystwyth to detect QTL for traits such as flowering time, leaf morphology, and stem traits.

Specific marker–trait associations can be developed based on association mapping in synthetic or natural experimental populations. A preliminary, proof-of-concept association study is currently underway using data from the replicated diversity trial in Aberystwyth (Figure 5.3). Based on analyses of molecular marker data, a subset of 144 *M. sinensis* genotypes were identified that are characterized by weak population substructure ($F_{ST} \leq 0.06$) and are therefore expected to represent an appropriate population for an association study. To assess levels of genetic variation in this population, we analysed phenotypic data using mixed linear models and used the resulting variance components to calculate broad-sense heritabilities ($H^2$; Falconer, 1989). All traits were moderately to highly heritable (as shown in Table 5.1), with values of $H^2$ being comparable to those calculated for the highly diverse nested association mapping population (NAM) in maize (Buckler et al., 2009; http://www.panzea.org/db/gateway?file_id=NAM_2006_trait_herit) and somewhat higher than those in experimental populations of sugarcane (O'Reilly et al., 1995; Jackson, 2005) and forage grasses (England, 1975; Humphreys, 1995; De Araujo and Coulman, 2002; Majidi et al., 2009). This suggests that the genetic variation of the germplasm captured in the field trial at Aberystwyth and the phenotyping protocols developed should allow the uncovering of the genetic basis for traits of interest through QTL and association mapping. Target traits include biomass productivity and cell wall characteristics which will accelerate plant breeding through genomic selection (Goddard and Hayes, 2009).

## 5.7 Summary

Biofuel production has recently generated controversy and misunderstanding over land-use requirements and greenhouse gas savings (Flavell et al., 2011). While food is clearly one of the necessities of life, modern civilizations also require energy for heating, lighting, cooking fuel, and transport. Biomass can be used as a replacement or in addition to fossil fuels for a multitude of energy applications. But if we grow both food and fuel, how can we minimize competition for land and what are the advantages of growing biomass for energy? Low input perennial energy crops, such as *Miscanthus*, have the advantage that they may utilize land unsuitable for high-input first-generation energy crops such as cereals and oilseeds. Despite this, yields of all crops will need to increase and it is likely that some degree of compromise in land use for food or fuel will have to be reached. The challenges of meeting demand for food and energy for a world population expected to reach 9 billion by 2050 are huge. Land is a finite resource and we must use it wisely. Valentine et al. (2012) set out the vision for bioenergy in terms of four major gains for society: (1) a reduction in carbon emissions by replacing fossil fuels with appropriate crops for energy generation; (2) a significant contribution to energy security by reductions in fossil fuel dependence, for example, to meet government targets; (3) new options that stimulate rural and urban economic development, and (4) reduced dependence of global agriculture on fossil fuels. Valentine et al. (2012) argue that, with improvements in food production systems (in particular reducing red meat consumption, particularly from grain-fed systems) and reductions in food waste (said to be 30–40% in both developed and developing countries), there are significant quantities of good- and medium-grade lands for bioenergy production. We believe our development of *Miscanthus* as a dedicated perennial lignocellulose energy crop can make an important contribution to the future renewable energy portfolio.

## Acknowledgments

Work reported in this manuscript was funded by the Department for Environment Food and Rural Affairs grant NF0426, the Biotechnology and Biological Sciences Research Council (BBSRC) Institute Strategic Programme Grant on Bioenergy and Biorenewables (BBS/E/W/00003134), the Engineering and Physical Sciences Research Council (EPSRC) Supergen—Bioenergy project (GR/S28204/01; EP/E039995/1), a BBSRC Institute career path fellowship (BB/E024319/1), and a BBSRC grant (BB/E014933/1). We wish to thank Roy Jones, Marc Loosley, Pete Roberts, and Chris Ashman for technical assistance on the breeding program at Aberystwyth.

## References

Adati S. Studies on the genus *Miscanthus* with special reference to the Japanese species useful for breeding purposes as a fodder crop. Bull Fac Agric Mie Univ, 1958; 17: 1–112.

Adati S, Shiotani I. The cytotaxonomy of the genus *Miscanthus* and its phylogenic status. Bull Fac Agric Mie Univ, 1962; 25: 1–24.

Allard RW. *Principles of Plant Breeding*. Wiley International Edition., 1960. John Wiley & Sons, Inc., New York.

Atkinson CJ. Establishing perennial grass energy crops in the UK: a review of current propagation options for Miscanthus. Biomass Bioenerg, 2009; 33: 752–759. DOI: 10.1016/j.biombioe.2009.01.005.

Brutnell TP, Wang L, Swartwood K, Goldschmidt A, Jackson D, Zhu X-G, Kellogg E, Ecka JV. Setaria viridis: a model for C4 photosynthesis. Plant Cell, 2010; 22: 2537–2544.

Buckler ES, Holland JB, Bradbury PJ, Acharya CB, Brown PJ, Browne C, Ersoz E, Flint-Garcia S, Garcia A, Glaubitz JC, Goodman MM, Harjes C, Guill K, Kroon DE, Larsson S, Lepak NK, Li HH Mitchell SE, Pressoir G, Peiffer JA, Rosas MO, Rocheford TR, Romay MC, Romero S, Salvo S, Villeda HS, da Silva HS, Sun Q, Tian F, Upadyayula N, Ware D, Yates H, Yu JM, Zhang ZW, Kresovich S, McMullen MD. The genetic architecture of maize flowering time. Science, 2009; 325: 714–718.

Christian DG, Yates NE, Riche AB. Establishing *Miscanthus sinensis* from seed using conventional sowing methods. Ind Crops Products, 2005; 21: 109–111.

Clifton-Brown JC, Lewandowski I, Andersson B, Basch G, Christian DG, Bonderup-Kjeldsen J, Jørgensen U, Mortensen J, Riche AB, Schwarz K-U, Tayebi K, Teixeira F. Performance of 15 *Miscanthus* genotypes at five sites in Europe. Agron J, 2001; 93: 1013–1019.

Clifton-Brown JC, Renvoize SA, Chiang Y-C, Ibaragi Y, Flavell R, Greef JM, Huang L, Hsu TW, Kim D-S, Hastings A, Schwarz KU, Stampfl P, Valentine J, Yamada T, Xi Q, Donnison I. Developing *Miscanthus* for bioenergy. In: Halford N and Karp A (eds) *Energy Crops*, 2011a. Royal Society of Chemistry, Cambridge.

Clifton-Brown JC, Robson P, Sanderson R, Hastings A, Valentine J, Donnison IS. Thermal requirements for seed germination in *Miscanthus* compared with switchgrass (*Panicum virgatum*), reed canary grass (*Phalaris arundinaceae*), maize (*Zea mays*) and rye grass (*Lolium perenne*). Global Change Biol Bioenerg, 2011b; 3: 375–386. DOI: 10.1111/j.1757-1707.2011.01094.x.

Daniels J, Roach BT. A taxonomic listing of *Saccharum* and related genera. Sugar Cane, 1987; Spring Supplement: 16–19.

De Araujo M, Coulman B. Genetic variation, heritability and progeny testing in meadow bromegrass. Plant Breed, 2002; 121:417–424.

Doebley J, Stec A, Gustus C. Teosinte branched1 and the origin of maize—evidence for epistasis and the evolution of dominance. Genetics, 1995; 141: 333–346.

Dohleman FG, Long SP. More productive than maize in the Midwest: how does Miscanthus do it? Plant Physiol, 2009; 150: 2104–2115. DOI: 10.1104/pp.109.139162.

England F. Heritabilities and genetic correlations for yield in Italian ryegrass (*Lolium multiflorum* Lam.) grown at different densities. J Agric Sci, 1975; 84: 153–158.

Falconer DS. *Introduction to Quantitative Genetics*, 1989. Longman Scientific and Technical Publications, London and New York.

Farrell AD, Clifton-Brown JC, Lewandowski I, Jones MB. Genotypic variation in cold tolerance influences yield of Miscanthus. Ann Appl Biol, 2006; 149: 337–345.

Flavell R, Cruz CHdB, Christie M, Allen J, Keller M, Gilna P, Kell DB. Moving forward with biofuels. Nat Outlook, 2011; 474: S44–S48.

Goddard M, Hayes B. Mapping genes for complex traits in domestic animals and their use in breeding programmes. Nat Rev Genet, 2009; 10: 381–391.

Greef JM, Deuter M. Syntaxonomy of *Miscanthus x giganteus* GREEF et DEU. Angew Bot, 1993; 67: 87–90.

Hirayoshi I, Mitsuishi S. Wild growing forage plants of the far east, especially Japan, suitable for breeding purposes. Part 1. Karyological study in *Miscanthus*. Bull Fac Agric Mie Univ, 1956; 12: 1–10 (in Japanese).

Hirayoshi I, Nishikawa K, Kato R, Kitagawa M. Cytogenetical studies on forage plants. (III) Chromosome numbers in Miscanthus. Jpn J Breed, 1955; 5: 49–50 (in Japanese).

Hodkinson TR, Chase MW, Lledo MD, Salamin N, Renvoize SA. Phylogenetics of *Miscanthus*, *Saccharum* and related genera (*Saccharinae, Andropogoneae, Poaceae*) based on DNA sequences from ITS nuclear ribosomal DNA and plastid trnL intron and trnL-F intergenic spacers. J Plant Res, 2002; 115: 381–392.

Humphreys M. Multitrait response to selection in *Lolium perenne* L. (perennial ryegrass) populations. Heredity, 1995; 74: 510–517.

Jackson P. Breeding for improved sugar content in sugarcane. Field Crops Res, 2005; 92: 277–290.

Jensen E, Farrar K, Thomas-Jones S, Hastings A, Donnison I, Clifton-Brown J. Characterization of flowering time diversity in *Miscanthus* species. Global Change Biol Bioenerg, 2011a; 3: 387–400.

Jensen E, Squance M, Hastings A, Jones S, Farrar K, Huang L, King R, Clifton-Brown J, Donnison I. (2011b) Understanding the value of hydrothermal time on flowering in *Miscanthus* species. Aspects Appl Biol, Accepted.

Lewandowski I, Clifton-Brown JC, Andersson B, Basch G, Christian DG, Jørgensen U, Jones MB, Riche AB, Schwarz K-U, Tayebi K, Teixeira F. Environment and harvest time affects the combustion qualities of *Miscanthus* genotypes. Agron J, 2003; 95: 1274–1280.

Leyser O, Ongaro V, Bainbridge K, Williamson L. Interactions between axillary branches of Arabidopsis. Mol Plant, 2008; 1: 388–400.

Majidi M, Mirlohi A, Amini F. Genetic variation, heritability and correlations of agro-morphological traits in tall fescue (Festuca arundinacea Schreb.). Euphytica, 2009; 167: 323–331.

O'Reilly K, Shanahan P, Levin J, Nuss K. Heritability and repeatability estimates of a sugarcane population grown under dryland conditions. Proceedings of the Annual Congress—South African Sugar Technologists' Association, 1995, pp. 9–13.

Robson PR, Mos M, Clifton-Brown JC, Donnison IS. Phenotypic variation in senescence in miscanthus: towards optimising biomass quality and quantity. Bioenerg Res, 2012. DOI 10.1007/s12155-011-9118-6.

Valentine J, Clifton-Brown J, Hastings A, Robson P, Allison G, Smith P. Food vs. fuel: the use of land for lignocellulosic 'next generation' energy crops that minimize competition with primary food production. Global Change Biol Bioenerg, 2012.

Wang DF, Portis AR, Moose SP, Long SP. Cool C-4 photosynthesis: pyruvate P-i dikinase expression and activity corresponds to the exceptional cold tolerance of carbon assimilation in Miscanthus x giganteus. Plant Physiol, 2008; 148: 557–567.

# Chapter 6
# Breeding Sorghum as a Bioenergy Crop

T.R. Stefaniak and W.L. Rooney

*Department of Soil and Crop Sciences, Texas A&M University, College Station, TX 77843-2474, USA*

## 6.1 Introduction

Sorghum (*Sorghum bicolor* L. Moench) is an important crop species in the United States and around the world. Because of the crops' innate heat and drought tolerance, sorghum production is traditional in semi-arid, subtropical, and tropical regions of the world. In addition to drought tolerance, sorghum is highly adaptable and very responsive to favorable conditions. The end uses of sorghum are varied such as cereal grain, forage, syrup, and more recently, as bioenergy production. In 2008, US farmers harvested about 5.9 million hectares of grain sorghum (USDA NASS). Worldwide, the top 20 sorghum-producing countries harvested 49 million metric tons of grain in 2007 (FAOSTAT, 2010). Concurrently, the United States and other countries are increasingly aware of the need of alternative energy sources to petroleum. For example, in 2004, the United States consumed about 140 billion gallons of gasoline for transportation. In that same year, approximately 3.4 billion gallons of ethanol were produced from starch crops, primarily corn (Renewable Fuels Association, 2012). This represented about 2% of gasoline consumption. Clearly, the development, deployment, and conversion of dedicated bioenergy crops to renewable fuel must occur if the United States is to meet the goals stated in the Energy Security Act of 2007; one of which is to replace 30% of the petroleum consumed with biofuels by 2030.

The most promising characteristics of a crop with high bioenergy potential include high-yield potential, wide adaptation, efficient use of resources, and adaptability to production systems (Perlack et al., 2005). Sorghum is a highly productive C4 photosynthetic species that is well adapted to warm growing regions and is one of the potential species for use as a bioenergy crop (Table 6.1). The best type of sorghum for biofuel production is dependent on the type of conversion process that will be used. Hybrid grain sorghum will provide starch for conversion, while sweet sorghum accumulates sugar in the stalk that can be directly fermented to ethanol. Cellulose is produced by all types of sorghum and specific genotypes are being developed to maximize this production. Each type thus fits in a different production system; no other species has the flexibility to produce large quantities of starch, sugar, or cellulose.

---

*Bioenergy Feedstocks: Breeding and Genetics*, First Edition. Edited by Malay C. Saha, Hem S. Bhandari, and Joseph H. Bouton.
© 2013 John Wiley & Sons, Inc. Published 2013 by John Wiley & Sons, Inc.

**Table 6.1.** Favorable characteristics of sorghum as a biofuel crop.

| As a Crop | As an Ethanol Feedstock | As a Bagasse |
| --- | --- | --- |
| Short growth period (3–4 mos) | Model bioenergy crop | High biological value |
| Drought tolerant | Carbon neutral process | Rich in micronutrients |
| Responds to inputs | Superior quality | Can be used to cogenerate electricity |
| Established production techniques | Less sulfur | Lignocellulosic feedstock for ethanol production |
| Versatile (feed, fodder, or juice) | High Octane | |
| Noninvasive | Suitable for automobile consumption | |
| Low soil $NO_2/CO_2$ emission | | |
| Established seed industry | | |
| Well-funded public research | | |
| Seed propagated | | |

Adapted from Reddy and Reddy (2003), Reddy et al. (2005), and (2009).

The versatility of sorghum makes it a promising dedicated bioenergy crop. Sorghum has an established production history as a crop in the United States and around the world. This history minimizes concerns about producer acceptance and adoption because it is a crop that producers are already cultivating. Second, in conjunction with the history of the crop, there is a well-established seed industry that is very efficient at sorghum improvement and seed production. Third, the annual nature of the crop, while a detraction to some, increases the speed and efficiency at which sorghum can be genetically improved. Finally, sorghum has evolved as a standard genetic model for the improvement of bioenergy crops. All of these factors contribute to the versatility and potential of sorghum (Table 6.1). No other crop combines the variation in harvestable biomass composition, accessible germplasm diversity, and robust research infrastructure. Therefore, sorghum will be an important energy and will likely play a pivotal role in the development and evolution of other energy crops. Herein, we describe the approaches that have been used for sorghum improvement and those approaches that will be used and adopted to improve and develop sorghum as a bioenergy crop.

## 6.2 Botanical Description and Evolution

### 6.2.1 Basic Characteristics

Although sorghum is a phenotypically diverse species, it is quite consistent genetically and botanically. Sorghum is a diploid with base chromosome number of $n = 10$ and $2n = 2x = 20$. There is some evidence that the crop is an ancient tetraploid, but functionally it behaves as a diploid (Kim et al., 2005). The genome size is larger than rice but substantially smaller than corn, and the complete genome has been sequenced with a genome size of about 730 Mb (Paterson et al., 2009). It is a member of the Panicoideae subfamily of grasses. It has a deep and fibrous root system, primary culm, and the capacity for both basal and axillary tillering. Sorghum has a complete flower that is exerted through the top leaf sheath prior to anthesis. Sorghum is predominantly self-pollinated with outcrossing rates reported between 2% and

30% and influenced by genotype (Sleper and Poehlman, 2006). The crop is amenable to the commercial production of hybrids utilizing cytoplasmic male sterility systems and exhibits sufficient heterosis to justify costs of seed production. It is usually grown and managed as an annual, but technically can be a perennial because once harvested, it will regrow unless environmental conditions (e.g., freezing temperatures) preclude further growth. Among commonly grown cereal crops, sorghum has a high water use efficiency and very good heat and drought tolerance. Plant maturity and height are highly variable and influenced by both genetics and the environment.

### 6.2.2 Evolution and Distribution

Sorghum evolved and was domesticated in arid areas of Northeastern Africa; it has been found in archaeological excavations dated to be over 6000 years ago (Kimber, 2000). After domestication, sorghum spread across Africa and into the continent of Asia through traditional trade routes. The species is relatively new to the Americas and Australia, arriving in the past 200–300 years. As a consequence of domestication and distribution, it is an extremely diverse species with a wide range of variation within domesticated sorghum. This variation has resulted in (or is the result of) many different end uses and environments in which the crop is produced. As the crop moved, new races were selected with specific adaptation to the region. Today, the species is divided into five primary races (Figure 6.1) and an array of secondary intermediates; these races provide the basis of many improved sorghums and a source of valuable traits for

**Figure 6.1.** Panicles of five primary race divisions of sorghum beginning in the lower left clockwise bicolor, kafir, durra, caudatum, and guinea. (*For color details, see color plate section.*)

further improvement (Rosenow and Dahlberg, 2000). The race Bicolor is considered the most primitive grain sorghum, having an open panicle and tightly clasped and covering glumes over small seed. The race Guinea is a grain sorghum commonly grown in West Africa; it has a loose panicle, smaller and open glumes, and very hard vitreous grain. The Caudatum race is the source for many of the yield attributes of modern grain sorghum hybrids. The Kafir race evolved in Southern Africa and has high-yielding semi-compact panicles; it has been an important source of alleles that maintain sterility in the cytoplasmic male sterility system. Finally, the Durra race has very tight compact panicles with large grain with good yield potential; this race is also the source of A1 cytoplasm that remains the primary CMS system used for hybrid seed production.

## 6.3 Traditional Breeding and Development

### 6.3.1 Initial Sorghum Improvement

Systematic genetic improvement of sorghum for both grain and forage production began in the early 20th century (Rooney and Smith, 2000). Up to this period and for generations prior, producers simply selected superior individual plants within heterogeneous populations of mostly homozygous landraces. Because sorghum is self-pollinated, it was quite simple to recover homozygous lines differing for specific traits. In temperate climates, early selections were primarily for early maturing types; sorghum is a tropical plant and early accessions were highly photoperiod sensitive. In the temperate climates, where most of the systematic breeding programs were developed, these photoperiod-insensitive selections formed the basis for much of sorghum improvement for many years. Once maturity was adapted to a region, variants for height were selected; both tall and short for use as forage and/or grain crops. In the mid-20th century, dwarfing genes ($Dw$) became important with the advent of mechanical harvest and the need for shorter and uniform height.

With rediscovery of Mendel's law, systematic crossing and selection in segregating progenies were initiated for sorghum improvement. Crossing in sorghum requires hand emasculation, or system of male sterility or techniques for limiting self-pollination. Pollination procedures are thoroughly described in Rooney (2004). Initial efforts were in the development of improved inbred cultivars. Selection was focused on the improvement of yield, quality, drought tolerance, disease, and pest resistance. This research was focused in public programs and resulted in an array of early publications on the genetic inheritance of many traits in sorghum as well as improved and important sorghum cultivars that have been grown around the world for grain or forage production.

### 6.3.2 Development of Hybrid Sorghum and Heterosis

The development of hybrid cultivars, or the third phase of sorghum improvement commenced when it was recognized that the heterogeneous populations were the result of segregation by H.N. Vinall and A.B. Cron in 1914 (Rooney and Smith, 2000). The success of hybrid corn in the early 1930s motivated sorghum breeders to assess the relative value of heterosis in sorghum. Stephens and Quinby (1952) documented the enhanced yield potential, reduced maturity and stability of hybrids that gave breeders incentive to identify production schemes for hybrid sorghum. Stephens and Holland (1954) solved the problem of self-fertility and large-scale hybrid seed production with the discovery and incorporation of a cytoplasmic male sterility

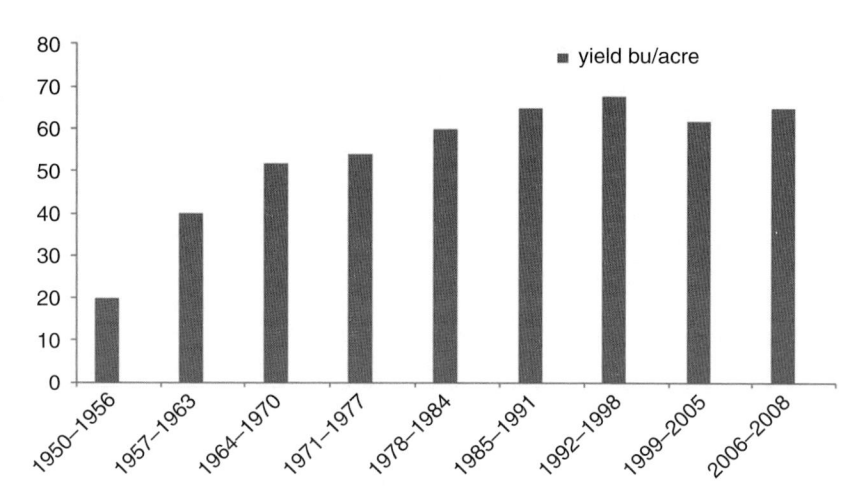

**Figure 6.2.** Average annual sorghum grain yield in the United States from 1950 to 2008. Data were compiled from the USDA National Agricultural Statistical Services and Smith and Frederiksen (2000).

system using milo cytoplasm and nonrestoring alleles from kafir sorghums. The system was rapidly incorporated and adopted, initially using standard cultivars as parental lines for both grain and forage sorghums. The first hybrid sorghums were commercially available in 1956; within 4 years, hybrid sorghum seed was planted on over 90% of the total area planted to grain sorghum in the United States. As a result, grain sorghum yields increased about 45% from 1956 to 1965 (Figure 6.2). It is also remarkable that sorghum hybrids maintained their yield levels after the end of the nineties when US sorghum production shifted to more arid land without irrigation. Later, forage sorghum hybrids for both hay and silage were developed and adopted by producers. While hybrid forage sorghums have increased yield, another important consideration has been in the enhanced logistics of seed production (higher seed yields, easier harvest, and better quality).

The development of hybrid sorghum was followed by the development of a hybrid seed industry. Within 10 years, sorghum breeding programs switched from a few publicly supported breeders to numerous privately funded breeders. This not only expanded the size and commitment to research, but also the scope of research as well. Traits such as yield under high inputs (water, fertilizer) became important as was the protection of yield potential from both biotic and abiotic stresses. These factors formed the basis of the modern sorghum breeding programs that are now working to improve the crop. These efforts have led to substantial improvements in yield potential, grain and forage quality, and tolerance to several abiotic and biotic stresses. While hybrid sorghums are commonly grown in many regions of the world, there are, for economic or cultural reasons, other regions that have not adopted hybrid sorghums. Consequently, the goals and products from sorghum breeding programs depend largely on the funding, goals, and the target area. More recent developments in biotechnology have also affected the implementation and breeding schemes used by sorghum breeders. While there are many approaches and methodologies that may be used for improvement of the crop, all phases of a sorghum breeding program begin with hybridization to create segregating populations that can be used for selection. In the case of hybrid sorghums, the method also ends with hybridization to create the sorghum hybrids that are produced.

### 6.3.3 Current Sorghum Breeding Approaches

Today's sorghum breeding programs integrate many traditional approaches with the tools and benefits of modern molecular genetics, agronomy, and agricultural engineering and analysis. Most of these approaches are currently used to improve energy sorghum. Whether breeding for cultivars or hybrids, each program develops inbred lines for *per se* testing (cultivars) or testcross evaluation (hybrids). Most of the sorghum breeding programs use some form of the pedigree selection method (Sleper and Poehlman, 2006). If the end product of the program is a cultivar, then replicated evaluation and yield testing of the lines *per se* may commence immediately after selection to some level of phenotypic and genotypic stability. If the end product of the program is a hybrid, then the parental lines must be evaluated for combining ability by crossing to established tester lines. Good general combining ability (GCA) is needed in the first testcross and further combinations and testing are required to identify those hybrids with excellent specific combining ability which is essential for commercialization. Predicting hybrid performance by evaluating the inbred lines is not typically an efficient use of resources (Rooney, 2004).

Sorghum germplasm has been classified into heterotic groups based on fertility restoration. While the different races of sorghum also represent logical heterotic groups, there is some justification for using fertility restoration as heterotic groups (Krishnamoorthy, 2005). Evaluation of new parental lines requires efficient use of testcross hybrids to identify the best parents because the correlation between inbred line and hybrid is of limited value (Maradiaga, 2003). As is the case, in maize good hybrid combinations are usually derived from elite and complementary inbred lines that maximize additive effects and effectively capture dominant gene action in the hybrid to maximize heterosis (Kambal and Webster, 1965; Miller and Kebede, 1981). However, not all quantitative characters have been shown to improve in hybrid combinations; in particular, stalk quality, cold tolerance, grain quality, tillering, A–R reaction, trueness to type, and some insect resistance (Maunder, 1983).

Many of the modern breeding tools available are being utilized by sorghum breeders. The sorghum genome has been sequenced (Paterson et al., 2009); two high-density genetic maps have been constructed (Bowers et al., 2003; Mace et al., 2009), as well as a cytogenetic map (Kim et al., 2005). Additionally, many genes have been cloned (Izquierdo and Godwin, 2005; Burow et al., 2009b; McIntyre et al., 2008) and a transformation system has been developed (Casas et al., 1997). Finally, the literature is replete with QTL studies in sorghum.

### 6.3.4 Germplasm Resources

The extensive array of genetic variability present within cultivated sorghum is captured in two major sorghum germplasm collections. The International Crops Research Institute for the Semi-Arid Tropics (ICRISAT), Hyderabad, India has 40,000+ accessions and the US Sorghum Germplasm collection (based in Griffin, GA, and Fort Collins, CO) holds over 42,000 accessions (Rosenow and Dahlberg, 2000). The passport data for the US collections are publicly available through the Germplasm Resource Information Network (GRIN). Many of the accessions in these collections have been classified into five primary races based on panicle characteristics (for review, see Rosenow and Dahlberg, 2000). About 50% of the accessions have categorized by 39 agronomic traits in GRIN. Additionally, many of the more popular breeding lines and their progenitors have been classified on the basis of stress and pest resistance. These collections represent a powerful tool for the breeders by giving them insight into sources of genetic variation for traits of interest for population development.

**Figure 6.3.** The effect of the sorghum conversion program on adaptation in more temperate climates (College Station, Texas). Genotype 1 is T×406 which is the donor parent for dwarfing and maturity insensitivity genes. Genotype 2 is the converted version of IS5670C. Genotype 3 is the unconverted IS5670. (*For color details, see color plate section.*)

Initial genetic variability in temperate sorghum breeding programs was limited to the sorghum germplasm that was day-length insensitive and would flower during the long summer days. Much of the world collection was photoperiod sensitive; much of this diversity was not accessible to many breeding programs. In order to make this diversity accessible, the sorghum conversion program was initiated in 1963 in the United States with its goal being to introgress the recessive alleles for day-length insensitivity and dwarfing, thus making this germplasm accessible to sorghum breeding programs worldwide. A converted line can be described as being a shorter version of the exotic line that flowers earlier, though to the layman a converted line does not much resemble its exotic parent (Figure 6.3). Breeding for conversion was a modified backcross strategy using the exotic germplasm as the donor parent and strategic environments in Puerto Rico and Texas (Stephens et al., 1967). During the time the conversion program was active (approximately 40 years), over 700 photoperiod-sensitive sorghum accessions were fully converted and released (Rosenow et al., 1997a, 1997b; Dahlberg et al., 1997). The lines have had a tremendous impact on sorghum improvements worldwide; it is difficult to find a sorghum grown anywhere (especially in temperate regions) that does not have sorghum conversion germplasm in its pedigree. In addition, conversion lines have been extremely important for genetic research and genomic analysis of sorghum for an array of traits ranging from adaptation, tolerance to biotic and abiotic stress, and the improvement of quality and composition of grain, forage, and eventually energy sorghums. The original conversion program was discontinued in 2003; more recently, there have been efforts to reprise the program, integrating marker-assisted selection (MAS) to facilitate the process (J. Dahlberg, personal communication, 2010). This is welcome news; there remain many valuable traits in exotic germplasm and significant genetic variation exists in the "unconverted" germplasm. This germplasm should be regarded as a repository for undiscovered beneficial alleles of interest to the bioenergy breeder.

## 6.4 Approaches to Breeding Sorghum as a Bioenergy Crop

The concept of sorghum as a bioenergy crop is not new; it was well documented that sorghum was a potential bioenergy crop during the energy crisis in the late 1970s and early 1980s (Monk et al., 1984). Initial efforts in the development of these types of sorghum were essentially the same as those developed for grain and forage sorghum improvement except that development and release focused primarily on line or cultivar development (Rooney, 2004). Current development focuses on hybrid development and some of the traits important to grain and forage sorghum breeders are also important to energy sorghum breeders. However, the importance of certain traits is very different between the sorghum types. For example, palatability is a critical component of a forage sorghum crop but it is unimportant in bioenergy sorghum. Thus, existing breeding strategies are well adapted for use in energy sorghum breeding but the traits of emphasis and how they are evaluated and selected will be adjusted.

The ultimate goal of the processor is to convert the biomass into a chemical or energy of value to the consumer. While the current emphasis focuses on ethanol production, breeding programs are already integrating with processors who are looking beyond ethanol to higher chain carbon compounds and higher value biochemical compounds. Therefore, exact traits of interest to the energy sorghum breeder and the relative priorities within that program are dependent upon the type of energy that is being produced. Given these factors, energy sorghums are divided into categories based primarily on the different portions of carbohydrate that they produce. These categories include grain sorghum, sweet sorghum, and biomass sorghum (Rooney et al., 2007).

### *6.4.1 Grain Sorghum*

Grain sorghum has been used to produce starch-based bioethanol in conjunction with maize for many years. In 2008, 29.7% of the US grain sorghum crop was used in ethanol production (Sorghum Grower, 2009). The primary goal of most grain sorghum breeding programs is to enhance the grain yield. Research in this area has been extensive quantifying genetic variation and/or identifying QTL for grain yield and its component traits (Gomez, 2002; Maradiaga, 2003; Murray et al., 2008; Ritter et al., 2008; Srinivas et al., 2009). While variation for seed composition (i.e., starch) exists in sorghum (Rami et al., 1998; Murray et al., 2008), simply increasing grain yield results in increased ethanol yield because of the increase in total starch per unit area. Existing improvement programs and their breeding goals which have a long and established history are well positioned to continue further enhancement of grain sorghums for feed and energy production.

### *6.4.2 Sweet Sorghum*

Traditionally, sweet sorghums have been used as a source of sugar; extracted juice is cooked and sugars are concentrated in syrup. Sweet sorghums are known in various parts of Africa and they became popular in Asia and North America in the seventeenth and eighteenth century (Smith and Frederiksen, 2000). In the early twentieth century, the United States was producing 20 million gallons of sorghum syrup annually (Oklahoma State Extension Service). Production dropped after World War II as crystal sugar became more available; today sorghum syrup production is essentially artisanal and it is concentrated within the Southeastern United States.

Sweet sorghum genotypes produce high quantities of simple sugars in juicy stalks. While some forage sorghum hybrids are occasionally portrayed as sweet sorghums, they lack one of these two traits and thus do not produce high yields of sugar. Sweet sorghums are usually tall, thick stalked, and have lower grain yields than grain sorghums. For energy production, sweet sorghum is processed in a similar manner to sugarcane; in fact, sugarcane processes and equipment provide a logical start point for utilizing sweet sorghum. Sweet sorghums have high biomass yield potential. Hunter and Anderson (1997) estimated that sweet sorghum has the potential to produce up to 8000 L ha$^{-1}$ ethanol or about twice as much as that of maize and 30% more than sugarcane ethanol in Brazil. Much of the carbohydrate content of the stalk juice of sweet sorghums is sucrose and/or glucose and it is fermentable without the need of starch hydrolysis. This has advantages and disadvantages because the fermentation process can proceed without any pretreatment (advantage) but fermentation must be initiated quickly because of the instability of the sugars in the stalks and/or juice (disadvantage). Preliminary results indicate that there can be a reduction of 16.8% sugar yield if juice extraction is delayed by 48 hours (Rao et al., 2009). Methods of stabilizing the sugar in the juice are either expensive or energy intensive.

The lack of stability may limit sweet sorghum production in temperate environments because the instability of the biomass limits the window for processing and the capital costs of building processing facilities require a longer season of use. Therefore, subtropical and tropical regions which have longer harvest seasons are the most likely production environments for sweet sorghum. In addition, there is a logical complementation between sweet sorghum and sugarcane; which can extend the harvest and milling season. Sweet sorghum compliments sugarcane in this scheme because it can be harvested twice in 1 year in tropical environments and of its enhanced drought tolerance and water use efficiency.

It was the interest in sweet sorghum in America that led to the development of existing sweet sorghum varieties. Breeding programs at several locations throughout the Southeastern United States developed and released well-adapted, high-yielding cultivars in the United States such as Dale (Broadhead et al., 1970) and Theis (Broadhead et al., 1974). Some important cultivars that have been released internationally are Rio, Theis, Roma, and Keller (M. Bitzer, personal communication, 2009). More recently, sweet sorghum varieties developed in India and China have impacted production in these regions. These varieties represent an array of different maturities, heights, and agronomic packages adapted to the regions as best identified by producers. They have become standards throughout the world and they form the basis for sweet sorghum breeding programs that are being established in many regions of the world.

In all cases, breeding focused on the development of inbred varieties. Sweet sorghums were selected from intentional crosses and advanced through the generations via self-pollination and selection to uniformity. Traits of importance were included but were not limited to maturity, height, sugar yield, sugar concentration, sugar quality, and agronomic adaptation traits such as stalk rot resistance, drought, and borer tolerance. Initial breeding efforts did not emphasize bagasse quality as this was a refuse by-product of little importance to the sweet sorghum producer. Naturally, the bagasse is substantially more important in energy situations because it is three to five times more concentrated in energy than the juice (Saballos, 2008). Bagasse can be burned to produce electricity for the ethanol facility or has the potential to be converted to ethanol itself in such time as lignocellulosic conversion processes become profitable.

While the sweet sorghum varieties are extremely valuable, they have limited capacity for large-scale production systems required by bioenergy programs. First, sweet sorghum varieties typically have very low seed yields. In addition, production of seed is difficult simply due to maturity and height of the varieties. Consequently, it is difficult, if not impossible, to produce quantities of seed required for planting bioenergy scale of land (Corn, 2009). The production of hybrid sweet sorghums provides the industry with a means of producing large quantities of seed while capturing heterosis that is inherent in a hybrid crop.

The ability to produce a true sweet sorghum hybrid was limited because sugar concentration is a primarily additive trait; since most seed parental lines have low sugar concentrations in the juice, their hybrids could only have intermediate sugar concentrations (Clark, 1981). Consequently, even though the juice volumes displayed significant heterosis, the overall sugar yields were reduced due to the lower sugar concentrations. Therefore, it is necessary to develop sweet sorghum germplasm, especially seed parents to produce true sweet sorghum hybrids. The development of pollinator parents is also important, but most sweet sorghum varieties can serve as pollinator parents as they typically restore fertility to their hybrids made using either A1 or A2 cytoplasm.

Sweet sorghum seed parental lines are being developed in several programs around the world (Rao et al., 2009). In studies of hybrids created from these seed parents and sweet sorghum cultivars as pollinators, the sweet sorghum hybrids are significantly superior to the seed parents (Corn, 2009). Parental selection emphasized high brix concentration, juicy stalk, and shorter height (to facilitate combine harvest of seed). While the first generation hybrids did not always outperform their respective pollinator parents, the seed production capacity of the hybrid was four to six times greater than the sweet variety and the seed can be mechanically harvested. In addition, these are first generation hybrids; it is logical to expect greater heterosis and higher yields in improved versions of these hybrids. In India, excellent progress has been made in developing hybrid sweet sorghum lines from sweet and grain parents. Reddy et al. (2005) described six hybrids significantly lower in brix than the control genotype that were nonetheless significantly higher in sugar yield than the control. These six lines also produced higher grain yields than the sweet sorghum genotype.

Hybrid sweet sorghum breeding methodologies follow traditional sorghum breeding approaches with modifications in the traits and selection protocol. For example, because of the additive effect of sugar concentration, it is critical to select for sugar concentration in both the seed and pollinator parents. However, juice volume is more of a dominant trait; consequently, it may be selected in either parent and it will be expressed in the hybrid. Most programs will use a pedigree breeding approach followed by sterilization of seed parent and testcrossing of both types on standard testers (Rooney, 2004).

Like yield in a grain crop, total sugar yield can be broken down into yield components. For sweet sorghum the main yield components are juice yield and soluble sugar concentration. Juice yields are directly related to biomass yield, and thus sweet sorghum breeding involves some level of selection for biomass yield. For efficiency, total sugar per unit area was calculated using total dry biomass yield and a coefficient for juice content, and hence juice yield is generally not reported. Soluble sugar concentration is typically measured by a handheld refractometer and is expressed in the unit brix. Care must be taken in evaluating brix concentrations in the field; they are highly dependent on environmental conditions and inconsistency in evaluation may result in bias if selection is based on brix alone (Corn, 2009). Lodging and stress tolerance are also traits of interest as they affect harvestibility, stability, and fermentable sugar yield. Finally, an important trait unique to sweet sorghum is the duration of optimal sugar yields in the hybrids. It is generally agreed that sugar yields peak in sorghum prior to physiological maturity and therefore, if that yield can be maintained for a longer period of time it could extend the economic harvest season.

### 6.4.3 Biomass Sorghum

Biomass sorghum generally describes genotypes with high-yield potential of primarily lignocellulosic biomass. Total yields of biomass sorghum may be very similar to sweet sorghums in certain production environments. However, these two sorghums are distinctly different for several traits. First, biomass sorghums are typically photoperiod sensitive. This characteristic results in an extended vegetative growing season, allowing the plant to capture and convert solar energy into biomass, provided that adequate moisture is available for growth. Biomass sorghums will reduce growth during periods of drought and then resume growth when moisture is available. Second, the composition of biomass sorghum is being developed specifically for the bioenergy market. Since biomass sorghum typically has dry stalks, the moisture content is lower than sweet sorghums and harvest systems and timing is based on balancing need, productivity, and composition.

The key to producing photoperiod-sensitive hybrids is based on the genetic control of photoperiod sensitivity and/or the target environments in which seed production is undertaken. In the winter months in tropical environments, daylengths are sufficiently short to allow commercial seed production of hybrids. The greatest challenge is planting seed and pollinator seed stock in time to ensure that both are in anthesis at the same time. In temperate environments, such production is not possible because of cool temperatures during the winter season. Therefore, seed production must rely on genetic systems that allow the production of a photoperiod-sensitive hybrid using two photoperiod-sensitive parental lines. Such a system was identified and characterized in forage sorghums (Rooney and Aydin, 1999) and it can be readily deployed within a bioenergy breeding program.

Breeding for biomass sorghum will follow the same strategies used for other hybrid sorghums. Vegetative biomass yield will be the most important trait and in most cases, biomass sorghums are being bred for the single harvest management scheme (Venuto and Kindiger, 2008). Seed parent inbred development will follow approaches similar to those used for grain parents. To adopt that system, most of the breeding efforts will be focused on the pollinator parent because existing seed parental lines possess most of the traits needed in a seed parent for biomass hybrids such as dry pithy stalk, low soluble sugar content, and acceptable parental line seed yields. There is also significant opportunity to develop pollinator parents. Potential pollinator parents range from existing elite sorghum germplasm that have excellent heterosis with existing seed parents to unique genotypes that maximize photoperiod sensitivity and that are derived from exotic sorghum accessions. Initial screening of potential parental lines for maturity, yield, composition, and agronomic desirability will be effective in identifying candidates for pollinator parents. Further breeding will be required to produce pollinator parents, optimally these lines will be optimized to complement existing seed parents at both maturity and dwarfing loci. This will allow for the relatively simple production of hybrids using lines that are moderately short and photoperiod insensitive, but that results in a hybrid that is tall and photoperiod sensitive (Quinby and Karper, 1948; Quinby, 1974; Rooney and Aydin, 1999).

## 6.5 Composition in Energy Sorghum Breeding

As is the case with forage sorghum, quality components of biomass are of concern to bioenergy breeders, but the definition of quality may be somewhat different. For biomass sorghums, lower protein content is desirable because less nitrogen is being removed from the production environment. Since bioenergy sorghum is grown for the production of carbohydrates, the composition of both nonstructural (sugar and starch) and structural carbohydrates (cellulose, hemicelluloses, and lignin) are important. The structural composition is critical for biomass sorghum while in sweet sorghum, nonstructural carbohydrates are of primary importance. In either case, a significant variation in both nonstructural and structural composition is observed (Corn, 2009; Packer, 2010). Stefaniak et al. (2012) reported a twofold range in variation among sorghum genotypes for lignin, hemicelluloses, and cellulose. While a proportion of this variation is environmental and maturity dependent (Gul et al., 2008), there is a genetic component that is heritable (Clark, 1981). Excessive ash content is strongly undesirable trait for biofuel processing and selection should also target for low ash content. Harvest approaches also influence this trait and any process that reduces ash uptake from the soil is good.

Research into the variation in forage composition and its effect on digestion in animal systems is of relevance to the bioenergy breeder because approaches to fermentation mimic a ruminant digestive system. There is considerable debate as to the net gain of energy using lignocellulosic ethanol conversion techniques (Patzek and Pimentel, 2005). However, the consensus is that this strategy of ethanol production from starch has a positive net energy gain utilizing current technologies (Shapouri et al., 2002; Shapouri and McAloon, 2004). Evidence suggests that the energy balance of lignocellulosic ethanol is far more favorable than starch-based ethanol (Hoover and Abraham, 2009). Additional improvements are needed to make lignocellulosic energy production more cost-effective as compared to fossil fuels or other renewable sources of energy (Vermerris et al., 2007, Saballos, 2008).

Lignin is a component in this process. The brown midrib mutation (*bmr*) in maize and sorghum has been identified as a model genotype for the evaluation of reduced lignin to cellulose/hemicellulose ratio (Porter et al., 1978). Oba and Allen (2000) reported a positive response to low lignin content when dairy cows were fed *bmr3* maize stover as compared to lines isogenic for normal midribs. The ADF concentrations, which are an estimate of cellulose plus lignin, were not statistically different between *bmr3* line and the control. The *bmr3* line had significantly lower lignin concentrations at both NDF levels evaluated. However, reduced lignin has been associated with reductions in biomass yield and increased lodging (Casler et al., 2002; Vogel et al., 2002; Pederson et al., 1982; Pederson et al., 2005; Coleman et al., 2008).

Researchers have begun to evaluate the efficiency of ethanol production from sorghum biomass directly. However, the exact method of cellulose to ethanol conversion has yet to be refined. The method of conversion will necessarily contain a pretreatment step that must efficiently release cellulose from the cell wall matrix, hydrolyze hemicelluloses, modify (syringinil/guaiacyl ratio) or remove lignin, and convert crystalline cellulose to an amorphous form (Saha, 2003;). Corredor et al. (2009) tested four forage sorghums known to differ in cell wall composition. Significant variation among genotypes existed for independent measures of cellulose, hemicellulose, and lignin. Lines with a low syringil/guaiacyl ratio in their lignin structures were easier to hydrolyze despite the initial lignin content. Additionally, the lines with the lowest crystallinity index (CrI) value were most efficiently hydrolyzed regardless of initial cellulose content. These results imply that further improvement of sorghum as a bioenergy feedstock can be accomplished both by increasing total structural carbohydrate yield per unit area, and by modifying carbohydrate and lignin structure as well. As with most crops, improving quality while maintaining yield is critical.

Systematic improvement in composition will require rapid and cost-effective screening methodologies. Currently, research to quantify the cellulose and hemicellulose fractions of sorghum biomass or to understand the underlying genetic mechanisms that influence their biosynthetic pathways has been limited. One reason for this is the lack of an inexpensive and reliable method for measuring these cell wall components in a breeding program. NIR technology has been used for years to measure forage quality in sorghum; this technology is readily available for use on bioenergy sorghum as well, once appropriate calibration curves are developed. A research project involving the National Renewable Energy laboratory (NREL), the National Sorghum Producers, and Texas A&M is developing these calibration curves (J. Dahlberg, personal communication, 2010).

## 6.6 Genetic Variation and Inheritance

To this point, this chapter has presented what makes sorghum a promising bioenergy crop, how developing cultivars for bioenergy would be similar and different to traditional sorghum breeding, and summarized breeding methodologies. Now it is necessary to step back and examine what we know concerning the nature of quantitative inheritance in sorghum. Estimating genetic variability, heritabilities, and the chromosomal locations of influential alleles are three powerful tools the modern breeder must use.

### *6.6.1 Grain Sorghum*

Historically, research into the genetic basis for yield in sorghum goes back to the late 1950s. Rao and Goud (1977) estimated variance components in a five-parent diallel. They reported highly significant general and specific combining affect variances for the yield components, indicating that additive and nonadditive components of variance were important for the traits measured (Murray et al., 2008). Significant genetic variance and high heritability (0.63) for grain yield were reported in this study in an RIL population of T×623/Rio. Given that this population was primarily selected for sweet sorghum work, and that the female parent is an excellent seed parent, these results are not overly surprising. In one pest-free environment, positive alleles for grain yield were inherited from both parents. This study also mapped the chromosomal locations of many QTLs. In water-stress environments, a major QTL was inherited from the grain parent and a minor QTL from the sweet parent. Ritter et al. (2008) also reported a very high broad-sense heritability indicating the presence of genetic variation for grain yield. Where significant genetic variances for grain yield exist, the potential for improvement in grain yield because genetic variance is a measure of allelic diversity. The reports of significant genetic variances give the breeder an insight into promising parents for inbred development.

The two main components of grain yield are seed size and seed number per unit area (Egli, 1998). Grain sorghum breeders have identified genetic variation for maturity traits affecting yield components (Hart et al., 2001; Oliver et al., 2005; Brown et al., 2006; Srinivas et al., 2009). As with most grain crops, the goal has generally been to select phenotypes that are late in flowering, but early enough for timely harvest and/or drought avoidance (Boyer, 1982). QTLs for days to anthesis have been identified (Brown et al., 2006, Ritter et al., 2008; Srinivas et al., 2009). These QTLs represented substantial amounts of the genetic variability for this trait and therefore are amenable to further manipulation.

QTLs for grain yield have been reported (Gomez, 2002; Maradiaga, 2003; Murray et al., 2008; Ritter et al., 2008; Srinivas et al., 2009), but the consistency of the QTLs across environments and populations is limited. Gomez (2002) constructed RIL and testcross populations and genotyped them with AFLP and SSR markers. In the RIL population, 14 genomic regions were identified that explained between 10% and 50% of the variation for panicle number, panicle exertion, panicle weight, panicle length, grain yield, maturity, and plant height. In the testcross population, 12 genomic regions were identified most of which were unique to the testcross population indicating the importance of QTL evaluation in hybrid combinations. Maradiaga (2003) also developed RIL and testcross populations for the mapping of grain agronomic traits. A total of 89 and 79 QTLs for seven traits were detected in the RIL and testcross populations, respectively. However, only a few of these QTLs were stable across environments and many were detected only when the logarithm of odds (LOD) was relaxed. However, in the RIL population one QTL for yield was detected in the combined environment analysis that had an additive effect of −0.21 and explained 8% of the variation in yield. Dividing yield into more heritable components offers breeders the possibility of obtaining more reliable selection tools than yield QTL themselves. For example, significant QTLs have been identified for panicle yield and seed size (Rami et al., 1998; Murray et al., 2008; Srinivas et al., 2009), as well as panicle length (Hart et al., 2001; Rami et al., 1998; Makanda et al., 2009). In Hart et al. (2001), Rami et al. (1998), and Srinivas et al. (2009), panicle length QTLs were mapped to the same location. The QTL for panicle length on chromosome 7 was reported to account for 8.6% of the additive variance in one location (Hart et al., 2001).

Tuinstra et al. (1997) mapped QTLs for grain yield and some of its components and evaluated their expression in lines with the stay-green phenotype in irrigated and drought-stressed environments. Six genomic regions, some being rather large, were associated with stay-green in this population. Two major QTLs were reported to have substantial effects on yield both in the rainfed and irrigated environments suggesting a pleiotropic effect under nondrought stress conditions as well. Eight QTLs for seed weight were detected and two were colocalized with yield QTLs under irrigated conditions. Significant variation within the RIL population along with the identification of beneficial alleles from both parents at QTLs suggest that in this population these yield and stay-green traits are amenable to improvement for biofuel through breeding.

### 6.6.2 Grain Quality/Starch Composition

The sorghum caryopsis is primarily starch and increasing starch yield per area can be accomplished by (1) increasing the grain yield per unit area and/or (2) increasing the starch content (on a percentage basis) of the sorghum seed. Increase in starch content improves ethanol fermentation efficiency (Wu et al., 2007). A substantial range of starch, 64.3–73.8%, was reported in 70 sorghum genotypes (Wu et al., 2007). The higher starch content generally translated into higher ethanol yield, but ethanol conversion efficiency did not correlate linearly with starch content ($R^2 = 0.04$), so other factors are also involved. Wu et al. (2007) also noted a weak and inverse correlation between protein content and ethanol conversion efficiency. They hypothesized that because grain starch is enmeshed within a protein matrix, interaction with water may be compromised when the starch-to-protein ratio is not optimal. Rami et al. (1998) reported a range of 3–5 and 2.5–5 for grain starch as expressed by a kernel flouriness rating in two RIL populations. QTLs for this trait were mapped to four linkage groups. The QTL mapped to chromosome F explained 54% of the phenotypic variance. This is a major QTL and a good candidate for MAS for ethanol conversion efficiency in a bioenergy breeding program. Increasing starch content without compromising yield potential might be an attainable goal for bioenergy breeders. Murray et al., (2008) estimated a significantly positive correlation between grain dry yield and starch. In maize a positive, albeit not significant, correlation between grain yield and starch content has been reported (Saleem et al., 2008).

Murray et al. (2008) also reported QTL locations for grain starch and grain crude protein content and the starch QTL were mapped in three linkage groups. The QTL on chromosome 2 could be the same QTL mapped to linkage group C (Rami et al., 1998). QTLs for crude protein were mapped to three chromosomes as well (Murray et al., 2008). The QTL on chromosome 1 may be the same as the protein QTL mapped to linkage group A (Rami et al., 1998). They also reported significant variation within their RIL population for starch with a range of 600–690 g kg$^{-1}$. The grain-quality QTL found in the literature could be useful to the breeder to initiate MAS programs.

The breeder now has powerful tools that can assist in resource allocation decisions for their program. Reliable QTLs that are tightly linked to genes affecting the trait of interest can allow nondestructive genotyping that can be accomplished easily at early seedling stage. This gives the breeder the opportunity to make selection decisions early in the season and could eliminate the need for early generation yield testing. Codominant markers such as simple sequence repeats (SSRs) could be useful selection tools for parents at the $F_2$ because they can identify individuals that are homozygous for the beneficial allele. Therefore, these markers could help improve selection early in the process of inbred development. Table 6.2 contains a list of references and summaries of results for inheritance studies in grain sorghum. A consensus on the naming of linkage groups (chromosomes) in sorghum was not reached until 2005, when Kim et al. (2005) presented standard designations agreed upon by the sorghum research community.

### *6.6.3 Dual Purpose—Grain and Stalk*

Murray et al. (2008) examined the relationship between grain and stalk quality variation and the possibility of improving both for bioenergy purposes. They concluded that total nonstructural carbohydrate yield (in grain and stalks) could be increased by selecting for stalk and grain yield traits rather than composition traits. For many years, silage sorghums have maximized biomass and grain sorghum yields (Bean and McCollum, 2006). Blummel et al. (2003) measured variability for stover and grain traits in 12 sorghum genotypes that showed that grain yield was not correlated with stover yield or quality under high fertilization. In fact, the cultivar with the highest grain yield also had the highest stover quality. In low fertility, grain and stover yield were found to be inversely correlated which is not surprising as the plant attempts to partition assimilate to the grain.

### *6.6.4 Soluble Carbohydrates*

Sweet sorghum has the potential to produce up to 8000 L ha$^{-1}$ ethanol which is about twice as much as that of maize and 30% more than sugarcane in Brazil (Hunter and Anderson, 1997). Ethanol production from sweet sorghum can be substantially less expensive than from sugarcane in India (Reddy et al., 2009). The juice of sweet sorghum is typically high in soluble sugars, with concentrations ranging from 8% to 30%, depending on the genotype, maturity, and environment. Most of the soluble sugars in sweet sorghum juice are sucrose with measurable levels of fructose. These sugars are readily fermentable without the need of hydrolysis using technology from the sugarcane process.

Genetic variation for biomass yield and sugar concentration has been observed (Clark, 1981; Natoli et al., 2002; Murray et al., 2008; Corn, 2009). Clark (1981) observed genetic variation for stem carbohydrates in a breeding population derived from the crosses between sweet and grain parents and seven of the hybrids exceeded the high parent values. Natoli et al. (2002) reported significant differences between lines for brix and biomass yield in an $F_3$ population that resulted in the recovery of superior individuals for total sugar yield. For these traits,

**Table 6.2.** Citation traits evaluated QTL names and QTL locations for genetic studies in grain sorghum.

| Study | Trait | QTL | Linkage Group |
|---|---|---|---|
| Rami et al., 1998[a] | Kernel flouriness | FL/11.1 | C |
| | | FL/15.6 | D |
| | | FL/17.6 | F |
| | | FL/55.1 | F |
| | | FL/53.6 | F |
| | Panicle length | PL/31.4 | A |
| | | PL/34.9 | A |
| | | PL/15 | B |
| | | PL/10.4 | B |
| | | PL/18 | F |
| | | PL/16.1 | F |
| | | PL/11.2 | G |
| | | PL/11.6 | H |
| | | PL/11.1 | K |
| | | PL/12 | K |
| | Kernel weight/panicle | KW/31.3 | A |
| | | KW/19 | F |
| | Protein content | PR/18.4 | A |
| | | PR/19.7 | C |
| | | PR/17.1 | C |
| | | PR/16.5 | C |
| | | PR/11 | D |
| | | PR/14.8 | H |
| | Seed size | TK/35.2 | A |
| | | TK/10.7 | C |
| | | TK/11.6 | G |
| Hart et al., 2001 | Days to anthesis | QMa.txs-F1 | 4 |
| | | QMa.txs-F2 | 7 |
| | | QMa.txs-G | 10 |
| | Panicle length | QPal.txs-E | 7 |
| | | QPal.txs-F | 7 |
| | | QPal.txs-G | 9 |
| Brown et al., 2006 | Days to anthesis | Flowering time | 8, 9 |
| Feltus et al., 2006 | Grain yield | GWT | 10 |
| | Days to anthesis | MA50 | 1, 3, 6, 8, 9 |
| | Panicle weight | HWT | 10 |
| | Seed size | KWT | 1, 2, 3, 4, 6, 87, 8, 10 |
| Srinivas et al., 2009 | Days to anthesis | QDan-sbi01-1,2 | 1 |
| | | QDan-sbi02-1,2 | 2 |
| | | QDan-sbi03 | 3 |
| | | QDan-sbi05 | 5 |
| | | QDan-sbi06 | 6 |
| | | QDan-sbi07 | 7 |
| | | QDan-sbi08 | 8 |
| | Grain yield | QGyl-sbi06 | 6 |
| | Panicle length | QPle-sbi02 | 2 |
| | | QPle-sbi06-1,2 | 6 |
| | | QPle-sbi07 | 7 |
| | Panicle weight | QPwe-sbi06 | 6 |
| | Seed weight | QSwe-sbi01 | 1 |
| | | QSwe-sbi04 | 4 |
| | | QSwe-sbi06 | 6 |

[a] Report published before 2005 containing nonstandard molecular linkage group (MLG) designations.

significant genetic variances were also reported. Relatively large heritabilities of 0.51, 0.62, and 0.53 were estimated for brix, dry matter yield, and total sugar yield, respectively. Murray et al. (2008) reported differences between individuals in an RIL population in grain, stem, and agronomic traits. Corn (2009) reported significant genotype effects for fresh biomass yield, brix and sugar yield in breeding population of hybrids, and their grain-type female and sweet-type male parents. Ritter et al. (2008) observed wider phenotypic ranges for glucose and fructose content; and dry matter and grain yield in RILs than in their parents indicating transgressive segregation for those traits.

Lodging is an important trait for sweet sorghum breeders that are inversely related to juice yield. Natoli et al. (2002) observed significant differences for this trait. The heritability for this trait was extremely high (0.96). Positive genotypic correlations were estimated between dry matter yield and brix; and dry matter yield, and sugar yield as well as a negative correlation between lodging and dry matter yield. Because of this, concurrent selection for both lodging resistance and biomass yield is essential to maximize productivity.

Makanda et al. (2009) observed heterosis for stalk sugar traits in hybrids between eight R-lines and eight A-lines. General combining effects were significant for all traits indicating primarily additive gene interaction. SCA effects were only significant for flowering, stem biomass, plant height, stem brix, and stem brix juice index. Improvement in this population could be achieved through hybridizations maximizing dominance and epistatic gene interactions Makanda et al. (2009). Twenty of the 64 hybrids showed better parent heterosis for brix, and 44 for stem biomass. One hybrid exhibited 250% heterosis over better parent and 20 hybrids were above 100% for stem biomass. The potential exists for developing dual-purpose sweet/grain sorghum for bioenergy hybrids between grains by sweet types using the grain type, as the female parent typically has more stalk sugar than their grain parent with similar grain yield (Hunter and Anderson, 1997). Cultivars with this dual crop capacity give the producer flexibility in coping with unexpected weather or economic events.

### 6.6.5 Breeding for Stress Tolerance

*Abiotic Stress*

Abiotic stress is defined as a yield-limiting factor caused by a nonbiological source. They include but are not limited to temperature, moisture, and fertility. The inheritance of tolerance to nearly all stresses has been described as quantitative in nature making QTL and inheritance studies important research resources for the employment of MAS and the identification of useful germplasm. The majority of sorghum grown worldwide is in dryland conditions. Boyer (1982) estimated that drought was the largest single factor in reducing grain sorghum yields. The same situation will be true in bioenergy sorghums, especially since production will be reliant on rainfall and not irrigation. Drought stress is a complicated condition that initiates the expression of numerous genes in a myriad of signal transduction pathways (Buchanan et al., 2000). Drought tolerance can be affected by the manipulation of all the plant organs and many physiological processes. Thusly, drought affects grain yield and quality, stalk juice yield and quality, and forage yield and quality.

*Drought*

Stay-green or delayed senescence is a drought tolerance trait that researchers have identified in sorghum, maize, and other grasses (Rosenow et al., 1983). Stay-green can mitigate the effects of terminal drought stress on the final processes of yield production such as grain fill and stalk development, and is of interest to breeders of dual-purpose sorghum (Haussmann et al., 2002). The stay-green trait includes modifications in charcoal stalk rot (Rosenow, 1984), lodging (Woodfin et al., 1988), and basal stem sugars (Duncan, 1984; van Oosterom et al., 1996). Exactly how, or even if these modifications represent overall improvement in the sorghum germplasm is unclear. Haussmann et al. (1999) suggested that a possible mechanism for the stay-green trait is reduced reproductive growth resulting in a poorly developed sink. For the grain breeder tolerance to drought means that a crop has acceptable yield under drought conditions.

Four or more putative stay-green QTLs have been reported (Tuinstra et al., 1997; Crasta et al., 1999; Xu et al., 2000; Subudhi et al., 2000; Tao et al., 2000; Kebede et al., 2001; Haussmann et al., 2002). It is not possible to ascertain whether these various QTLs are associated with the same or different genes due to the polygenic nature of the trait, variation in mapping populations, mapping functions used, and testing environments. In any case breeders, always try to pyramid multiple beneficial alleles for quantitative traits in order to ameliorate genetic vulnerability. Xu et al. (2000) constructed an RIL population and mapped four stay-green QTLs; *stg1* and *stg3*, mapped to linkage group A; *stg2*, mapped to linkage group D; and *stg4* to linkage group J. The QTL *stg1*, *stg2*, and *stg3*, accounted for 46% of the phenotypic variability for stay-green. The chromosomal regions containing *stg1* and *stg2* are known to contain important heat shock proteins, key photosynthetic enzymes, and an abscisic responsive genes indicating that stay-green is a phenotype influenced by an array of biosynthetic pathways (Rochester et al., 1986). Subudhi et al. (2000) reconfirmed the stability and genomic locations of four QTLs identified by Xu et al. (2000).

Haussmann et al. (2002) mapped QTL for drought tolerance, and estimated their effect on grain and stover yield, total biomass, and harvest index under terminal drought-stress conditions using two RIL populations with a common stay-green parent E36-1. The different parents for these populations were IS9830 and N13. They estimated relatively high repeatabilities ranging from 0.55 to 0.95 in RIL population 1 and 0.49 to 0.94 in RIL population 2. Grain yield was correlated with all the stay-green traits in RIL population 2, but no correlation was detected in population 1, possibly due to different environmental conditions in the evaluation of population 1 or genetic background. High stover yield was correlated with stay-green at 15 days post flowering in both populations, at 30 days in RIL population 2, and at 45 days in RIL population 1. Five QTLs for stay-green were detected in RIL population 1 and eight in RIL population 2. Three of these were mapped to the same position in the two populations grown in different years. This validation across backgrounds and environments indicates that these chromosomal regions are good targets for introgression into a breeding population using these parents.

Given all the research into QTLs for drought tolerance in sorghum, it is clear that MAS for drought tolerance is nearing reality. As genotyping technology continues to improve, breeders are interested in using these markers. It is impossible to discern whether the drought tolerance QTL, identified by Xu et al. (2000), Haussmann et al. (2002), or Subudhi et al. (2000), consist of the same genetic elements because they were mapped using different markers and populations. However, QTLs from Xu et al. (2000) and Subudhi et al. (2000) have been fine mapped; the opportunity for MAS using these QTLs is possible (P. Klein, personal communication, 2009).

*Temperature*

By virtue of its C4 photosynthetic pathway and its tropical origin, sorghum cultivation has been limited to relatively warm climates. Consequently, breeders interested in broadening sorghum's area of temperate adaptation have been focusing on tolerance to cold temperatures.

The sorghum plant is sensitive to cold temperatures from the very beginning of its life cycle with germination being inhibited at temperatures as below 15°C and completely halted at 7°C (Rooney, 2004). Early growth is compromised by a retarding of the chlorophyll biosynthetic pathway (McWilliams et al., 1979). Genetic variation for cold tolerance has been identified in sorghum (Singh et al., 1985; Knoll and Ejeta, 2008; Burow et al., 2009a). In Mexico, sorghum genotypes were screened in a cold tolerance nursery at high elevations (Singh, 1985). This early work has identified germplasm tolerant to cold temperatures from anthesis to maturation and this germplasm has been a valuable source of cold tolerance. Sorghum breeders have utilized this germplasm and developed breeding lines that are adapted to the highlands of Mexico, Eastern, and Southern Africa (Peacock, 1982).

At more temperate latitudes, the effects of cold temperatures are manifested earlier in the plant's life cycle. For cold temperatures, seedling emergence and seedling vigor genetic variation have also been observed (Bacon et al., 1985; Knoll and Ejeta, 2008). Bacon et al. (1985) completed four cycles of recurrent selection for seedling cold tolerance and observed a 2.8% improvement per cycle. These demonstrate that this trait is heritable and amenable to selection. Beneficial alleles for cold tolerance have also been identified in genotypes of Chinese origin (Smith and Frederiksen, 2000). Knoll and Ejeta (2008) developed two mapping populations variable for cold tolerance and were able to map a QTL that had positive effects for seedling vigor in both populations (Table 6.3). This QTL represents a tool for MAS that could augment phenotypic selection.

Table 6.3. Citation traits evaluated QTL names and QTL locations for genetic studies in forage and sweet sorghum.

| Study | Trait | QTL | Linkage Group |
|---|---|---|---|
| Feltus et al., 2006 | Leaf length | LLN | 4, 6, 8, 9, 10 |
| | Leaf width | LWD | 2, 3, 6, 7, 8, 9 |
| Brown et al., 2006 | Dwarf | Dw3 | 7 |
| | Dwarf | dw3 | 9 |
| Murray et al., 2008 | Crude protein | Crude_Protein_Grain | 1, 3, 4, 6 |
| | Days to anthesis | Flowering_Time | 6, 9 |
| | Grain starch | Starch_Grain | 1, 3, 4, 6 |
| | Grain yield | Grain_Dry_Yield | 4, 6, 9 |
| | Juice yield | Juice_Yield_1st_Press | 1, 6, 7, 9 |
| | Panicle yield | Fresh_Panicle_Yield | 4, 6, 7, 9 |
| | Plant height | Height | 7, 9 |
| | Seed size | Thousand_Seed_Weight | 6, 8, 9 |
| | Soluble Solids | Brix | 6 |
| | Stem diameter | Mean_Stem_Thickness | 4, 6 |
| | Stem yield | Stem_Fresh_yield | 6, 7, 9 |
| | Sucrose concentration | Sucrose_Juice | 3 |
| | Total sugar yield | Sugar_Yield | 1, 6, 7, 9 |
| Natoli et al., 2002[a] | Biomass yield | Dry matter yield | J |
| | Days to anthesis | Flowering time | A |
| | Plant height | Plant height | A, J |
| | Total sugar yield | Sugar yield | J |
| Ritter et al., 2008 | Biomass yield | Total dry matter | 1, 6, 10 |
| | Days to anthesis | Flowering Time | 1, 4, 6, 10 |
| | Grain yield | Grain yield | 2, 3, 10 |

[a]Report published before 2005 containing nonstandard MLG designations.

*Soil Fertility/Toxicity*

Aluminum toxicity can limit sorghum production in acidic soils (Anas and Yoshida, 2000), which are common in the humid tropics (Sanchez, 1976). Excessive aluminum in soils inhibits root growth by damaging the root tip (Delhaize and Ryan, 1995). Genetic variability for aluminum tolerance has been observed (Boye-Goni and Marcarian, 1985; Foy et al., 1993; Anas and Yoshida, 2004). Boye-Goni and Marcarian (1985) used a half-diallel mating design to estimate inheritance parameters for aluminum tolerance. Aluminum tolerance was measured as a function of root growth in a nutrient solution. Highly significant GCA and SCA were estimated with the GCA effects being nine times greater. Narrow-sense heritability was 0.78 suggesting this trait is amenable to improvement using recurrent or pedigree selection. However, progress for improving this trait might be slower than the results obtained by Boye-Goni and Marcarian (1985). Anas and Yoshida (2004) reported low realized heritability for aluminum tolerance measured as a complex of stress response traits in two populations of 0.43 and 0.35, respectively. Magalhaes (2007) identified and cloned an aluminum tolerance gene $Alt_{SB}^5$. Increased root citrate exudation was associated with expression of this gene and was proposed as a mechanism to identify aluminum tolerant haplotypes with this gene.

Nitrogen use efficiency (NUE) in cereal crop production worldwide has been estimated to be at about 33% (Raun and Johnson, 1999). With respect to sorghum, the bioenergy breeder will need to be especially concerned with improving NUE, as sorghum requires higher rate of nitrogen relative to other proposed biofuel feedstocks such as switchgrass and miscanthus. While research into improving NUE in sorghum through breeding is limited, research in rice (Bufogle et al., 1997) and wheat (Kanampiu et al., 1997) indicates that NUE could be improved through the improvement of harvest index. However, for lignocellulosic bioenergy production, breeding should emphasize improving the NUE for biomass yield rather than grain. Currently, experiments are being conducted using sweet sorghum to identify genotypes and QTL that can improve NUE (DOE, 2010).

Saline soils are a major constraint to crop productivity especially in many areas that rely on irrigation. About 10% of the world's soil is saline (Szabolcs, 1994), and that percentage is expected to grow (Ghassemi et al., 1995). Sorghum productivity is reduced in saline soils primarily due to the reduction of seedling emergence (Igartua et al., 1994) and biomass yield (Almodares et al., 2008). Igartua et al. (1994) developed a population variable for seedling salt tolerance using an incomplete 6 × 6 factorial mating design. They reported large difference among hybrids for salt tolerance at germination and emergence stages which was attributable to SCA and female GCA for emergence, and female GCA for germination. The male GCA was also significant for both traits. Azhar and McNeilly (1988) estimated the genetic effects of salinity on sorghum seedlings in 8 × 8 diallel cross hybrids. Their results indicated that both dominance and additive effects were important sources of variation, with the dominance effects being more substantial. Almodares et al. (2008) identified genotypes that varied in stem yield in both high and low salinity. They also found that stem sucrose tended to be inversely correlated with stem yield. At the highest salt concentration, the genotype with the lowest reduction in stem yield (relative to the yield at the low salt concentration) also had the highest sucrose concentration. This indicates that the inverse relationship of sucrose on stem yield does not hold for genotypes that are not severely reduced in stem yield at very highly saline soils. Two implications of these results are that it might be possible to improve both salt tolerance and juice traits in sweet sorghum simultaneously, and that stem sucrose content might be used as a selectable trait for the improvement of salt tolerance.

*Biotic Stress*

Biotic stresses are defined as yield-limiting factors of a biological source, typically either an insect pest or a disease pathogen. There are numerous pathogens and pests that reduce the yield and quality of every type of sorghum. However, the relative importance of each biotic stress differs dependent on the type of sorghum and the location of production. For example, bioenergy sorghums will be managed to eliminate or minimize reproductive growth. Consequently, the pests and pathogens of reproductive growth are not important. The pests and pathogens described herein are focused on their predicted importance to energy crops (bioenergy and sweet sorghum).

*Insect Pests*

Traditionally, important insect pests of sorghum include the sorghum midge *Stenodiplosis sorghicola* (Coquillett), greenbug *Schizaphis graminum* (Rondani), shoot fly *Atherigona soccata* (Rondani); and stalk boring insects such as sugarcane borer *Diatraea saccharalis* (Fabricius), southwestern corn borer *D. grandiosella* Dyar, spotted stem borer *Chilo partellus* (Swinhoe), and maize stalk borer *Busseola fesca* (Fuller). In bioenergy crops, the insects of greatest interest to the breeder are those that affect the harvested plant organs. For lignocellulosic bioenergy, the greenbug and the borers are of particular importance because they affect the leaves and stalks, whereas midge and shoot fly are important for starch crops because they reduce grain yield.

For grain sorghum, sorghum midge is the most widespread and most damaging insect in the southwestern United States (Teetes and Pedleton, 2000). When susceptible varieties are grown, yield losses of 890.8 kg ha$^{-1}$ have been reported (Damte et al., 2009). Genetic variation for sorghum midge has been reported (Boozaya-Angoon, et al., 1984; Sharma, 1985; Agrawal et al., 1988; Nwanze et al., 1991). Several groups observed ovipositional antixenosis or the reduction in egg deposition in sorghum genotypes (Franzmann, 1993; Sharma et al., 2002 Tao et al., 2003). Actual antibiosis or the death of larvae in response to feeding on resistant genotypes has also been observed (Sharma et al., 2002; Tao et al, 2003; Waquil et al., 1986). Tao et al. (2003) mapped two QTLs in sorghum for resistance to sorghum midge. The allele from the resistant parent accounted for a 12–15% of the phenotypic variability for resistance. One genomic region was identified for antibiosis resistance and the favorable alleles at the two flanking QTLs accounted for a 34.5% of the phenotypic variability. These results indicate that pyramiding QTLs using two separate mechanisms (ovipositional antixenosis and antibiosis) could be possible.

The shoot fly is a pest of sorghum in most of its production areas, but is not found in the Americas or Australia (Teetes and Pendleton, 2000). Losses from this insect can be substantial. In India, it has been estimated that annual losses can be up to 5% (Jotwani, 1983). Larvae of this insect cut the growing point of the apical shoot resulting in a deadheart symptom. Genetic variation for resistance to shoot fly has been observed (Nimbalkar and Bapat, 1988; Nwanze et al., 1991; Dhillon et al., 2006; Satish et al., 2009). Nimbalkar and Bapat (1988) studied the inheritance of shoot fly resistance in 8 × 8 diallel. They found highly significant SCA and GCA for all resistance traits measured additive and dominance effects were significant for all traits except yield per plant suggesting that this resistance is amenable to recurrent or pedigree selection procedures. Researchers have identified and mapped 29 QTLs for five components of shoot fly resistance (Satish et al., 2009). The QTLs of related component traits were colocalized suggesting pleiotropy or tight linkage. Linkage of genes for this trait would facilitate the introgression of multiple sources of resistance using these QTLs.

The mechanisms of injury due to stalk borer are generally indirect in nature. Tunneling activity can disrupt the vasculature causing nutrient or water deficiency in addition to the general weakening of the stalk which can cause lodging (Teetes and Pendleton, 2000). The wound due to stalk borer damage provides entry points for infections of stalk rot pathogens (Reagan and Flynn., 1986). Sorghum crops in the United States are not seriously infested with stalk boring insects (Teetes and Pendleton, 2000). However, if bioenergy sorghums are widely grown these may become very serious pests of the crop. In Africa and India, where losses have been estimated to be $260 million annually (ICRISAT, 1992), the spotted stem borer is a particular threat and can cause damage to all aboveground organs (Kumar et al., 2006). Genetic variation for resistance to spotted stem borer has been identified (Singh and Rana, 1989; Nwanze et al., 1991; Sharma et al., 2007). Sharma et al. (2007) reported significant genetic variation for deadheart formation, leaf feeding, and overall resistance to this insect. The sources of genetic variation for the resistance parameters rated were both additive and additive × dominant and therefore, this trait is amenable to selection using recurrent or pedigree selection procedures.

*Sorghum Diseases*

Many biotic stresses limit sorghum production and will be of interest to bioenergy breeders. Among the most important are anthracnose (*Colletotrichum sublineolum*), downy mildew (*Peronosclerospora sorghi* (Weston and Uppal) C. G. Shaw), and head smut (*Sporisorium reilianum* (Kuhn) Langdon and Fullerton) (Table 6.4). Considerable effort has been put into developing cultivars that are resistant to these pathogens. Anthracnose is one of the most economically important diseases of sorghum and can affect all aboveground plant organs (Cardwell et al., 1989; Rosewich et al., 1998). This disease is likely to become more important in energy sorghums (Rooney et al., 2000). Reported yield losses due to anthracnose have been in excess of 50% (Harris et al., 1964). Gene-for-gene relationship has not been conclusively demonstrated and therefore, the assignment of races or pathotypes has been difficult (Rosewich et al., 1998). Genetic variation for resistance has been reported (Rooney et al., 2000; Mehta et al., 2005; Perumal et al., 2009). Both dominant and recessive alleles for resistance have been reported (Mehta et al., 2005). Perumal et al., (2009) used a mapping population to identify gene (*Cg1*) for anthracnose resistance. Two markers, *Xtxa* 6227 and *Xtxp* 549, were found to be tightly linked to *Cg1*. These markers were used to genotype 13 resistant lines of which 12 had the resistant allele and were positive at both loci indicating the markers were robust and could be used for marker-assisted selection.

Another pathogen that affects sorghum production worldwide is downy mildew. Genetic variation for resistance to this pathogen has been reported (Craig and Shertz, 1985; Sifuentes and Frederiksen, 1988; Reddy et al., 1992). Multiple races of downy mildew have been identified and consequently resistance genes have come from multiple sources (Sifuentes and Frederiksen, 1988). Resistance to single races is described as oligogenic in most cases (Rooney et al., 2000). The dominant alleles at both loci confer resistance and the resistant genotype was designated $Pl_aPl_aPl_bPl_b$. Pyramiding resistance genes for multiple races will be possible.

Head smut affects sorghum crops in Africa, Asia, Australia, Europe, and North America (Rooney et al., 2000). Because head smut cannot be effectively controlled by seed treatment or other management practices (Smith and Frederiksen, 2000), genetic resistance is crucial to the control of this disease. Several races of head smut have been identified (Cao et al.,

**Table 6.4.** Citation traits evaluated QTL names and QTL locations for genetic studies stress tolerance traits.

| Study | Trait | QTL | Linkage Group | Study |
|---|---|---|---|---|
| Tuinstra et al., 1997[a] | Drought tolerance | Seed weight stability | | A, E |
| | Drought tolerance | Stay-green | | B, F, G, H, I |
| | Drought Tolerance | Yield stability | | B, C, F |
| Crasta et al., 1999[a] | Days to anthesis | Mat | | B, G |
| | Drought tolerance | Stay-green | | A, B, D2, G, I |
| Subudhi et al., 2000[a] | Drought tolerance | SG97DLL | | A, D, E |
| | Drought tolerance | SG98HW | | A, D |
| | Drought tolerance | SG3478 | | A, D, J |
| Tao et al., 2000 | Midge resistance | Antibiosis | | 7 |
| | Midge resistance | Antixenosis | | 3, 9 |
| Xu et al., 2000[a] | Drought tolerance | Stay-green HL93 | | A, D |
| | Drought tolerance | Stay-green LL93 | | A, D |
| | Drought tolerance | Stay-green HL94 | | A, D, J |
| | Drought tolerance | Stay-green LL94 | | A, D |
| | Drought tolerance | Stay-green LD94 | | A, J |
| Kebede et al., 2001[a] | Drought tolerance | Stay-green | | A, B, C, D, E, F, J |
| | Drought tolerance | Preflowering drought tolerance | | C, E, F, G |
| Haussmann et al., 2002 | Drought tolerance | GL15-1 | | 3 |
| | Drought tolerance | GL15-2 | | 4 |
| | Drought tolerance | GL15-3 | | 7 |
| | Drought tolerance | GL15-4 | | 10 |
| | Drought tolerance | GL15-5 | | 10 |
| | Drought tolerance | GL30-1 | | 1 |
| | Drought tolerance | GL30-2 | | 3 |
| | Drought tolerance | GL30-3 | | 4 |
| | Drought tolerance | GL30-4 | | 10 |
| | Drought tolerance | GL30-5 | | 10 |
| | Drought tolerance | GL30-6 | | 8 |
| | Drought tolerance | GL30-7 | | 8 |
| | Drought tolerance | GL45-1 | | 1 |
| | Drought tolerance | GL45-2 | | 1 |
| | Drought tolerance | GL45-3 | | 4 |
| | Drought tolerance | GL45-4 | | 4 |
| | Drought tolerance | GL45-5 | | 10 |
| | Drought tolerance | GL45-6 | | 8 |
| | Drought tolerance | GL45-7 | | 8 |
| Perumal et al., 2009 | Anthracnose | Xtxa 6227 | | 5 |
| | Anthracnose | Xtxp549 | | 5 |
| Knoll and Ejeta, 2008 | Cold tolerance | Xtpxp43 | | 1 |
| Satish et al., 2009 | Shoot fly resistance | QEg21.dsr-1 | | 1 |
| | Shoot fly resistance | QEg21.dsr-7 | | 7 |
| | Shoot fly resistance | QEg21.dsr-9 | | 9 |
| | Shoot fly resistance | QEg21.dsr-10 | | 10 |
| | Shoot fly resistance | QEg28.dsr-5 | | 5 |
| | Shoot fly resistance | QEg28.dsr-7 | | 7 |
| | Shoot fly resistance | QEg28.dsr-10 | | 10 |
| | Shoot fly resistance | QDh.dsr-5 | | 5 |
| | Shoot fly resistance | QDh.dsr-9 | | 9 |
| | Shoot fly resistance | QDh.dsr-10.1 | | 10 |
| | Shoot fly resistance | QDh.dsr-10.2 | | 10 |
| | Shoot fly resistance | QDh.dsr-10.3 | | 10 |
| | Shoot fly resistance | QDh.dsr-10.4 | | 10 |
| | Shoot fly resistance | QTdu.dsr-10.1 | | 10 |
| | Shoot fly resistance | QTdu.dsr-10.2 | | 10 |
| | Shoot fly resistance | QTdl.dsr-1.1 | | 1 |
| | Shoot fly resistance | QTdl.dsr-1.2 | | 1 |
| | Shoot fly resistance | QTdl.dsr-4 | | 4 |
| | Shoot fly resistance | QTdl.dsr-6 | | 6 |
| | Shoot fly resistance | QTdl.dsr-10.1 | | 10 |
| | Shoot fly resistance | QTdl.dsr-10.2 | | 10 |

[a]Report published before 2005 containing nonstandard MLG designations.

1988; Magill et al., 1997). In Mexico and China the race dynamics are similar to those in the United States and therefore, management in South Texas where four races have been identified can be used as a model for those regions (Smith and Frederiksen, 2000). Early efforts in breeding for resistance in Texas resulted in the hybrid RS626. The fact that this hybrid's resistance lasted only a few years highlights the need for pyramiding resistance to multiple races. Markers linked to resistance to this pathogen have been identified. Oh et al. (1994) identified two RFLP markers and one RAPD marker linked to the gene *Shs* which provides resistance to race 5. Evidence indicates that resistance to this pathogen follows quantitative as well as qualitative inheritance depending on the race evaluated (Cao et al., 1988). Selection procedures for quantitative resistance will need to be either of the pedigree or recurrent type, while genotypic selection should be effective for improving resistance to head smut that is qualitative.

## 6.7 Wide Hybridization

### 6.7.1 Interspecific Hybridization

There is potential to improve sorghum for bioenergy using the undomesticated relatives of *S. bicolor* in which there resides an array of potentially useful traits. One possible phenotype would be perenniality. There are several sorghum species that are perennial. *Sorghum propinquum*, a fully fertile relative has a perennial growth both with and without rhizomes. *Sorghum halepense*, commonly known as johnsongrass is well known to producers as a noxious weed capable of vegetative reproduction through rhizomatous growth. While there are significant and legitimate concerns regarding perennial sorghum and their potential weediness, there are several groups researching various possibilities for the production of perennial sorghums that minimize or eliminate the potential for weediness (Glover et al., 2010).

*Sorghum bicolor x S. halepense* hybrids are known to occur naturally. Fertile hybrids then must be either diploid or tetraploid for stable bivalent pairing. Triploid hybrids from this cross could arise from unreduced gametes in the sorghum parent (Hadley, 1953, 1958). Diploid hybrids result from the formation of monohaploid gametes from the johnsongrass parent as has been documented in millet (Hanna 1990), or possibly zygomatic elimination of one genome from johnsongrass (Dweikat, 2005). Clearly, elucidating the inheritance of quantitative traits in a hybrid with such variability in genome structure could be a formidable challenge. However, Dweikat (2005) reported the transfer of the rhizomatous growth habit from johnsongrass into a fertile diploid that had added beneficial alleles from johnsongrass for resistance to greenbug and chinchbug, and improved cold tolerance.

A mapping population from the interspecific cross *Sorghum bicolor x S. propinquum* was used for mapping the sorghum genome (Feltus et al., 2006). Some of the progenies from this mapping population were observed to have perennial characteristics (Paterson et al., 1995). These progeny displayed variation for rhizomatousness between 46% and 92% of the $BC_1$ and $F_2$ progeny surviving a mild winter in south Texas. Results suggested that selecting for perenniality would result in a yield penalty (Cox et al., 2002). Such a yield penalty would need to be offset by the reductions in production costs associated with reduced planting and tillage costs. Other indirect benefits of the development of perennial cultivars include the preservation of soil structure, the reduction of exhaust pollutants associated with reduced equipment usage, and less impact on ground water resources.

The use of other *Sorghum* sp. had been limited by the inability to recover hybrids. Hodnett et al. (2005) reported that a primary barrier to hybridization in these crosses was pollen tube growth inhibition of alien (non-*S. bicolor*) pollen. Using a specific mutant allele designated *iap* that was described by Laurie and Bennett (1989), Price et al. (2006) made interspecific crosses between *S. bicolor* and *S. angustum,* S. bicolor and *S. nitidum,* and *S. bicolor* and *S. macrospermum.* The recovery of viable hybrids from this work suggested that large genome segments can be transferred within Poaceae species, which was confirmed in the reports by Kuhlman et al. (2008) and 2010).

### 6.7.2 Intergeneric Hybridization

Additional genetic diversity for sorghum can be found in closely related genera such as *Saccharum* species. Hodnett et al. (2010), used *iap* mutant female sorghum in crosses with sugarcane. The long-term objectives of this study were to assess the feasibility of developing hybrids that combine the perennial growth habit and juice quality of sugarcane, with seed propagation, drought tolerance, and wide adaptation from sorghum. The near-term objectives of this study were to quantify to what degree sorghum germplasm containing *iap* can increase hybrid formation and what effect different sugarcane pollinators had on seed set and progeny viability. Pollen parents were highly variable in their ability to complete fertilization and produce viable seed. Some parent combinations were reasonably prolific. In fact, this group recovered 1371 sorghum × sugarcane hybrids which is a sufficiently large population for selection of agronomically superior genotypes. Selection within the sugarcane germplasm could identify male parents that could further improve hybridization potential between these two species (Hodnett et al., 2010).

## 6.8 Conclusions

Sorghum is among the most versatile of crop species due to its potential for wide adaptation and diversity of end uses. The extensive and successful breeding history gives the bioenergy breeder of sorghum an excellent head start for cultivar development. The genetic resources available to sorghum breeders, contained in both the currently utilized breeding germplasm as well as the extensive public germplasm collections, represent a rich repository of genetic variation which is necessary for breeding progress. Genetic variability from other related genera and species has already been introgressed into viable hybrid lines. Much research has been published regarding the inheritance and genetic location of beneficial alleles for an array of traits in sorghum. Historical breeding progress indicates current breeding methodology combined with modern techniques of molecular breeding that would enhance cultivar development of sorghum for bioenergy production.

## References

Almodares A, Hadi MR, Ahmadpour H. Sorghum stem yield and soluble carbohydrates under different salinity levels. Afr J Biotech, 2008; 7: 4051–4055.

Anas A, Yoshida T. Screening of Al-tolerant sorghum by hematoxylin staining and growth response. Plant Prod Sci, 2000; 3: 246–253.

Anas A, Yoshida T. Heritability and genetic correlation of Al-tolerance with several agronomic characters in sorghum assessed by hematoxylin staining. Plant Prod Sci, 2004; 7: 280–282.

Agrawal BL, Abraham CV, House LR. Inheritance of resistance to midge, *Contarinia sorghicola* Coq in sorghum *Sorghum bicolor* (L.) Moench. Insect Sci Appl, 1988; 9: 43–45.

Azhar FM, McNeilly T. The genetic basis of variation for salt tolerance in *Sorghum bicolor* (L.) Moench seedlings. Plant Breed, 1988; 101: 114–121.

Bacon RK, Cantrell RP, Axtell JD. Selection for seedling cold tolerance in grain sorghum. Crop Sci, 1985; 26: 900–903.

Bean B, McCollum T. Summary of six years of forage sorghum variety trials. Pub. SCS-2006-04, Texas Cooperative Extension and Texas Agricultural Experiment Station, College Station, TX, USA, 2006.

Blummel M, Zerbini E, Reddy BVS, Hash CT, Bidinger F, Khan AA. Improving the production and utilization of sorghum and pearl millet as livestock feed: progress towards dual-purpose genotypes. Field Crops Res, 2003; 84: 143–158.

Bowers JE, Abbey C, Anderson S, Chang C, Drave X, Li Z, Lin Y, Liu S, Marler BS, Ming R, Mitchell SE, Qiang D, Reischmann K, Schulze SR, Skinner DN, Wang Y, Kresovich S, Schertz KF, Paterson AH. A high density genetic recombination map of sequence-tagged sites for Sorghum, as a framework for comparative structural and evolutionary genomics of tropical grains and grasses. Genetics, 2003; 165: 367–386.

Boye-Goni SR, Marcarian V. Diallel analysis of aluminum tolerance in selected lines of grain sorghum. Crop Sci, 1985; 25: 749–752.

Boyer JS. Plant productivity and the environment. Science, 1982; 218: 444–448.

Brown PJ, Kein PE, Bortiri E, Acharya CB, Rooney WL, Kresovich S. Inheritance of inflorescence architecture in sorghum. Theor Appl Genet, 2006; 113: 931–942.

Boozaya-Angoon D, Starks KJ, Weibel DE, Teetes GL. Inheritance of resistance in sorghum (*Sorghum bicolor*) to the sorghum midge *Contarinia sorghicola* (Diptera: Cecidomyiidae). Environ Entomol, 1984; 13: 1531–1534.

Broadhead DM, Coleman OH, Freeman KC. Dale, a new variety of sweet sorghum for sirup production. Miss State Exp Sta Inform, 1970; 1099.

Broadhead DM, Freeman KC, Coleman OH, Zummo N. Theis a new variety of sweet sorghum for syrup production. Miss State Exp Sta Inform, 1974; 1236.

Buchanan BB, Gruissem W., Jones RL. *Biochemistry and Molecular Biology of Plants*, 2000. American Society of Plant Physiologists, Rockville, MD.

Bufogle A, Bollich PK, Kovar JL, Macchiavelli RE, Lindau CW. Rice variety differences in dry matter and nitrogen accumulation as related to stature and maturity group. J Plant Nutr, 1997; 20: 1203–1224.

Burow GB, Xin Z, Burke JJ, Franks CD. Genetic enhancement of cold tolerance in sorghum: Mapping of QTL's for early season cold tolerance[abstract]. Great Plains Sorghum Conference, 2009a; Amarillo, Texas, CDROM.

Burow GB, Franks CD, Acosta-Martinez V, Xin Z. Molecular mapping and characterization of *BLMC*, a locus for profuse wax (bloom) and enhanced cuticular features of sorghum (*Sorghum bicolor* (L.) Moench. Theor Appl Genet, 2009b; 118: 423–431.

Cao RH, Wang XL, Ren JH, Nan CH. The resistance of sorghum to head smut and its inheritance. In: Suzuki S (ed.) *Crop Genetic Resources of East Asia*, 1988, pp. 121–124. International board for Plant Genetic Resources, Rome, Italy.

Cardwell KF, Hepperly PR, Frederiksen RA. Pathotypes of *Colletotrichum graminicola* and seed transmission of sorghum anthracnose. Plant Disease, 1989; 73: 255–257.

Casas AM, Kononowicz AK, Haan TC, Zhang L, Tomes DT, Bressan RA, Hasegawa PM. Transgenic sorghum plants obtained after microprojectile bombardment of immature inflorescences. Plant, 1997; 33: 92–100.

Casler MD, Buxton DR, Vogel KP. Genetic modification of lignin concentration affects fitness of perennial herbaceous plants. Theor Appl Genet, 2002; 104: 127–131.

Clark JW. The inheritance of fermentable carbohydrates in stems of *Sorghum bicolor* (L.) Moench. 1981. Ph.D. dissertation, Texas A&M University, TX.

Coleman HD, Samuels AL, Guy RD, Mansfield SD. Perturbed lignifications impacts tree growth in hybrid poplar – a function of sink strength, vascular integrity, and photosynthetic assimilation. Plant Physiol, 2008; 148: 1229–1237.

Corredor DY, Salazar JM, Hohn KL, Bean S, Bean B, Wang D. Evaluation and characterization of forage sorghum as feedstock for fermentable sugar production. Appl Biochem Biotechnol, 2009; 158: 164–179.

Corn R. Heterosis and composition of sweet sorghum. 2009. Ph.D. dissertation, Texas A&M University, TX.

Cox TS, Bender M, Picone C, Van Tassel DL, Holland JB, Brummer EC, Zoeller BE, Paterson AH, Jackson W. Breeding perennial grain crops. Crit Rev Plant Sci, 2002; 21: 59–91.

Craig J, Shertz KF. Inheritance of resistance in sorghum to three pathotypes of Peronosclerospora. Genetics, 1985; 75: 1077–1078.

Crasta OR, Xu WW, Rosenow DT, Mullet J, Nguyen HT. Mapping of post-flowering drought resistance traits in grain sorghum: associations between QTLs influencing premature senescence and maturity. Mol Gen Genet, 1999; 262: 579–588.

Dahlberg JA, Rosenow DT, Peterson GC, Clark LE, Miller FR, Sotomayor-Rios A, Hamburger AJ, Madera-Torres P, Quiles-Belen A, Woodfin CA. Registration of forty converted sorghum germplasm. Crop Sci, 1997; 38: 564–565.

Damte T, Pendleton BB, Almas LK. Cost benefit analysis of sorghum midge, *Stenodiplosis sorghicola* (Coquillett)-resistant sorghum hybrid research and development in Texas. Southwest Entomol, 2009; 34: 395–405.

Delhaize E, Ryan PR. Aluminum toxicity and tolerance in plants. Plant Physiol, 1995; 107: 315–321.

Dhillon MK, Sharma HC, Reddy BVS, Singh R, Naresh JS. Inheritance of resistance to sorghum shoot fly, *Atherigona soccata*. Crop Sci, 2006; 46: 1377–1383.

Duncan RR. The association of plant senescence with root and stalk diseases in sorghum. In: Mughogho LK (ed.) *Sorghum Root and Stalk Rots, A Critical Review*. Proc Workshop. Bellagio, Italy. 1984, pp. 99–110.

Dweikat I. A diploid, interspecific, fertile hybrid from cultivated sorghum, *Sorghum bicolor*, and its common johnsongrass weed *Sorghum halepense*. Mol Breeding, 2005; 16: 93–101

Egli DB. Seed biology and the yield of grain crops, 1998. CAB International. New York.

Feltus FA, Hart GE, Schertz KF, Casa AM, Kresovich S, Abraham S., Klein PE, Brown PJ, Paterson AH. Alignment of genetic maps and QTLs between inter and intra-specific sorghum populations. Theor Appl Genet, 2006; 112: 1295–1305.

Food and Agricultural Organization of the United Nations (FAOSTAT). Agricultural data, 2010 http://faostat.fao.org/site/567/DesktopDefault.aspx?PageID=567#ancor. Accessed on 6 May 2010.

Foy CD, Duncan RR, Waskom RM, Miller DR. Tolerance of sorghum genotypes to an acid, aluminum toxic soil. J Plant Nutr, 1993; 16: 97–127.

Franzmann BA. Ovipositional antixenosis to *Contarinia sorghicola* (Coquillett) (Diptera: Cecidomyaiidae) in grain sorghum. J Aust Entomol Soc, 1993; 32: 59–64.

Ghassemi F, Jakerman AJ, Nix HA. Salinisation of Land and Water Resources: Human Causes, Extent, Management and Case Studies, 1995, p. 526. CAB International, Wallingford, UK.

Glover JD, Reganold JP, Bell LW, Borevitz J, Brummer EC, Buckler ES, Cox CM, Cox TS, Crews TE, Culman SW, DeHann LR, Eriksson D, Gill BS, Holland J, Hu F, Hulke BS, Ibrahim AMH, Jackson W, Jones SS, Murray SC, Paterson AH, Ploschuk E Sacks EJ, Snapp S, Tao D, Van Tassel DL, Wade LJ, Wyse DL, Xu Y. Increased food and ecosystem security via perennial grains. Science, 2010; 328: 1638–1639.

Gomez AS. Identification of quantitative trait loci for grain yield in a recombinant inbred B-line population in sorghum, 2002. Ph.D. dissertation. Texas A&M University, TX.

Gul I, Demial R, Kilieal N, Sumerli M, Kilic L. Effect of crop maturity stages on yield, silage composition and *in vivo* digestibilities of the maize, sorghum, and sorghum-sudangrass hybrids grown in semi-arid conditions. J Anim Vet Adv, 2008; 7: 1021–1028.

Hadley HH. Cytological relationships between *Sorghum vulgare* and *S. halepense*. Agron J, 1953; 45: 139–143.

Hadley HH. Chromosome numbers, fertility and rhizome expression of hybrids between grain sorghum and johnsongrass. Agron J, 1958; 50: 278–282.

Hanna WW. Transfer of germplasm from the secondary to the primary gene pool in pennisetum. Theor Appl Genet, 1990; 80: 200–204.

Harris HB, Johnson BJ, Dobson JW, Luttrell ES. Evaluation of anthracnose on grain sorghum. Crop Sci, 1964; 4: 460–462.

Hart GE, Shertz KF, Peng Y, Syed NH. Genetic mapping of *Sorghum bicolor* (L.) Moench QTLS that control variation in tillering and other morphological characters. Theor Appl Genet, 2001; 103: 1232–1242.

Haussmann BIG, Mahalakshmi V, Reddy BVS, Seetharama N, Hash CT, Geiger HH. QTL mapping of stay-green in two sorghum recombinant inbred populations. Theor Appl Genet, 2002; 106: 133–142.

Hodnett GL, Burson BL, Rooney WL, Dillon SL, Price HJ. Pollen–pistil interactions result in reproductive isolation between Sorghum bicolor and divergent sorghum species. Crop Sci, 2005; 45: 1403–1409.

Hodnett GL, Hale AL, Packer DJ, Stelly DM, da Silva J, Rooney WL. Elimination of a reproductive barrier facilitates intergeneric hybridization of Sorghum bicolor and Saccharum. Crop Sci, 2010; 50: 1188–1145

Hoover F, Abraham J. A review: comprehensive comparison of corn-based and cellulosic-based ethanol as a biofuel source. Nanotech Conference and Expo, 2009; 3: 30–33.

Hunter E, Anderson I. Sweet Sorghum. In: Janick J. (ed.) *Horticultural Reviews*, 1997, pp. 73–104. Wiley, New York.

International Crops Research Institute for the Semi-Arid Tropics (ICRISAT) The Medium Term Plan. Part II. International Crops Research Institute for the Semi-Arid Tropics, 1992. Patancheru 502 324, Andhra Pradesh, India.

Igartua E, Gracia MP, Lasa JM. Characterization and genetic control of germination-emergence responses of grain sorghum to salinity. Euphytica, 1994; 76: 185–193.

Izquierdo L, Godwin ID. Molecular characterization of a novel methioneine-rich [delta]-kafirin seed storage protein gene in sorghum (*Sorghum bicolor* L.). Cereal Chem, 2005; 82: 706–710.

Jotwani MG. Losses due to shoot fly in high yielding sorghum. In: Krishnamurthy Rao BH, Murthy KSRK (ed.) *Crop Losses due to Insect Pests, Special Issue of Indian J. Entomol.* 1983, pp. 213–220. Entomological Society of India, Rajendranagar, Hyderabad, Andhra Pradesh, India.

Kambal AE, Webster OJ. Estimates of general and specific combining ability in grain sorghum, *sorghum vulgare* pers. Crop Sci, 1965; 5: 521–523.

Kanampiu FK, Raun WR, Johnson GV. Effect of nitrogen rate on plant nitrogen loss in winter wheat varieties. J Plant Nutr, 1997; 20: 389–404.

Kebede H, Subudhi PK, Rosenow DT, Nguyen HT. Quantitative trait loci influencing drought tolerance in grain sorghum (Sorghum bicolor L. Moench). Theor Appl Genet, 2001; 106: 133–142.

Kim J-S, Klein PE, Klein RR, Price HJ, Mullet JE, Stelly DM. Chromosome Identification and nomenclature of *Sorghum bicolor*. Genetics, 2005; 169: 1169–1173.

Kimber C. Origins of Domesticated Sorghum and Its Early Diffusion to India and China. In: Wayne SC, Frederiksen RA (ed.) *Sorghum Origin, History, Technology and Production*, 2000. John Wiley and Sons, Inc., New York.

Knoll J, Ejeta G. Marker-assisted selection for early-season cold tolerance in sorghum: QTL validation across populations and environments. Theor Appl Genet, 2008; 116: 541–553.

Krishnamoorthy G. A study of heterotic relationships in sorghum, 2005. Ph.D. dissertation, Texas A&M University, TX.

Kuhlman LC, Burson BL, Klein PE, Klein RR, Price HJ, Rooney WL. Genetic recombination in *Sorghum bicolor* x *S. macrospermum* interspecific hybrids. Genome, 2008; 51: 749–756.

Kuhlman LC, Burson BL, Klein PE, Klein RR, Price HJ, Rooney WL. Introgression breeding using *S. macrospermum* and analysis of recovered germplasm. Genome, 2010; 53: 419–429.

Kumar VK, Sharma HC, Reddy KD. Antibiosis mechanism of resistance to spotted stem borer, *Chilo partellus* in sorghum, *Sorghum bioclor*. Crop Prot, 2006; 25: 66–72.

Laurie DA, Bennett MD. Genetic variation in sorghum for the inhibition of maize pollen tube growth. Ann Bot, 1989; 64: 675–681.

Mace ES, Rami J, Bouchet S, Klein PE, Klein RR, Kilian A, Wenzl P, Xia L, Halloran K, Jordan DR. A consensus genetic map of sorghum that integrates multiple component maps and a high-throughput diversity array technology (DArT) markers. BMC Plant Biol, 2009; 9: 13.

Magalhaes JV, Liu J, Claudia T Guimarães CT, Lana UGP, Alves VMC, Wang YH, Schaffert RE, Hoekenga OA, Piñeros MA, Shaff JE, Klein PE, Carneiro NP, Coelho CM, Trick HN, Kochian LV. A gene in the multidrug and toxic compound extrusion (MATE) family confers aluminum tolerance in sorghum. Nat Genet, 2007; 39: 1156–116.

Magill CW, Boora KS, Kumari RS, Osorio J, Oh BJ, Gowda BS, Cui Y., Frederiksen RA. Tagging sorghum genes for disease resistance: expectations and reality. In: *Proceedings of the International Conference on Genetic Improvement of Sorghum and Pearl Millet*, 1997, pp. 316–325. INTSORMOL. Lubbock, TX.

Makanda I, Pangirayi T, Derera J. Combining ability and heterosis of sorghum germplasm for stem sugar traits under off-season conditions in tropical lowland environments. Field Crop Res, 2009; 114: 272–279.

Maradiaga JL. Quantitative trait loci affecting the agronomic performance of a *Sorghum bicolor* (L.) Moench recombinant inbred restorer line population, 2003. Ph.D. dissertation. Texas A and M University, TX.

Maunder AB. Development and perspectives of the hybrid sorghum seed industry in the Americas. In: Ejeta G (ed.) *Proceedings of Hybrid Sorghum Seed for Sudan Workshop*, 1983, pp. 39–48. W. Lafayette. IN.

McIntyre CL, Drenth J, Gonzales N, Henzell RG, Jordan DR. Molecular characterization of the waxy locus in sorghum. Genome, 2008; 51: 524–533.

McWilliams, J.R., W. Manokaran, T. Kipnis. Adaptation to chilling stress in sorghum. In: Lyons M, Graham D, Raison JR (eds) *Low Temperature Stress in Crop Plants*, 1979. New York.

Mehta PJ, Wiltse CC, Rooney WL, Collins SD, Frederiksen RA, Hess DE, Chisis M, TeBeest DO. Classification and inheritance of genetic resistance to anthracnose in sorghum. Field Crops Res, 2005; 93: 1–9.

Miller FR, Kebede Y. Genetic contributions to yield gains in sorghum 1950–1980. In: Fehr WR (ed.) *Genetic Contributions to Yield Gains in Five Major Crop Plants*. CSSA Spec. Publ. 7. 1981. CSSA, ASA, Madison, WI.

Monk RL, Miller FR, McBee GG. Sorghum improvement for energy production. Biomass, 1984; 6: 145–385.

Morgan PW, Finlayson SA. Physiology and Genetics of Height. In: Smith CW, Frederiksen RA (eds) *Sorghum Origin, History, Technology, and Production*, 2000, pp. 227–260. John Wiley and Sons, New York.

Murray SC, Sharma A, Rooney WL, Klein PE, Mullet JE, Mitchell SE, Kresovich S. Genetic improvement of sorghum as a biofuel feedstock: I QTL for stem sugar and grain nonstructural carbohydrates. Crop Sci, 2008; 48: 2165–2178.

Natoli A, Gorni C, Chegdani F, Ajmone Marsane P, Colombi C, Lorenzoni C, Morocco A. Identification of QTLs associated with sweet sorghum quality. Maydica, 2002; 47: 311–322.

Nimbalkar VS Bapat DR. Inheritance of shoot fly (*Antherigona soccata*) resistance in F2 diallel crosses of sorghum (*Sorghum bicolor*). Indian J Agr Sci, 1988; 58: 955–957.

Nwanze KF, Reddy YVR, Taneja SL, Sharma HC, Agrawal BL. Evaluating sorghum genotypes for multiple insect resistance. Insect Sci, 1991; 12: 183–188.

Oba M, Allen MS. Effects of brown midrib 3 mutation in corn silage on productivity of dairy cows fed two concentrations of dietary neutral detergent fiber: 1. Feeding behavior and nutrient utilization. J Dairy Sci, 2000; 83: 1333–1341.

Oh BJ, Frederiksen RA, Magill CW. Identification of molecular markers linked to head smut resistance gene (*Shs*) in sorghum by RFLP and RAPD analyses. Phytopathology, 1994; 84: 830–833.

Oklahoma State University Extension http://oklahoma4h.okstate.edu/aitc/lessons/extras/facts/milo.html.

Oliver AL, Pederson JF, Grant RJ, Klopfenstein TJ, Jose HD. Comparative effects of the sorghum *bmr-6* and *bmr-12* genes II. Grain yield stover yield and stover quality in grain sorghum. Crop Sci, 2005; 45: 2240–2245.

Packer D. High-biomass sorghums for biomass biofuels, 2010. Ph.D. Dissertation. Texas A and M University, TX.

Paterson AH, Schertz KF, Lin Y, Liu S, Chang Y. The weediness of wild plants: molecular analysis of genes influencing dispersal and persistence of Johnsongrass, *Sorghum halepense* (L.) Pers. Proc Natl Acad Sci, 1995; 92: 6127–6131.

Paterson AH, Bowers JE, Bruggmann R, Dubchak I, Grimwood J, Gundlach H, Haberer G, Hellsten U, Mitros T, Poliakov A, Schmutz J, Spannagl M, Tang H, Wang X, Wicker T, Bharti AK, Chapman J, Feltus FA, Gowik U, Grigoriev IV, Lyons E, Maher CA, Martis M, Narechania A, Otillar RP, Penning BW, Salamov AA, Wang Y, Zhang L, Carpita NC, Freeling M, Gingle AR, Hash CT, Keller B, Klein P, Kresovich S, McCann MC, Ming R, Peterson DG, Mehboob-ur-Rahman, Ware D, Westhoff P, Mayer KF, Messing J, Rokhsar

DS. The Sorghum bicolor genome and the diversification of grasses. Nature, 2009; 457: 551–556.

Patzek TW, Pimentel D. Thermodynamics of energy production from biomass. Crit Rev Plant Sci, 2005; 24: 327–364.

Peacock, J.M. Response and tolerance of sorghum to temperature stress. In: House LR, Mughogho LK, Peacock JM (eds) *Orghum in the Eighties: Proceeding of the International Symposium on Sorghum Patancheru, India*, 1982, pp. 143–159.

Pederson JF, Gorz HJ, Haskins FA, Ross WM. Variability for quality and agronomic traits in forage sorghum hybrids. Crop Sci, 1982; 22: 853–856.

Pederson JF, Vogel KP, Funnell DL. Impact of reduced lignin on plant fitness. Crop Sci, 2005; 45: 812–819.

Perlack RD, Wright LL, Turhollow AF, Graham RL, Stokes BJ, Erbach DC. Biomass as a feedstock for bioenergy and bioproducts industry: the technical feasibility of a billion-ton annual supply, 2005. Prepared by Oak Ridge National Laboratory. http://feedstockreview.ornl.gov/pdf/billion_ton_vision.pdf. Accessed on 20 July 2010.

Perumal R, Menz MA, Mehta PJ, Katilé S, Gutierrez-Rojas LA, Klein RR, Klein PE, Prom L.K., Schlueter JA, Rooney WL, Magill CW. Molecular mapping of *Cg1*, a gene for resistance to anthracnose (*Colletotrichum sublineolum*) in sorghum. Euphytica, 2009; 165: 597–606.

Porter KS, Axtell JD, Lechtenberg VL, Colenbrander VF. Phenotype, fiber concentration, and in vitro dry matter disappearance of chemically induced brown midrib (bmr) mutants of sorghum. Crop Sci, 1978; 28: 205–208.

Price JH, Hodnett GL, Burson BL, Dillon SL, Stelly DM, Rooney WL. Genotype dependent interspecific hybridization of Sorghum bicolor. Crop Sci, 2006; 46: 2617–2622.

Quinby JR. *Sorghum Improvement and the Genetics of Growth*, 1974. Texas A&M University Press, College Station, TX.

Quinby JR, Karper RE. The effect of different alleles on the growth of sorghum hybrids. Agron J, 1948; 40: 255–259.

Quinby JR, Karper RE. Inheritance of Height in Sorghum. Agron J, 1954; 46: 211–216.

Rami JF, Dufour P, Trouche G, Fliedel G, Mestres C, Davrieux F, Blanchard P, Hamon P. Quantitative trait loci for grain quality, productivity, morphological and agronomical traits in sorghum (*Sorghum bicolor* L. Moench). Theor Appl Genet, 1998; 97: 605–616.

Rao MJV, Goud JV. Inheritance of grain-yield and its components in sorghum. Indian J Genet Pl Br, 1977; 37: 31–39.

Rao SP, Rao SS, Seetharama N, Umakanth AV, Reddy PS, Reddy BVS, Gowda CLL. Sweet sorghum for biofuel and strategies for its improvement, 2009. ICRISAT Information bulletin No. 77.

Raun WR, Johnson GV. Improving nitrogen use efficiency for cereal production. Agron J, 1999; 91: 357–363.

Reagan TE, Flynn JL. Insect pest management of sweet sorghum in sugarcane production systems of Louisiana: problems and integration. In: Smith WH (ed.) *Biomass Energy Development*, 1986, pp. 227–239. Plenum, New York.

Reddy, B.V.S., L.K. Mughogho, Y.D. Narayana, K.D. Nicodemus, and J.W. Stenhouse. Inheritance pattern of downey mildew resistance in advanced generations of sorghum. Ann Appl Biol, 1992; 121: 249–255.

Reddy BVS, Reddy PS. Sweet sorghum: characteristics and potential. Int Sorghum Millets Newsl, 2003; 44: 26–28.

Reddy BVS, Ramesh S, Reddy PS, Ramaiah B, Salimath PM, Rajashekar K. Sweet sorghum – a potential alternative raw material for bioethanol and bio-energy. Int Sorghum Millets Newsl, 2005; 46: 79–86.

Reddy BVS, Rao PS, Kumar AA, Reddy PS, Rao PP, Sharma KK, Blummel M, Reddy CHR. Sweet sorghum as a biofuel crop: where are we now?. In: Proceedings of the Sixth Int. Biofuels Conference, 2009, pp. 191–202. March 2009. New Delhi.

Ritter KB, Jordan DR, Chapman SC, Godwin ID, Mace ES, McIntyre CL. Indentification of QTL for sugar-related traits in a sweet x grain sorghum (*Sorghum bicolor* L. Moench) recombinant inbred population. Mol Breeding, 2008; 22: 367–384.

Ritter KB, Jordan DR, Chapman SC, Godwin ID, Mace ES, McIntyre CL. Identification of QTL for sugar-related traits in a sweet x grain sorghum (*Sorghum bicolor* L. Moench) recombinant inbred population. Mol Breeding, 2008; 22: 367–384.

Rochester DE, Winer JA, Shah DM. The structure and expression of maize genes encoding the major heat shock protein, *hsp* 70. EMBO J, 1986; 5: 451–458.

Rooney WL, Aydin S. The genetic control of photoperiod sensitive response in sorghum. Crop Sci, 1999; 39: 397–400.

Rooney WL, Smith CW. Techniques for developing new cultivars. In: Smith CW, Frederiksen RA (eds) *Sorghum Origin, History, Technology, and Production*, 2000, pp. 309–328. John Wiley and Sons, New York.

Rooney WL, Collins SD, Klien RR, Mehta PJ, Frederiksen RA, Rodriguez-Herrera R. Breeding sorghum for resistance to anthracnose, grain mold, downy mildew, and head smuts. In 3rd Global Conference on Sorghum and Millet Diseases. Guanajuato, Mexico, 2000. pp. 273–279.

Rooney WL. Sorghum Improvement – Integrating Traditional and New Techniques to Produce Improved Genotypes. Adv Agron, 2004; 83: 37–109.

Rooney WL, Blumenthal J, Bean B, Mullet JE. Designing sorghum as a dedicated bioenergy feedstock. Biofuels Bioprod Bioref, 2007; 1: 147–157.

Rosenow DT, Quisenberry JE, Wendt CW, Clark LE. Drought tolerant sorghum and cotton germplasm. Agr Water Manage, 1983; 7: 207–222.

Rosenow DT. Breeding for resistance to root and stalk rots in Texas. In: Mughogho LK (ed.) *Sorghum Root and Stalk Rots, A Critical Review*. Proc Workshop. Bellagio, Italy, 1984. pp. 209–217.

Rosenow, D.T., J.A. Dahlberg, J.C. Stephens, F.R. Miller, D.K. Barnes, G.C. Peterson, J.W. Johnson, and K.F. Shertz. Registration of 63 converted sorghum germplasm from the sorghum conversion program. Crop Sci, 1997a; 37: 1399–1400.

Rosenow DT, Dahlberg JA, Peterson GC, Clark LE, Miller FR, Sotomayor-Rios A, Hamburger AJ, Madera-Torres P, Quiles-Belen A, Woodfin CA. Registration of fifty sorghums from the sorghum conversion program. Crop Sci, 1997b; 31: 1397–1398.

Rosenow DT, Dahlberg JA. Collection, conversion, and utilization of sorghum. In: Smith CW, Frederiksen RA (eds) *Sorghum Origin, History, Technology, and Production*, 2000, pp. 309–328. John Wiley and Sons, New York.

Rosewich UL, Pettway RE, McDonald BA, Duncan RR, Frederiksen RA. Genetic structure and temporal dynamics of a colletotrichum graminicola population in a sorghum disease nursery. Phytopathology, 1998; 88: 1087–1093.

Saballos A. Development and utilization of sorghum as a bioenergy crop. In: Vermerris W (ed.) *Genetic Improvement of Bioenergy Crops*, 2008, pp. 211–248. Springer, New York.

Saha BC. Hemicellulose bioconversion. J Ind Microbiol Biotechnol, 2003; 30: 279–291.

Saleem M, Ahsan M, Aslam M., Majeed A. Comparative evaluation and correlation estimates for grain yield and quality attributes in maize. Pak J Bot, 2008; 40: 2361–2367.

Sanchez PA. *Properties and Management of Soils in the Tropics*, 1976, John Wiley and Sons, New York.

Satish K, Srinivas G, Madhusudhana R, Padmaja PG, Reddy RN, Mohan SM, Seetharama N. Identification of quantitative trait loci for resistance to shoot fly in sorghum [*Sorghum bicolor* (L.) Moench]. Theor Appl Genet, 2009; 119: 1425–1439.

Shapouri H, Duffield JH, Wang M. The energy balance of corn ethanol, an update, 2002. Agric Econ Rep. USDA Office of Energy Policy and new Uses, No. 813.

Shapouri H, McAloon A. The 2001 net energy balance of corn-ethanol www.usda.gov/oce/reports/energy/net_energy_balance.pdf. (verified 15 July 2010), 2004.

Sharma HC. Breeding for sorghum midge resistance and resistance mechanisms. In Proc Int Workshop Sorghum Insect Pests, 1985, pp. 275–292.

Sharma HC, Fanzmann, BA, Heenzell, RG. Mechanisms and diversity of resistance to sorghum midge *Stenodiplosis sorghicola* in *Sorghum bicolor*. Euphytica, 2002; 124: 1–12.

Sharma, B.C., Dillon, MK, Pampapathy, G., Reddy BVS. Inheritance of resistance to spotted stem borer, *Chilo partellus*, in sorghum, *Sorghum bicolor*. Euphytica, 2007; 156: 117–128.

Sifuentes J, Frederiksen, RA. Inheritance of resistance to pathotypes 1, 2, and 3 of *Peronoscerospora sorghi* in sorghum. Plant Dis, 1988; 72: 332–333.

Singh SP. Sources of cold tolerance in grain sorghum. Can J Plant Sci, 1985; 65: 251–257.

Singh BU, Rana BS. Resistance in sorghum to spotted stem borer, *Chilo partellus* (Swinhoe). Insect Sci Appl, 1989; 10: 3–27.

Sleper DA, Poehlman JM. *Breeding Field Crops*, 5th edn, 2006. Blackwell Publishing, Ames, IA.

Smith, CW, Frederiksen R. History of Cultivar Development in the United States: From "Memoirs of A.B. Maunder – Sorghum Breeder". In: Smith CW, Frederiksen RA (eds) *Sorghum Origin, History, Technology, and Production*, 2000, pp. 309–328. John Wiley and Sons, New York.

Sorghum Grower. Sorghum markets change over time. Sorghum Grower, 2009; 3(13): 13.

Srinivas G, Satish K, Madhusudhana R, Reddy RN, Mohan SM, Seetharama N. Identification of quantitative trait loci for agronomically important traits and their association with genic-microsatellite markers in sorghum. Theor Appl Genet, 2009; 118: 1439–1454.

Stefaniak TR, Wolfrum E, Dahlberg JA, Bean BR, Dighe N, Rooney WL. Variation in biomass composition components among forage, biomass, and sweet sorghums. Crop Sci, 2012; 52: 1949–1954.

Stephens JC, Quinby JR. Yield of hand-produced hybrid sorghum. Agron J, 1952; 4: 231–233.

Stephens JC, Holland, RF. Cytoplasmic male sterility for hybrid sorghum seed production. Agron J, 1954; 46: 20–23.

Stephens JC, Miller FR, Rosenow DT. Conversion of alien sorghums to early combine genotypes. Crop Sci, 1967; 7: 396.

Subudhi PK, Rosenow DT, Nguyen HT. Quantitative trait loci for stay-green trait in sorghum (Sorghum bicolor L. Moench): consistency across genetic backgrounds and environments. Theor Appl Genet, 2000; 101: 733–741.

Szabolcs I. Soils and salinization. In: Pessarakali M (ed.) *Handbook of Plant and Crop Stress*, 1994, pp. 3–11. Marcel Dekker, New York.

Tao YZ, Henzell RG, Jordan DR, Butler DG, Kelly A.M., McIntyre CL. Identification of genomic regions associated with stay-green in sorghum by testing RILs in multiple environments. Theor Appl Genet, 2000; 100: 1225–1232.

Tao YZ, Hardy A., Drenth J, Henzell RG, Franzmann BA, Jordan DR, Butler DG, McIntyre CL. Identifications of two different mechanisms for sorghum midge resistance through QTL mapping. Theor Appl Genet, 2003; 107: 116–122.

Teetes GL, Pendleton BB. Insect pests of sorghum. In: Smith CW, Frederiksen RA (eds) *Sorghum Origin, History, Technology, and Production*, 2000, pp. 309–328. John Wiley and Sons, New York.

Tuinstra MR, Grote EM, Goldsbrough PB, Ejeta G. Genetic analysis of post-flowering drought tolerance and components of grain development in *Sorghum bicolor* (L.) Moench. Mol Breed, 1997; 3: 339–343.

U.S. Department of Agriculture (USDA) NASS http://usda.mannlib.cornell.edu/MannUsda/viewDocumentInfo.do;jsessionid=5B408B60A71527B1968E5565B47CDA9A?documentID=1593. Accessed on May 6, 2010.

U.S. Department of Energy (DOE) Office of Science. 2010. http://www.genomicscience.energy.gov/research/DOEUSDA/abstracts/2009dweikat_abstract.shtml. Accessed on December 13, 2012.

van Oosterom EJ, Jayachandran R, Bidinger FR. Diallel analysis of the stay-green trait and its components in sorghum. Crop Sci, 1996; 36: 549–555.

Venuto B, Kindiger B. Forage and biomass feedstock production from hybrid forage sorghum and sorghum-sudangrass hybrids. Grassland Sci, 2008; 54: 189–196.

Vermerris W, Saballos A, Ejeta G, Mosier N, Ladisch MR, Carpita NC. Breeding to enhance ethanol production from corn and sorghum stover. Crop Sci, 2007; 47(S3): S142–S153.

Vogel KP, Hopkins AA, Moore KJ, Johnson KD, Carlson IT. Winter survival in switchgrass populations bred for high IVDMD. Crop Sci, 2002; 42: 1857–1862.

Waquil, J.M., Teetes JL, Peterson GC. Adult sorghum midge *Contarinia sorghicola* (Coquillett) (Diptera: Cecidomyyaiidae) nonpreference for a resistant hybrid sorghum. J Econ Entomol, 1986; 79: 883–837.

Woodfin CA, Rosenow DT, Clark LE. Association between the stay-green trait and lodging resistance in sorghum. Agronomy abstracts, 1988. ASA, Madison, WI.

Wu X, Zhao R, Bean SR, Seib PA, McLaren JS., Madl RL, Tuinstra M, Lenz MC, Wang D. Factors impacting ethanol production from grain sorghum in the dry-grind process. C Chem, 2007; 84: 130–136.

Xu W, Subudhi PK, Crasta OR, Rosenow DT, Mullet JE, Nguyen HT. Molecular mapping of QTLs conferring stay-green in grain sorghum (Sorghum bicolor L. Moench). Genome, 2000; 43: 461–469.

# Chapter 7
# Energy Cane

Phillip Jackson

*CSIRO Plant Industry, Australian Tropical Science Innovation Precinct, Townsville, QLD 4814, Australia*

## 7.1 Introduction

"Energy canes" are defined for the purpose of this chapter as comprising sugarcane genotypes and other related species or interspecific hybrids which may be crossed with sugarcane that are suitable for existing or future bioenergy production systems (e.g., biofuel and biopower). Sugarcane is an economically important crop in many tropical and subtropical regions of the world, and is the leading industrial crop with very large scale production (>1.5 billion tonnes processed annually). Sugarcane currently supplies about 65% of the world production of sugar, with the remainder being supplied by sugar beet. It is also being increasingly used for renewable biofuel production, currently supplying about 40% of the world's ethanol, mainly from Brazil. Its use for biopower, particularly for electricity generation in many countries, is growing rapidly. Energy cane has a number of potential advantages for bioenergy production in relation to some other candidate crop species, most notably:

(i) Production of a high proportion of biomass (to over 50% of dry matter) in the form of readily fermentable sugars (sucrose, glucose, fructose), allowing for easy and commercially viable production of biofuels, especially ethanol.
(ii) It is perennial, that is, it regrows vigorously after harvesting to produce a series of annual "ratoon" crops (i.e., crops produced from regrowth of the underground stems following harvesting). Depending on harvesting conditions and the environment, from two to over ten ratoon crops may arise from an original planting before yield gradually declines to a point where it is profitable to plough out and replant. The perennial character greatly reduces costs of production, and also reduces potential environmental damage and energy usage otherwise associated with regular ground disturbance and planting.
(iii) It has an efficient photosynthetic process, utilizing the $C_4$ pathway that supports efficient radiation and water use efficiency, especially under high temperatures.

---

*Bioenergy Feedstocks: Breeding and Genetics*, First Edition. Edited by Malay C. Saha, Hem S. Bhandari, and Joseph H. Bouton.
© 2013 John Wiley & Sons, Inc. Published 2013 by John Wiley & Sons, Inc.

(iv) It can produce high biomass yields, a function of point (iii) and also its 12-month growth duration per harvest cycle and quick canopy development in ratoon crops, which maximize proportion of incident radiation intercepted by leaves.

Collectively, these features provide both economic and energy balance advantages to sugarcane as a source of bioenergy in relation to other candidate crops. Energy output/input (O/I) ratios have been reported for sugarcane in different parts of the world. These vary widely depending on methodology and assumptions used, but there is a consensus that the ratio is greater than 1, and greater than that reported for grain crops, sugar beet, and cassava, amongst others (e.g., de Vries et al., 2010). The main advantage in sugarcane production systems compared with corn and some others is associated with the availability of bagasse (i.e., cane residue following crushing of cane, consisting of around 50% fiber and 50% water) as an energy source for processing, and for exporting excess energy as electricity. Macedo et al. (2004) reported an O/I ratio of 8:1 for ethanol production from sugarcane in Brazil. Pimental and Patzek (2007) report a less favorable, but still positive, ratio (1.38). Renouf et al. (2008) compared Australian sugarcane production with corn production in the United States and sugar beet production in the United Kingdom, and reported a negative energy input of about $-8$ MJ kg$^{-1}$ of monosaccharide for sugarcane, compared with approximately $+6$ MJ kg$^{-1}$ of monosaccharide for the other two crops.

In this chapter, potential energy production systems, some of which have already been developed, are briefly described. Information about such production systems is relevant for understanding breeding targets for energy cane, and how these may differ to those used to date for raw sugar production systems and sugarcane. An overview of genetic improvement of sugarcane is provided, along with a review of the potential role of closely related species in development of energy cane. Initial efforts and research relating to breeding of energy cane are reviewed, along with possible future directions of these programs and priority questions to be resolved through research that will affect these directions.

## 7.2 Sugar and Energy Production Systems

### 7.2.1 Current Global Sugarcane Production

Sugarcane is a crop of major international importance, and is harvested in greater quantities than any other crop globally. Over 1.5 billion tons of sugarcane is harvested and processed at sugar mills annually (Table 7.1 ). Approximately 100 countries produce significant quantities of commercial sugarcane (http://faostat.fao.org), but production is dominated by a few countries, with the top three being Brazil, India, and China (Table 7.1). World sugarcane production has increased significantly over the last 20 years, although the increase has been driven primarily by Brazil and to a lesser extent, China (Table 7.1). The expansion of sugarcane production has been supported by: (i) a steady increase in global demand for sugar, and (ii) the biofuel program in Brazil (described below). There is a long-term incremental increase in sugar consumption of around 2% per annum, associated with increased affluence of people in developing countries.

In nearly all countries in the past, except Brazil, sugarcane has been grown almost exclusively for raw sugar production, with the fiber component of harvested cane burned to provide steam needed for extracting juice, evaporation of water from the juice, and production of crystal raw sugar within the sugar mill. However, sugarcane is now being increasingly used for bioenergy production, including ethanol production from fermentation of sugar, and electricity production

**Table 7.1.** Production of sugarcane in top producing countries and total world production of cane-delivered mills. Weights are on a fresh weight basis.

| Country | 1990—total production of cane (million tons) | 2000—total production of cane (million tons) | 2009—total production of cane (million tons) | 2009—cane yield (t ha$^{-1}$) |
|---|---|---|---|---|
| Brazil | 262 | 327 | 689 | 79 |
| India | 225 | 300 | 285 | 69 |
| China | 63 | 69 | 114 | 71 |
| Thailand | 33 | 54 | 67 | 71 |
| Pakistan | 35 | 46 | 50 | 51 |
| Mexico | 40 | 44 | 51[a] | 76 |
| Colombia | 28 | 35 | 38 | 100 |
| Australia | 24 | 38 | 31 | 85 |
| Argentina | 16 | 18 | 30 | 84 |
| Indonesia | 28 | 24 | 27 | 62 |
| Philippines | 25 | 24 | 26 | 66 |
| Cuba | 81 | 36 | 15 | 41 |
| USA | 25 | 36 | 27 | 71 |
| Others | 163 | 204 | 261 | |
| Total world | 1052 | 1257 | 1736 | |

*Source*: http://faostat.fao.org
[a] 2008 data.

from turbines powered by steam generated from burning bagasse. Production of ethanol is most notable in Brazil, where juice extracted from cane is partitioned approximately 50:50 between production of raw sugar and ethanol. In nearly all other countries, juice is used almost exclusively for production of raw sugar, while ethanol production, if it occurs, is only from molasses, the by-product of sugar manufacture. Capital investment in sugar mills to produce either sugar or ethanol is driven largely by commercial decisions. In most situations, there is little incentive to sugar milling businesses to produce ethanol from cane juice instead of sugar. For example, at present, sugar prices in the open market were about US$550 ton$^{-1}$. A reasonable estimate of cost of production of ethanol from cane juice is often used by applying the sugar price value and converting sugar to ethanol according to the stoichiometric conversion rate (1 ton sucrose = 537 kg ethanol = 4.27 barrels of ethanol). This conversion is reasonable given the identical cost of feedstock (cane) and costs of juice extraction, and similar processing costs from juice of either raw sugar or ethanol. According to this conversion, the cost of ethanol production is US$550 (ton$^{-1}$ of sucrose)/4.27 (barrels ton$^{-1}$ sucrose) = US$128 barrel$^{-1}$. Considering ethanol needs to compete with oil, and has about two-thirds the energy content, this cost does not provide an incentive to make longer-term investments to divert juice to ethanol in most mills. By contrast, in Brazil, government mandates (commencing in the 1970s) to blend ethanol from domestic sugarcane production with gasoline created the necessary driver for diversion of sugar in juice to ethanol production, and facilitated greatly expanded sugarcane production in this country (Goldemberg, 2008). The Brazilian experience is described in more detail below.

Currently, sugarcane is the second most important crop producing bioethanol, after corn, with the vast majority of this from Brazil (Balat and Balat, 2009). The cost of production of sugarcane ethanol is widely acknowledged as being the lowest out of the major crops currently used or considered for ethanol production (e.g., Balat and Balat, 2009), with corn used on a large scale mainly because of U.S. government domestic market protective mechanisms and the high domestically supported price of sugar, especially when world prices were lower.

### 7.2.2 Bioenergy Production from Sugarcane in Brazil

The development of sugarcane-based industries in Brazil deserves special mention because of its dominating share in global sugarcane production (Table 7.1), and the leading role it played in developing biofuel production on an industrial scale. Use of sugarcane for ethanol in Brazil is widely regarded as a world-leading example of biofuel production.

The Brazil ethanol story has been reviewed in detail elsewhere (e.g., Andrietta et al., 2007; Goldemberg et al., 2008; Matsuoka et al., 2011), and is outlined only briefly here. In 1975, the Brazilian government created a national alcohol program called ProAlcohol in response to threats posed by sharp world oil price rises, to reduce the country's dependence on imported oil, and to support the sugar industry which was facing global overproduction at that time. The first phase of this program involved creating a requirement to blend gasoline with 5% anhydrous ethanol. Then in the early 1980s, the government subsidized the local car manufacturing industry to produce cars fuelled by 100% hydrous ethanol. The sales of ethanol-fuelled cars peaked around 1985–1986, with around 75% of cars sold. Around the mid-1980s to 1990s, the world oil price decreased and ethanol prices rose to be significantly higher than gasoline. Also at this time, ethanol supply became unreliable, further contributing to a loss of consumer confidence in ethanol-fuelled cars, resulting in a drastic reduction in sales which dropped to around 1% by the mid-to-late 1990s. This loss in confidence in the ProAlcohol program continued until 2003. At this time, flex fuel vehicles began to be sold, which could run on any mix of ethanol or gasoline, and this completely changed consumer demand for ethanol. Flex fuel cars have since reached around 86% of new car sales (IBGE, 2006; cited in Matsuoka et al., 2011). Currently, about half the cane juice produced in Brazil is used to produce ethanol (with the remainder used to produce raw sugar) and about 90% of the ethanol production is being consumed domestically (IBGE, 2006).

### 7.2.3 Overview of Main Components in Existing Sugarcane Production Systems

*Growing*

Sugarcane is initially established from vegetative propagation using pieces of stem called setts. The first crop harvested after planting is called the plant crop, and crops subsequently produced from regrowth after harvesting, on an annual basis, are called ratoon crops. Because significant ground preparation is needed for good germination of setts, and because of the physical bulk of the planting material (in contrast to most other crops in which seeds are sown), planting is an expensive component of a sugarcane production system (Wilcox et al., 2000). The ability to produce a number of relatively high-yielding ratoon crops has a large impact on reducing the production cost of sugarcane. Generally, cane yield declines by 5–15% on average in each successive ratoon crop, depending on the variety, management, and environmental conditions. Typically, a plant crop and three to four ratoon crops would be grown before cane is ploughed out and replanted, but this number depends on environmental conditions, incidence of diseases, and crop management practices. The reliance on heavy machinery for harvesting and associated damage to the plant stems and roots beneath the soil surface, particularly in wet harvesting conditions, is detrimental to yields of ratoon crops (Swinford and Boevey, 1984; Braunack et al., 1993). One of the key contributions by breeding programs historically has been to improve ratooning ability of cultivars (Jackson, 2005).

A wide range of land preparation and row spacing configurations are used in sugarcane production. Traditionally, a fine tilth of soil to about 30 cm has been sought to facilitate satisfactory germination. There has been progressive experimentation and adoption in some countries of approaches using less aggressive tillage (Wilcox et al., 2000). Row spacing of between about 0.8 m and 1.8 m are used in different countries and regions, with narrower row spacing used where manual harvesting is practiced and wider spacing used to accommodate mechanized cultivation and harvesting.

Nutrient management of sugarcane has been extensively studied and reviewed (e.g., Calcino et al., 2000). High nutrient use efficiency is expected to become increasingly important for environmentally sustainable and energy-efficient production systems. Like with any cropping system, initial alleviation of any micronutrient or secondary macronutrient (Ca, S, Mg), and near-optimal annual application of N, P, and K is a prerequisite to a profitable production system. In sugarcane, there are large annual applications of N and K because of the high biomass yields and associated high levels of removal of these elements in harvested cane (typically 100–200 kg ha$^{-1}$ year$^{-1}$) and losses to leaching or gasification. Losses of N due to denitrification and leaching may be particularly important in high-rainfall, tropical environments. Improved nutrient use efficiency may become a target of breeding programs in future if costs of fertilizer application (from either an economic or environmental perspective) become more important. There has been extensive research on biological N fixation in sugarcane, but there is uncertainty about the level of contribution (Herridge et al., 2008). However, increased rates of biological N fixation in sugarcane may also be a potential future breeding target for sustainable bioenergy production systems.

Sugarcane is affected by a range of pests and diseases (Ricaud et al., 1989; Allsopp et al., 2000; Rott et al., 2000). The major diseases causing most losses worldwide include ratoon stunting disease (*Leifsonia xyli*), smut (*Ustilago scitaminea*), sugarcane mosaic (caused by sugarcane mosaic virus), red rot (*Colletotrichum falcatum*), and rusts (*Puccinia* spp.). Larvae of numerous species of beetles and moths cause damage to roots or stalks of sugarcane in many countries. Most pests and diseases are managed through the use of resistant varieties. Disease resistance has been a key objective of sugarcane breeding programs (Walker, 1987; Ricaud et al., 1989; Machado, 2001). Ongoing control of disease incursions through resistant varieties has been possibly the most important contribution by sugarcane breeding programs to sustainable sugar production systems in the past, and this vital contribution would be expected to remain the case for bioenergy production systems in the future.

## Harvesting and Transport

Harvesting and transport from farm gate to sugar mill represents a proportionally large component cost in sugarcane production systems compared with some other crops because of the bulk and weight of the harvested fresh cane. Because of the high labor requirements in manually harvested systems, or alternatively high capital costs of harvesting and transport equipment, the harvesting period in sugarcane usually extends between 5 and 9 months. In most sugarcane growing regions in the world, harvesting occurs during the winter and spring periods, which coincides with drier and cooler weather. Dry, cool weather is conducive to both field access for harvesting and to increased sucrose content in cane, as the plant reduces the rate of stalk elongation proportionally more than photosynthesis reductions, and partitions more assimilate to sucrose storage rather than new growth (e.g., Singels et al., 2005). Highest possible sucrose content on a fresh weight basis remains one of the highest priorities for sugarcane breeding programs. Genetic improvement in sugar yields through increased sucrose content is more beneficial than the same proportional increase in cane yield, because it does not attract additional harvesting, transport, or milling costs arising with cane yield increases (Jackson et al., 2000). This is likely to remain the case with future bioenergy production systems based on cane, but the emphasis toward sugar content in selection indices in breeding programs, while remaining, may be less extreme than with production systems targeting sugar alone (see Section 4).

Sugarcane has been harvested mechanically in some countries since the late 1960s, but is still harvested manually in countries where labor costs are low. In both manual and mechanized harvesting, a key task up to now is to separate the stalk from green and dead leaves, prior to transport to the mill. Typically, trash represents about 15% of the aboveground biomass (Robertson and Thorburn, 2000). Cane can be either burnt or harvested green, but there is an increasing trend toward the latter because of air pollution concerns in some regions, and a growing awareness of the agronomic value of the unburnt trash left in the field. Under mechanical harvesting systems, around 75% of the trash is left in the field, with the rest (termed extraneous matter) transported to the mill (Macedo et al., 2001). Typically, trash contains lower sucrose and higher of impurities than cane stalks, and is highly undesirable for sugar production (Ivin and Doyle, 1989; Crook et al., 1999). In future energy production systems, the net benefits from separation of trash in the field at harvesting may diminish since the trash may be used to produce electricity or biofuels.

In future energy production systems, there are three main options for processing trash: (i) collection from the field following harvesting, (ii) harvesting and transport together with cane stalks from the field, and separation upon delivery at the mill, and (iii) same as (ii) but without separation. The costs of handling trash for these options have been considered (e.g., Macedo, 2001; Moller et al., 2010; McGuire et al., 2011) and as expected are sensitive to product value. However, the issue is complicated because of benefits that may arise through retaining trash in the field. The agronomic benefits of retaining trash in the field on soil fertility and moisture retention have been widely documented (e.g., Wood, 1991; Denmead et al., 1997; Chapman et al., 2001; Thorburn et al., 2004). In developing cultivars for future energy production systems, the yield of trash may need to be considered in assessing relative economic value of candidate genotypes under selection. This would contrast to most existing sugarcane breeding programs in which only stalk weights and composition are considered during selection.

*Milling*

Detailed descriptions of sugarcane processing procedures currently used for sugar manufacture are provided in the landmark texts Chen and Chou (1993) and Rein (2006). Future factories for producing energy from cane are likely to build on these traditional processes, although some major changes and innovations are also expected, especially in relation to improving overall energy efficiency and processing bagasse (Leal, 2007; Oliverio et al., 2010). The key steps in processing of cane are shown in a highly simplified format in Figure 7.1. These steps are: extraction of juice (usually via mechanical shredding and crushing cane through rollers), purification of juice (called clarification), evaporation of water from juice by boiling under partial vacuum, crystallization of sucrose through further boiling and crystal nuclei seeding, and separation (by centrifuging) of the crystals from the remaining molasses. In some mills, notably in Brazil, the juice is also partitioned for ethanol production via fermentation and distillation processes. Energy for mechanical power (for juice extraction) and heat (for

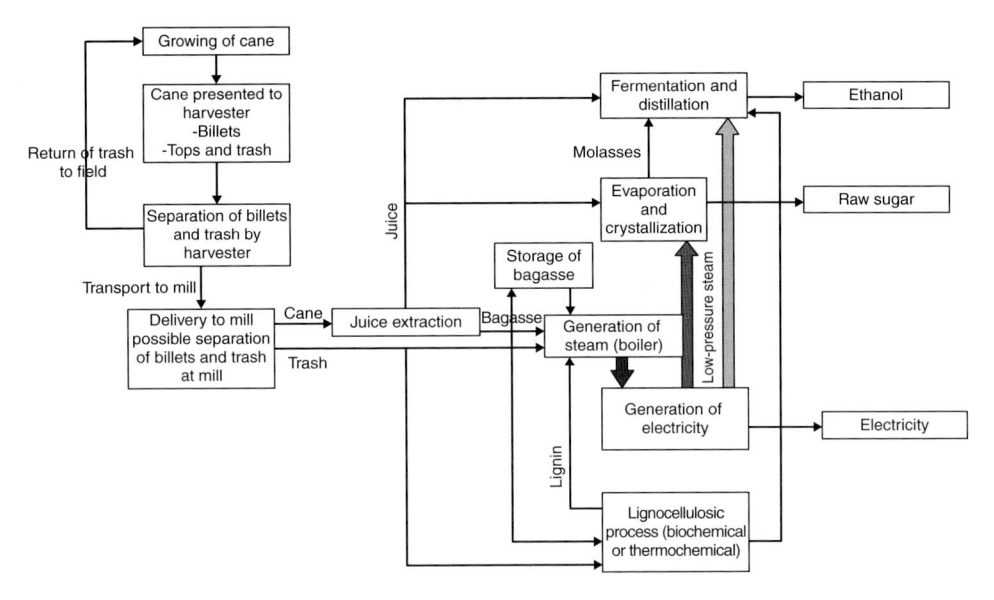

**Figure 7.1.** Generalized steps in current or future sugar and energy cane production systems.

evaporation and crystallization) is supplied from burning bagasse (the residue comprising mostly fiber and water remaining after the juice extraction process) and generation of steam.

In nearly all sugar mills, all of the bagasse is burnt in boilers. In the past, sugar mills were not designed to achieve high energy efficiency, because a concurrent goal had been to ensure that all the bagasse is disposed through burning, to avoid bagasse disposal costs. However, sugar mills are now looking increasingly at using bagasse for electricity production (e.g., Beeharry, 2001; Albert-Thenet, 2003; Autrey and Tonta, 2005; Walter and Ensinas, 2010), and in many countries, sugar mills are obtaining increased revenue from this source. While this is known to be occurring widely and increasing in response to increasing electricity prices and incentives provided for renewable energy, global statistics relating to this could not be readily collated during this review. Figure 7.1 indicates the possibility of using lignocellulosic processes to produce a combination of biofuel and electricity. While there is considerable investment into research and development of these processes globally, it is unclear if and to what extent this is being applied commercially at this stage. However, given trends in improved technology development and increases in biofuel prices, there is a high chance of such processes being part of sugar mills in future.

### 7.2.4 Overview and Potential Trends

Development of energy production systems involving sugarcane will continue to evolve from existing sugar production systems because of the large investment of capital already available in these systems and the additional marginal benefits that can be attained from enhancing these assets and processes for production of electricity and biofuels. However, it is also possible that new greenfield factories could be constructed focusing purely on biofuel and electricity production, especially if prices for energy products increase significantly, and/or if further progress is made in reducing costs of production of lignocellulosic derived biofuels. The main areas of change regarding future production systems compared with current ones based on sugar production would appear to be:

- The treatment of trash: to what extent this will be removed from the field, whether it will be separated at the factory or not, and relative costs and benefits of these options.
- Improvements of energy efficiency in mills to obtain greater economic value from energy content of cane and increase the proportion of bagasse used for electricity export and/or biofuels.
- An increased proportion of sugar or molasses which is diverted to ethanol.
- The inclusion of lignocellulosic processing to produce high-value biofuels from bagasse.
- The possibility of expansion into more marginal land if future larger scale bioenergy production develops.

Future directions will depend on economic drivers and process technology development. Breeding programs targeting different systems are likely to have different weightings for different traits, and the implications are considered below.

## 7.3 Sugarcane Improvement

### 7.3.1 Taxonomy and Crop Physiology

The taxonomy and physiology of sugarcane and its relatives, and the history of sugarcane breeding, is important in future breeding for energy production. Future energy cane breeding efforts are likely to build on progress made in sugarcane breeding to a large degree. It is also likely that future joint sugar plus energy production systems, or energy production systems, may be able to take greater advantage of favorable characteristics of the wild relatives of sugarcane, particularly *Saccharum spontaneum*. The optimal balance in commercial cultivars between *Saccharum officinarum* and *S. spontaneum* may shift toward the latter, for reasons discussed later.

The taxonomy of sugarcane, and particularly energy cane, is somewhat complex and controversial. Sugarcane cultivars belong to the genus *Saccharum,* within the tribe *Andropogoneae*. The *Andropogoneae* is frequently polyploid, and the evolution, speciation, and taxonomy are complex and unclear in some cases, with the monophyletic status of many genera being in doubt (Spangler et al., 1999; Hodkinson et al., 2002). In addition, it is difficult to assign taxonomic boundaries because of frequent interspecific and intergeneric hybridization. Sugarcane breeders and geneticists have used the term "Saccharum complex," a term first used by Mukherjee (1957) to describe an interbreeding group linked with the origin of sugarcane. Daniels and Roach (1987) provided a comprehensive review of the members of this complex.

Most sugarcane specialists in the past have recognized six species within the genus *Saccharum* (Daniels and Roach, 1987), two of which are wild (*S. spontaneum* and *Saccharum robustum*), and the other four being cultivated forms. *S. officinarum* is characterized by thick stalks, broad leaves, high sugar content, low fiber content, and a chromosome number usually of $2n = 80$, $x = 10$. *S. officinarum* clones are often called the "noble" canes. *S. spontaneum* ($2n = 36–128$) is a highly variable species that is found in the wild and distributed widely in the tropics to the subtropics in Asia and Africa in a wide diversity of habitats. Plants vary in appearance from short bushy types to large stemmed clones over 5 m in height. Stalk diameter and leaf width also vary but most *S. spontaneum* clones have thin stalks and leaves. Sugar content is usually very low but some *S. spontaneum* clones have been reported as having brix levels of up to 17% (Irvine, 1999) and 20% (He et al., 1999). *S. robustum* has characteristics similar to *S. officinarum,* but with more variability in most characteristics reported, and has been suggested as being the wild ancestor of *S. officinarum* (Brandes, 1958). *Saccharum barberi and Saccharum sinense* comprise the ancient landraces of India and China, respectively, with characteristics generally between *S. officinarum* and *S. spontaneum*, but with greater overall similarity to the former. Recent genomic *in situ* hybridization suggests that *S. barberi* and *S. sinense* were derived from interspecific hybridization between *S. officinarum* and *S. spontaneum* (D'Hont et al., 2002). *Saccharum edule* is a species characterized by an aborted inflorescence and, therefore, cannot be used for breeding.

Irvine (1999) challenged this traditional division of *Saccharum* into six species arguing that there is little basis for the separation. All six species overlap in most phenotypic characters and not only are the species interfertile but chromosomal pairing and recombination have been demonstrated (D'Hont et al., 1996). Molecular marker and other evidence suggest that the six species should more properly consist of two: one being *S. spontaneum*, and the other, comprising the five other species to be called *S. officinarum* (Irvine, 1999).

Several other closely related genera are also potentially important in sugarcane breeding, and these, together with the *Saccharum* genus are regarded as being part of the "*Saccharum* complex." The related genera include *Erianthus, Miscanthus, Narenga,* and *Sclerostachya*. These genera within the *Saccharum* complex are considered to be relevant to sugarcane improvement because: (i) they can interbreed, sometimes generating fertile $F_1$ progeny, (ii) they all have some origins around the frontiers of India, Yunnan, and Burma, and (iii) they have been implicated in the evolution of sugarcane (Guimaraes and Sobral, 1998).

Sugarcane and its relatives are perennial $C_4$ grasses, a class of plants ideally suited to efficient production of biomass (e.g., Carruthers, 1994). A widely accepted and useful framework for determining biomass accumulation in field crops was developed by Monteith (1977) as follows:

$$W_h = S \cdot \varepsilon_1 \cdot \varepsilon_2 \cdot \eta / k$$

where $W_h$ is the dry matter weight at harvest (g m$^{-2}$), $S$ is the incident solar radiation (MJ m$^{-2}$), $\varepsilon_1$ is the fraction of the incident radiation intercepted by the crop canopy, $\varepsilon_2$ is the efficiency that the intercepted radiation is converted into biomass, $\eta$ is the fraction of total biomass which is harvested, and $k$ is the energy content of the biomass (MJ g$^{-1}$). Perennial $C_4$ plants offer several advantages over other groups in relation to two of these components. First, $C_4$ plants have maximum conversion efficiency, $\varepsilon_2$, that is, about 40% higher than $C_3$ plants (Monteith, 1978). This advantage is regarded as being greatest under tropical conditions, although advantages are also realized under cool climates (Beale and Long, 1995). Second, perennial grasses such as sugarcane have an additional advantage of achieving high rates of canopy development after harvesting (e.g., Robertson et al., 1996a). An additional economic and environmental advantage of a perennial growth habit is that yearly expensive ground preparation and planting does not need to occur as frequently as with annual crops, reducing input costs, energy use, and potential for controlling soil erosion and water runoff.

Sugarcane yields are commonly reported on a per area basis of the weight of sugarcane stalks and trash (leaves that are not separated from the stalks at harvest) that are delivered to sugar mills for processing. Dry matter content varies according to environmental conditions, but is typically 30–35% of the fresh weight of delivered cane. Reducing moisture content of harvested cane is desirable for commercial production of either sugar or energy products, since harvesting, transport and some milling costs are in near proportion to the total weight of cane. One of the attractions of sugarcane as a bioenergy crop is the relatively high dry matter yields obtained in relation to some other crops. Such high yields may be attributed to its efficient $C_4$ photosynthetic system, the long duration of commercial crop cycles (typically 12 months between harvests), the high proportion of this time that the crop has a canopy intercepting over 90% of incoming solar radiation, and the rapid rate of canopy development in regrowth after harvesting. Commercial yields of up to 260 t ha$^{-1}$ for a 13-month old crop have been cited (Waclawovsky et al., 2010), but such reports need to be treated cautiously. Maximum experimental yield averaging 212 t ha$^{-1}$ year$^{-1}$, and a theoretical maximum of harvested sugarcane stalks assuming highly favorable environmental conditions, a harvest index (i.e., ratio of stalks to total biomass) of 0.8 and 30% dry matter, of 381 t ha$^{-1}$ were reported (Waclawovsky et al., 2010). Actual commercial maximum yields under favorable, irrigated, environments of about 160–190 t ha$^{-1}$ year$^{-1}$ are observed in some regions (e.g., Burdekin region in Australia and Colombia). In practice, however, most commercial yields are less than these maxima due to a range of environmental (e.g., water deficits or water logging, diseases) or management constraints (e.g., harvesting at nonoptimal times of year). In breeding programs, genetic response to these constraints, rather than achieving maximum yields under optimal conditions, will remain a key objective.

Mature sugarcane stalks are capable of accumulating high concentrations of sucrose, with levels of up to 650–700 mM in stalk tissue of being reported (Welbaum and Meinzer, 1990; Moore, 1995). The physiological and biochemical pathways of sucrose accumulation and the genes associated with sucrose accumulation have been extensively studied (e.g., Rae et al., 2005; Papini-Terzi et al., 2009), but the mechanisms and potential genetic control points are still not well understood.

A major issue for genotypes of all crops suited to future renewable bioenergy production will be adaptation to dry environments and achieving high levels of water use efficiency where irrigation is available. This arises due to increasing demands for land and freshwater resources for crops for food, and concerns about additional pressures arising if plants are additionally used for bioenergy (e.g., Berndes, 2002; Food and Agriculture Organisation, 2008; Gerbens-Leenes et al., 2009). In most sugarcane growing regions of the world, water is already seriously limiting production (Inman-Bamber and Smith, 2005). It is occasionally erroneously inferred that sugarcane has lower water use efficiency than other crops because it is often grown commercially in high-rainfall zones or under irrigation. In fact, like other $C_4$ species, sugarcane has relatively high water use efficiency relative to $C_3$ species. Tanner and Sinclair (1983) reported results showing an approximately twofold increase in transpiration efficiency on $C_4$ species (corn, sorghum) compared with $C_3$ species (potato, alfalfa, soybean). Robertson and Muchow (1994) reported a range in water use efficiency in sugarcane of between 4.8 and 12.1 tons of cane per mL of water used. This was attributed mainly to variations in vapor pressure deficit and stalk dry matter content. A range of physiological factors may influence crop response to water stress or improve crop water use efficiency (i.e., yield per water used), including access of water through deep roots, regulation of water loss through stomatal closure in response to water stress or other mechanisms such as leaf rolling or leaf shedding, inherent differences in transpiration efficiency, and ability of plants to tolerate water deficits and heat (Inman-Bamber and Smith, 2005). To date, there is no sugarcane breeding program that has managed to take advantage of physiological understanding or use physiological traits as indirect selection criteria for improving yields under dry environments. Globally, limited water availability (in both rain-fed and irrigated regions) will remain the number one biophysical constraint to higher productivity in sugarcane. While there may be opportunities to help speed up breeding sugarcane for water-limited environments through physiological understanding, there are also challenges facing this approach, particularly related to large genotype × environment interactions in relation to plant response to varying water deficits (e.g., Condon et al., 2004).

### 7.3.2 History of Sugarcane Breeding

Improvement of sugarcane through breeding has been conducted as a deliberate, directed activity since the late 1800s following the observation that sugarcane produced viable seed (Heinz, 1987). Historically, three main phases have been identified in modern (post 1900) sugarcane improvement (Roach, 1989). The first was intraspecific hybridization of selected *S. officinarum* clones. Generally, the goal was more disease-resistant forms with high commercially extractable sucrose levels. "Noble" cultivars are characterized by high commercially extractable sugar content, low fiber content, but very poor ratooning ability, poor adaptation to nontropical climates, and susceptibility to some diseases and insect pests (Stevenson, 1965). Famous noble cane varieties include "Otaheiti" from Java, "H109" from Hawaii, "B716" in Barbados, and Badila from New Guinea. Industries around the world in the early 1900s were based on noble cultivars.

The second phase was interspecific hybridization involving mainly *S. officinarum* × *S. spontaneum* clones, and led by breeders in Java and India. One key motivation for this was the threat of diseases to which some *S. officinarum* clones were susceptible. However, it was also found that resulting hybrids were generally vigorous, tolerant to a range of environmental stresses, and were strong ratooning. Following initial interspecific hybridization, it was necessary to backcross the hybrids to *S. officinarum* to "dilute" the undesirable characters in the wild canes, particularly low sugar content. This backcrossing process is termed "nobilization" by sugarcane breeders (Bremer, 1961). One of the most famous varieties produced as part of these early nobilization programs was the cultivar "POJ2878" in 1921 in Java that became not only an important cultivar, but is also in the ancestry of most commercial varieties grown today. In Coimbatore, India, production of the "Co" clones also became internationally important, especially in subtropical regions, and these also feature strongly in the ancestry of most sugarcane breeding programs.

The third phase, beginning around the 1930–1940s and continuing until today involved exploiting the material produced in the second phase (Roach, 1989). This has involved intercrossing among the original hybrids, and recurrent crossing and selection among progeny with increasingly larger populations. Most crosses made in breeding programs throughout the world today are based on a relatively small number of ancestors derived from the early interspecific hybrids produced in the second phase. Significant efforts have been made since the 1960s by sugarcane breeding programs to broaden the genetic base of modern sugarcane breeding programs including in China (Chen et al., 1985; Deng et al., 2002), Australia (Roach, 1989; Symington, 1989), Barbados (Rao and Kennedy, 2004), and Louisiana (Tew, 2003; Hale et al., 2010). However, modern cultivars in all parts of the world today trace back to the interspecific hybrids produced in Indonesia and India. Genomic *in situ* hybridization (GISH) indicates that modern cultivars typically contain 10–20% of their 100–130 chromosomes from *S. spontaneum*, 70–80% from *S. officinarum*, and about 5–20% being from recombination of chromosomes from both species (D'Hont et al., 1996, Piperidis and D'Hont, 2001).

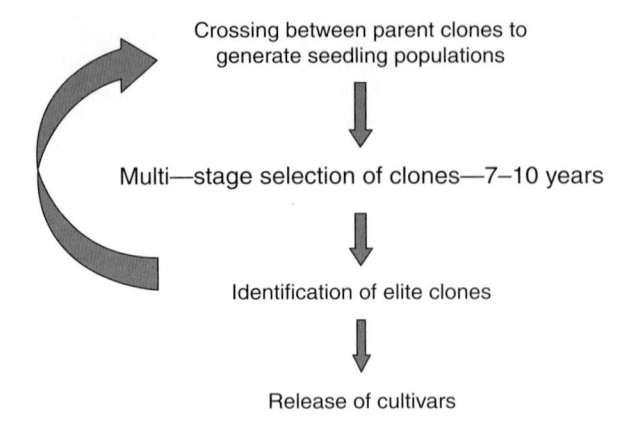

```
                    Crossing between parent clones to
                    generate seedling populations
                                ↓
        Multi—stage selection of clones—7–10 years
                                ↓
                    Identification of elite clones
                                ↓
                        Release of cultivars
```

**Figure 7.2.** General schematic of sugarcane breeding programs.

## 7.3.3  Basic Features of Sugarcane Breeding Programs

A detailed account of some important aspects of sugarcane breeding, including important practical aspects was provided by Heinz (1987). All sugarcane breeding programs around the world have their own particular features, but most follow the general scheme outlined in Figure 7.2. This general scheme would be expected to be followed in breeding programs that develop or evolve to become focused on energy production systems. Parent clones are initially selected on the basis of good phenotypic characteristics or commercial performance, and these are propagated at crossing stations for the purpose of producing flowers. Not all sugarcane clones flower and photoperiod manipulation is used in some programs to help induce and synchronize flowering (Moore and Nuss, 1987). Crosses between parents are chosen using criteria that vary between programs. However, specific combining ability effects in sugarcane are important (e.g., Miller, 1977; Hogarth, 1987) and unpredictable and the general strategy of most programs has been to make a large number of crosses to find the relatively small number from which commercial cultivars arise. This is done also to obtain estimates of breeding values (or general combining abilities) of parents (Heinz and Tew, 1987). Depending on breeding strategy used, clones found to have high breeding value (based on progeny performance) may be retained for further use, along with new ones brought in annually from the recurrent selection process, and those with poor breeding value discarded.

Following crossing, seed is sown in germinating cabinets and individual seedlings raised in glasshouses, before being transplanted into the field (Breaux and Miller, 1987). Most breeding programs transplant thousands of seedlings to the field each year (typically, 20,000 to over 1,000,000) (Loureiro et al., 2011), arising from less than one hundred to over a thousand different crosses. The rationale is that new cultivars are rare individual clones and large populations are needed to identify these individuals. The large numbers of clones generated also allow for high selection intensities to increase genetic gains from selection (Falconer and Mackay, 1996). Planting seedlings to the field marks the start of the major effort in sugarcane breeding: the multistage selection process. The number of stages of selection generally varies between 4 and 6. In the first stage, selection is conducted usually on the basis of visual assessment of individual clones, and in some programs using family selection as an indirect selection criterion for selecting superior clones (Skinner et al., 1987; Jackson et al., 1996).

During each stage the aim is to select a much smaller number of clones that appear to have the highest economic value. In subsequent selection stage, more accurate evaluation (through larger plots, more replicates, more measurements, more sites or years) is conducted on the smaller set of superior clones. For each stage of selection, parameters for selection trials such as numbers of clones evaluated, plot size, numbers of replicates per clone, numbers of sites and years in which the trial is conducted over, and measurements made need to be traded off while aiming to achieve maximum gains from selection per unit cost. Such decisions are made intuitively by the breeder, or more appropriately through application of appropriate objective statistical genetics principles (Skinner et al., 1987; Hu et al., 2008).

Selection criteria applied in each stage are a critical feature and are expected to be the major focus of changes in sugarcane breeding as it shifts in emphasis in future from sugar only to combined sugar plus energy products, or energy-only products. Selection criteria applied in each stage varies from program to program, but usually the most commonly applied selection criteria in the past have been strongly related to commercial sucrose yield (tons sucrose per hectare) with a bias toward commercial sucrose content (Wei et al., 2008), and resistance to diseases. Other criteria such as ratooning performance, adaptation to particular environmental stresses, and traits affecting harvestability (size of stalks, trash, lodging propensity) are also applied in varying degrees in different programs (e.g., Skinner et al., 1987). Often in the first one or two stages, visual appearance has been applied as an indirect selection criterion for cane yield and for disease resistance. Selection for sugar content is relatively more expensive than visual selection, and in some programs measurement of sugar content does not commence until the later stages.

Treatment of diseases and pests in most sugarcane breeding programs is done by: (i) considering resistance of parents to important local diseases and avoiding crossing susceptible pairs of parents, and (ii) selection for apparent resistance of clones in field trials throughout the multiple stages of selection. Many breeding programs also conduct separate inoculation trials for important diseases near the final stages of selection to ensure clones being considered for release, to ensure clones have been exposed to important diseases, and to determine accurate resistance ratings (relative to known check cultivars in the same trials exposed to identical pest or disease pressures).

There is usually a heavy weighting applied to commercially extractable sugar content in sugarcane breeding programs in the past, and this can be justified for two reasons. First, sugar content can be reliably measured in early stage selection trials because it has a high degree of genetic determination in relation to genotype × environment interaction variance and experimental error variance (Skinner et al., 1987; Jackson and McRae, 2001). Second, any given incremental contribution of higher sugar content on sugar yields is economically more valuable than the same contribution by cane yield (Jackson et al., 2000). This is because increases in cane yield also comes with higher harvesting, transport, and milling costs, in contrast to minimal marginal costs associated with increased sugar content. Genetic correlations between sucrose content and cane yield are low, indicating no overall genetic trade-offs in selecting for these two traits (e.g., Brown et al., 1969; Milligan, 1990).

Increased fiber content directly decreases sucrose content through displacing dry matter that could exist as sugar. In most genetic populations generated in breeding programs, negative genetic correlation is observed between fiber and sucrose content and commercially extractable sucrose (Milligan, 1990; Jackson, 1994; Berding and Pendrigh, 2009), and so strong selection pressure for high sucrose will tend to result in indirect selection pressure for reduced fiber content. In parallel with this effect is a requirement in most sugar industries for cultivars to maintain moderate fiber levels. High fiber content in cane impacts adversely on sugar production systems through increased loss of sucrose in bagasse and slowing down of milling rate. Quantifying the economic impact is complex but models of the extraction process have been developed in mills (e.g., Kent, 1997), and generally each 1% unit increase in fiber (e.g., moving from 12% to 13% on a fresh weight basis) reduces extraction of sugar by approximately 0.5% of the total sugar present in cane. If the bagasse is burnt then the value of this lost sugar is only realized for its contribution via heating, which is much less than its value as raw sugar crystal. However, in future lignocellulosic processes, if sugar in bagasse is converted to ethanol then the economic impact on sugar loss may not be as great.

### 7.3.4 Composition of Cane for Sugar or Energy Production

The composition of cane in modern cultivars of sugarcane used primarily for sugar production has been strongly influenced by breeding and selection for varieties which have high levels of commercially extractable sucrose levels. This is favored by high sucrose content, low levels of soluble impurities (i.e., nonsucrose dissolved solids in juice), and low fiber content. Typical composition of modern sugarcane cultivars is given in Table 7.2. The economic impact of both soluble impurities and fiber on the economics of sugar production and how this would change in energy production systems is explained below.

A high level of impurities in cane affects sugar production through increased loss of sugar in molasses. Molasses is the final product from the sugar mill containing residual sucrose which cannot profitably be crystallized in the raw sugar manufacturing process. Unless the molasses is directly used for ethanol production, increased loss of sucrose via molasses has a negative impact on industry profitability because molasses is a much lower value product than raw sugar. The method used to determine commercial cane sugar content (CCS, %) in sugarcane in Australia (BSES, 1984) illustrates the impact of changing levels of impurities on sugar production. The CCS formula is based on the assumption that in the raw sugar manufacturing processes in sugar mills 75% of soluble impurities are eliminated via molasses and that molasses can be produced at 40% purity (i.e., 40% of soluble solids exist as sucrose).

Table 7.2. Composition of sugarcane and juice from mature stalks of a typical modern cultivar.

| Millable stalks | Percentage of cane weight |
| --- | --- |
| Water | 73–76 |
| Fiber | 11–16 |
| Soluble solids (brix) | 10–16 |
| Juice constituents | Percentage of soluble solids |
| Sucrose | 72–92 |
| Glucose | 2–4 |
| Fructose | 2–4 |
| Salts | 3–4.5 |
| Organic acids | 1.5–5.5 |
| Protein | 0.5–0.6 |
| Starch | 0.001–0.1 |
| Gums, waxes, fats | 0.35–0.75 |
| Others | 3–5 |

From Clark (1993).

It can be readily deduced that if X is the amount of impurities entering the sugar mill, 0.75X will carry with it 0.5X of sucrose at 40% purity. Therefore:

$$\text{Recoverable sucrose} = \text{sucrose content} - \text{impurities}/2$$

In practice, the recovery of sucrose is also affected by the nature of the impurities. Increasing the ratio of reducing sugars to ash lowers the purity of molasses and thus decreases loss of sucrose via molasses (Miller et al., 1998). However, these impacts are relatively minor and the above formula is considered to provide an adequate approximation for cane payment purposes.

In production of ethanol, impurities present as fermentable sugars, most likely as glucose and fructose, will contribute to ethanol yield. In some reports pertaining to developing sugarcane for energy production, it is inferred that brix (percentage of total soluble solids) is closely related to total fermentable sugars. However, this may not necessarily be the case, particularly for early generation interspecific hybrids which have low purity levels (e.g., Wang et al., 2008). There appear to be few reports documenting genetic variation in proportions of impurities present as fermentable sugars versus other constituents, and therefore, it is difficult to comment at this stage on the importance of this trait in selecting profitable varieties for ethanol production. However, Robertson et al. (1996b) reported on significant variation between commercial cultivars for glucose and fructose levels in relatively immature cane, but these differences became smaller in mature cane. Unpublished data from Australia have shown large variation in proportion of impurities existing as glucose and fructose in genotypes in $F_1$ and $BC_1$ generations derived from *S. spontaneum*, ranging from 20% to 80%. Such variation may be particularly important from an economic perspective for ethanol production in relatively immature cane when impurity levels are high.

In production of liquid or gas biofuel from fiber (lignocellulosic biomass), efficiency of conversion and yields may be affected by fiber composition (Himmel et al., 2007; Vermerris, 2008). Lignin, cellulose, and hemicelluloses are the main components of cell walls and their production is mediated by complex pathways. However, identifying targets for breeding seem to be complicated by uncertainty about which processes will be widely deployed in commercial practice. There appear to be few studies estimating economic impact of changing particular characteristics of fiber on production costs for different conversion technologies, and so gauging the relative importance of manipulating fiber composition as a breeding target in relation to other traits such as total biomass yield, is difficult. There is limited knowledge of variation of fiber composition in sugarcane. Lingle et al. (2008) analysed fiber in a range of *Saccharum* germplasm, but no interesting variants were noted (cited by Berding and Pendrigh, 2009).

### 7.3.5 Application of Molecular Genetics in Developing Energy Cane

There has been considerable investment in molecular genetics applications to sugarcane improvement over the last 20 or more years, following pioneering research such as that of Bower and Birch (1992), Wu et al. (1992), D'Hont et al. (1996) amongst many others. However, to date all cultivars grown commercially have been developed through application of traditional crossing and selection methods, and it is still unclear when first examples of cultivars arising with clear contributions from molecular genetics will appear in farmer's fields. For sugar industries and investors interested primarily in maximizing rates of genetic improvement for industry profitability, the issue of balance of investment in traditional breeding programs versus molecular genetics research and application is an important, but difficult, question. On the one hand, based on their past and current achievements, traditional breeding programs provide an almost sure bet avenue for incremental improvements in productivity and control of disease outbreaks. On the other hand, progress in molecular genetics technologies, including in sugarcane, is occurring at a rapid rate, and for many scientists it seems surely inevitable that practical impacts will arise within the next decade or two.

In large advanced sugarcane industries, investment in molecular genetics and leveraging of developments being made in other crop species as much as possible is, generally, a high priority. However, of some concern to many "traditional" sugarcane breeders is: (i) the high (and possibly increasing) proportion of R&D funds allocated to molecular genetics projects which anticipate potential practical outcomes which do not materialize, (ii) the susceptibility of governments or other R&D investors in bioenergy, to proposals about "transformational" molecular genetics science, in which the levels of technical risks of delivering commercial outcomes within a reasonable timeframe are high but not accurately disclosed by proponents, and (iii) the relatively small number of the best and brightest young scientists interested in and undertaking training in plant breeding. This leads to a range of concerns about whether an optimal proportion of R&D devoted to investment in breeding programs delivering new cultivars for industries will occur in the future.

As with other crop species, one focus of molecular genetics research programs in many countries has been adding or manipulating expression levels of specific genes or metabolic pathways. This includes targeting traits such as disease or pest resistance (e.g., Joyce et al., 1998; McQualter et al., 2004), sucrose metabolism (e.g., Ma et al., 2000; Botha et al., 2001), herbicide resistance (e.g., Gallo-Meagher and Irvine, 1996; Leibbrandt and Synman, 2003), and alternative high-value products (e.g., McQualter et al., 2005). Specifically, for application to bioenergy production in sugarcane, Sainz and Dale (2009) reported on research based on a transgene induction and amplification system aiming to express celluloses in bagasse to very high levels upon induction, and help enable commercially viable production of lignocellulosic derived ethanol.

Progress in developing transgenic sugarcane appears to have been much slower than what the early proponents were hoping for nearly two decades ago. Important limitations include availability of suitable gene expression elements, effects of transgene silencing, level of understanding of sugarcane metabolism, and limited knowledge about heritability of transgenes in sugarcane (Lakshmanan et al., 2005; Botha, 2010). An emerging issue is also government regulatory and market challenges to overcome, in common with all GM products (Bonnett et al., 2010).

The first commercial applications of transgenic plants are expected to arise from herbicide resistance, with large (but to date unpublished) current investments in this trait in sugarcane by multinational companies. This level of investment is being driven by the large and growing market for new sugarcane varieties in Brazil. As with other crops, herbicide resistance traits are expected to improve the ease and effectiveness of weed control through the ability to spray with environmentally benign broad-spectrum herbicides, and without the sometimes narrow window of suitable times with conventional weed control methods. This allows for more flexible management operations, reduced need for expensive and energy-consuming soil tillage, and may provide for a large positive impact on the economic, environmental, and social aspects of sugarcane production. These impacts are expected to be particularly important for large-scale bioenergy production systems in tropical regions where wet weather may frequently interfere with timely control operations. Herbicide resistance traits in sugarcane would be expected to be eagerly embraced by farmers, with the major risk in final adoption perhaps being government regulatory requirements. If the transgenic sugarcane is used for sugar production, buyer's acceptance of GM-derived sugar will be an important issue.

The complexity of the sugarcane genome has provided an ongoing challenge in genetic and genome analysis of sugarcane and in development of DNA marker applications (Grivet and Arruda, 2001; Alwala and Kimbeng, 2010; D'Hont et al., 2010; Pastina et al., 2010). The main progenitor of sugarcane, *S. officinarum*, is an octoploid with a basic chromosome number of x = 10 (D'Hont et al., 1996, 1998). Clones labeled as *S. officinarum*, but with higher chromosome numbers, are generally regarded as interspecific hybrids. The other main contributor to the sugarcane genome, *S. spontaneum*, has a chromosome number ranging from 2n = 40 to 2n = 128, with five major cytotypes of 2n = 64, 80, 96, 112, and 128 (Panjie and Babu, 1960), and a basic chromosome number of x = 8 (D'Hont et al., 1996, 1998). Modern sugarcane cultivars which are derivatives of interspecific hybridization involving these two species are complex aneu-polyploids, comprising 70–80% *S. officinarum*, 10–20% *S. spontaneum*, and 5–20% recombinant chromosomes (D'Hont et al., 1996; Piperidis and D'Hont, 2001).

Progress in sugarcane genetics and genomics clearly lags behind other crops of comparable economic value largely because of its genome complexity. A large effort is being maintained in this area and progress has been reviewed elsewhere (see chapters in Henry and Kole, 2010). Genetic linkage maps are difficult to construct because of the high level of polyploidy and random pairing of homologous chromosomes. The usage of single-dose markers (i.e., markers present in a single copy, and, therefore, segregate 1:1 in gametes) provided an avenue for construction of linkage maps (Wu et al., 1992). Sugarcane maps containing in excess of 1000 linked markers are available (Hoarau et al., 2001; Aitken et al., 2005) but are still incomplete and unsaturated. Presence of multiple copies of some linkage groups may make completion of maps using genetic mapping alone difficult because of the dependence on use of low copy number markers (single- or double-dose markers) using current approaches.

A range of QTL mapping studies have been done in sugarcane (listed and reviewed by Pastina et al., 2010). Most studies have been based on about 100–300 progeny derived directly from biparental populations using a pseudo-testcross strategy. Only a small number of association studies across a broader range of genotypes have been reported to date (Pastina et al., 2010). Single marker associations or interval mapping are applied to marker data obtained from each parent separately. Analyses on a range of traits, including brix, sucrose content, fiber, cane yield, and disease resistance have been conducted. The general result reported is of multiple, small QTL explaining proportions of variation ($r^2$) from 2% to 22%. However, the values in the higher ranges are in studies with relatively small numbers of genotypes and are most likely overestimates, considering the methods used.

There is a high level of linkage disequilibrium (LD) observed in sugarcane breeding populations, due to the limited number of key ancestral clones used to found commercial breeding programs, and limited number of meiotic events (usually less than 10) since foundation (Jannoo et al., 1999; Raboin et al., 2008). Under these circumstances, association mapping would be expected to be a useful approach for identifying markers and genome regions linked to QTL causing variation in breeding populations. Some evidence for this has been reported (Wei et al., 2006, 2010). However, the large size and complexity of the sugarcane genome means that very large numbers of markers are required for adequate genome coverage, despite the high LD.

The limitations to QTL mapping and subsequent development of marker-assisted breeding approaches in sugarcane include availability of low-cost marker technologies to simultaneously screen very large numbers (100,000 or more) of markers across large numbers of genotypes, as well as statistical methods that can cope with the complex segregation patterns and interallelic and interlocus interactions possible in a high-level polyploid. The ability to accurately score dosage levels of alleles (not just presence versus absence) is expected to be important in sugarcane both for initial QTL mapping and subsequent breeding applications. While such tools do not yet appear to have been reported, it is likely that they will arise at some stage in the next decade or so, given the rapid rate of progress in DNA screening technology. The sequencing of sorghum, a close relative of sugarcane will provide an initial framework for understanding the sugarcane genome and developing marker applications. The international sugarcane genetics community is currently discussing potential sequencing strategies for sugarcane. In the meantime, the almost exponential reduction in costs of resequencing will provide for more cost effective and deep analysis of allelic variation, which could be particularly important in sugarcane. The prospects of rapid progress in DNA marker screening technologies provide attraction for further investment in this area. However, apart from challenges related to large-scale, cost-effective marker screening and QTL mapping, other challenges will subsequently arise in identifying effective breeding strategies utilizing markers that enhance genetic gains per unit cost invested of existing breeding programs.

The other "omics" tools—transcriptomics, proteomics, and metabolomics—have also been a major area of investment in sugarcane over the last decade, and there has been an extremely rapid progress in the supporting technologies (Casu et al., 2010; Watt et al., 2010). It is anticipated that these tools will provide insights into factors controlling yield and composition of sugarcane in response to environmental factors. However, key challenges remain interpretation of the complex data sets which arise, including understanding the myriad of complex interactions which occur at the levels being examined, and relating these to higher hierarchical levels of plant physiology, and ultimately to performance across a whole growing cycle in commercial production environments (Moore, 2005).

## 7.4 Selection of Sugarcane Genotypes for Energy Production

### 7.4.1 Overall Directions

Exact directions of breeding programs targeting future development of "energy cane" is currently an important but potentially complex and confusing issue for most sugarcane breeding programs. Uncertainty about future renewable energy (electricity and biofuel) demand and prices and rates of commercialization of lignocellulosic process technology creates uncertainty about optimal selection indices in breeding programs. Uncertainty about future sugarcane industry needs is not aided by these often contrasting viewpoints among industry participants and investors. This, in turn, creates further uncertainty about appropriate objectives of sugarcane breeding programs. Given the long lead times in developing and commercializing cultivars, breeding programs ideally should anticipate industry needs and opportunities well in advance if the best possible crosses are to be made and if optimal selection criteria are to be applied in early stages of selection. To address this problem, several reviewers have suggested "Type 1" and "Type 2" energy cane definitions (e.g., Alexander, 1985; Tew and Cobill, 2008).

Type 1 energy cane is one selected for both its sugar content as well as fiber. Alexander (1985) suggested that greater gains for energy production could be attained by relaxing constraints against high fiber content. Type 2 energy cane has been defined as one which is selected specifically for high fiber yields. Such varieties would be used for generation of electricity or production of cellulosic ethanol. It is conceivable that germplasm improvement may focus on maximizing energy yield per unit area per time (Botha, 2009). Tew and Cobill (2008) reported that proponents of the existing Brazilian sugar industry are opposed to low-sugar, high fiber varieties on the grounds that this reduces flexibility of shifting production between sugar and ethanol depending on relative prevailing economics. They also cite views that any decision to develop high-fiber energy canes would mean dilution of an ongoing breeding program targeting sugar production. In relation to breeding sugarcane for the United States, Tew and Cobill (2008) have a two-way bet, suggesting the best pathway forward is to continue to produce sugar from sugarcane, but also explore opportunities for producing ethanol from sugarcane including from bagasse.

In practice, actual industry demand for varieties with particular characteristics will be led primarily by commercial considerations, driven by the relative costs and revenues of generating sugar and bioenergy products from the cane feedstock with different composition (in terms of proportion of fiber, sucrose, other fermentable sugars, and other constituents). In many existing sugarcane production systems, there are two reasons why Type 2 sugarcane may not be as economically competitive as either Type 1 energy cane or existing sugarcane, at least in the foreseeable future. Firstly, in these systems, there exists a large sugar milling infrastructure already producing raw sugar. This, together with reasonable parity expected between sugar prices and liquid biofuel prices mean that even if biofuel prices soar (favoring the economics of Type 2 sugarcane), returns from raw sugar production would be expected to improve in similar proportion. The argument that the world raw sugar price will remain in reasonable parity (on average over a long-term period) with the liquid biofuel price is due to the flexibility of many sugar mills in Brazil, producing a large proportion of the global sugar supply, to swing a proportion of its cane juice into either raw sugar or ethanol, depending on price. Second, the cost of producing liquid biofuel from easily extracted fermentable sugars from cane is less, and is likely to remain so, than the more technically demanding process of lignocellulosic production of biofuel from fiber.

For selecting clones for potential future production systems involving joint sugar + bioenergy, the main differences compared with selecting clones for a sugar-only production system (as has occurred in sugarcane breeding programs up to now) are expected to be changes in the relative weightings given of the four traits: (i) sucrose content, (ii) cane yield, (iii) fiber content, and (iv) other fermentable sugars and impurities.

In a "sugar-only" production system, gain in fiber attracts a negative weighting (i.e., high fiber penalized during selection) because high fiber can both slow down crushing rate and increase sucrose losses, and these impacts can represent major costs. By contrast, for production of either electricity or ethanol from fiber, the fiber content will have some positive value since it can be converted to those energy products, and so long as the price of the energy products is high enough, this positive value could outweigh the negative impact on sugar extraction. It is likely that the relative weightings given to sugar content versus cane yield would also shift in balance in a sugar + bioenergy production system toward increased weighting favoring cane yield.

It is believed that weightings given to disease resistance levels will remain largely unchanged across any production system. For example, it would not be possible to release or grow varieties that are susceptible to important diseases, regardless of the products produced. In any production system producing biofuels from juice, nonsucrose fermentable sugars would also have more value than in those based just on sugar. In those based just on sugar, nonsucrose sugars have a negative effect (not just a neutral effect) because they increase sucrose loss in molasses.

For optimal selection, relative weightings given to different traits need to be quantified as precisely as possible, not just qualitatively described as above. Conceptually, it is easy to appreciate the importance of knowing something about the relative economic value of different traits for breeding programs. Obviously traits that are important economically in the industry targeted by the breeding program should be selected for, if possible. Also, greater attention or weighting should normally be given to those which are most important economically compared to those that are least important. However, quantification of the relative economic value of traits in optimizing selection is important because this enables a selection index to be applied to optimize gains toward a given goal. The theory of index selection is not new and was developed previously by Hazel (1943) and Smith (1936). A selection index is in the form of a linear equation of observed trait values for each genotype with a coefficient for each trait value:

$$\text{Selection index}_i = v_1 * GV_{i1} + v_2 * GV_{i2} + v_3 * GV_{i3} + \cdots + v_n * GV_{in}$$

where selection index$_i$ is the index for genotype $i$, $v_n$ is the economic weighting of trait $n$, and $GV_{in}$ is the genetic value of genotype $i$ for trait $n$. The economic weighting is defined as the economic benefit (on the basis of a suitable common unit, e.g., US\$ ha$^{-1}$) arising from an increase in one unit of the trait (e.g., the economic benefit from an increase in cane yield by 1 t ha$^{-1}$, or from an increase in fiber by 1%). Genetic values of a set of genotypes considered for selection are defined as the performance of genotypes relative to each other when measured in a common experiment or set of experiments. Commonly, best linear unbiased predictors (BLUPs) are considered to provide the best prediction of genetic values for selection purposes (Cotterill and Dean, 1990). All genotypes are ranked according to the index for the purpose of selection, and the index is considered to be synonymous with the breeding objective (Cotterill and Dean, 1990; Wei et al., 2008). In Australia, the index "Relative economic genetic value" (rEGV, Wei et al., 2008) was developed to serve as the primary selection criterion for selecting sugarcane clones with highest expected profitability for a sugar-only production system.

### 7.4.2 Example of Economic Weightings for Selecting Sugarcane for Energy Products

The first step in developing an appropriate selection index is to estimate the economic gain from changes in different traits. This is normally based on an economic model of the targeted production system, which captures estimates of the major costs and revenues that are affected by changing characteristics of the crop supply. While there is a lot of uncertainty in many assumptions, the use of best estimates from experts could provide a better direction for selection than arbitrary selection criteria without best expert inputs.

In the following analysis, three different production systems were considered: (i) sugar-only production, (ii) sugar produced from juice, and electricity produced from available bagasse, and (iii) sugar produced from juice, and ethanol produced from bagasse. The overall system depicted in Figure 7.1 was used as a framework, and it was assumed that the production process was set up using current best practice for efficient energy balance, based on values of parameters in Australia. The specific costs and details of different regions and factors may vary, but the assumptions used in this analysis were considered broadly representative of an internationally competitive sugarcane-based production system, producing either or both energy and sugar products, and should provide a useful guide to how economic weightings for different traits will vary generally for different types of production systems. Key inputs and cost assumptions included: base cane yield, sucrose content, fiber content, other fermentable sugar content at 80 t ha$^{-1}$, 13%, 12%, and 2%, respectively; growing costs of US\$1000 ha$^{-1}$; and harvesting

**Table 7.3.** Estimated impact (in US$ ha$^{-1}$) of one unit changes in sucrose, other sugars, fiber, and cane yield for four different production systems each producing different products from cane and for different prices of sugar and ethanol (two combinations of price levels, given a likely parity linkage between ethanol and sugar prices), and electricity (two levels). Key assumptions are given in the text.

| Products from processing cane[a] | Trait | Sugar and ethanol price of US$400 ton$^{-1}$ and US$0.65 L$^{-1}$ respectively Electricity price 0 | 100 | 200 | Sugar and ethanol price of US$600 ton$^{-1}$ and US$1 L$^{-1}$ respectively Electricity price (US$/MW-hr) 0 | 100 | 200 |
|---|---|---|---|---|---|---|---|
| Sugar + cogeneration | Sucrose (%) | 303 | 303 | 307 | 448 | 452 | 456 |
|  | Other sugars (%) | −79 | −68 | −58 | −166 | −155 | −145 |
|  | Fiber (%) | −19 | 76 | 170 | −27 | 66 | 160 |
|  | Cane yield (t ha$^{-1}$) | 27 | 41 | 52 | 57 | 69 | 80 |
| Ethanol from sugar + cogeneration | Sucrose (%) | 274 | 277 | 281 | 432 | 436 | 440 |
|  | Other sugars (%) | 245 | 255 | 266 | 391 | 401 | 411 |
|  | Fiber (%) | −14 | 80 | 173 | −21 | 73 | 167 |
|  | Cane yield (t ha$^{-1}$) | 49 | 60 | 72 | 86 | 98 | 110 |
| Sugar + lignocellulosic process | Sucrose (%) | 304 | 305 | 311 | 449 | 455 | 460 |
|  | Other sugars (%) | −79 | −65 | −57 | −165 | −157 | −149 |
|  | Fiber (%) | 44 | 106 | 166 | 71 | 131 | 192 |
|  | Cane yield (t ha$^{-1}$) | 42 | 50 | 56 | 80 | 86 | 93 |
| Ethanol from sugar + lignocellulosic process | Sucrose (%) | 274 | 277 | 281 | 433 | 438 | 444 |
|  | Other sugars (%) | 245 | 255 | 266 | 391 | 399 | 407 |
|  | Fiber (%) | 49 | 80 | 173 | 77 | 138 | 198 |
|  | Cane yield (t ha$^{-1}$) | 63 | 60 | 72 | 109 | 115 | 122 |

[a]Production of sugar is also associated with production of molasses, with an assumed value of US$65 ton$^{-1}$. Cogeneration refers to export of excess electricity generated from burning bagasse. In all cases, it is assumed that the unit changes in traits depicted do not impact on required level of capital investment.

and transport costs of US$11 ton$^{-1}$ of cane. Electricity outputs were based on assumptions used by Hobson et al. (2006), including 1.27 MW-hr of electricity generated per ton dry fiber from excess bagasse, steam supplied at 520°C and 65 bar (abs), 40% steam on cane required for processing needs. Yields of ethanol and electricity from lignocellulosic processing were based on choosing conservative values within ranges reported in the literature in studies such as Laser et al. (2009) of 350 L ethanol and 0.175 MW-hr per dry ton bagasse. Different prices for sugar, electricity, and ethanol were assumed, as indicated in Table 7.3. Of key interest is the relative weighting of the four traits: cane yield, fiber content, sucrose content (measured by pol), and nonsucrose fermentable sugars. For example, consider a production system producing sugar + cogeneration, assuming US$400 ton$^{-1}$ for sugar price and US$0/MW-hr for electricity (similar to the situation for many sugar mills currently in the world). For this system, a one unit increase in sucrose content of cane (i.e., going from 12% to 13% fresh weight) would provide for a US$303 ha$^{-1}$ industry gain in profit. For the same system, a one (%) unit increase in nonsucrose sugars would provide a US$79 ha$^{-1}$ loss in profit, while a one (%) unit increase in fiber content would provide a US$19 ha$^{-1}$ loss, and a 1 t ha$^{-1}$ increase in cane yield would provide a US$27 ha$^{-1}$ gain in profit. By contrast, in a production system producing ethanol from sucrose, with lignocellulosic processing and assumed ethanol price of US$0.65 and electricity price of US$100/MW-hr, similar one unit changes in sucrose, nonsucrose sugars, fiber, and cane yield would deliver profits of US$277 ha$^{-1}$, US$255 ha$^{-1}$, US$80 ha$^{-1}$, and US$60 ha$^{-1}$, respectively.

Some key points in the results include:

- Nonsucrose sugars have a large negative value on sugar-only production systems, due to impact on sugar loss through molasses. Purity (ratio of sucrose to nonsucrose soluble matter) of molasses is approximately 40% (varying slightly depending on other factors) and therefore, increased nonsucrose sugars will result in reduced sugar production. If ethanol is manufactured from molasses or a high price is attained for molasses, then this impact will be reduced.
- If no income is received from selling electricity (scenarios assuming US$0/MW-hr for electricity price), and there is no lignocellulosic processing of fiber, then increased fiber has a small negative impact on profitability. This corresponds to the situation with most sugar mills in the past, and still currently in many cases. This negative impact is due to increased sugar loss in bagasse and/or impact on milling throughput rates.
- In all cases involving nonzero values received for electricity and/or where lignocellulosic processes are involved, increasing fiber content has a positive weighting. However, in all cases, the value of an increase in sucrose content is greater than a corresponding unit increase in fiber content. The highest value of increased fiber in relation to sucrose is about 45%.
- The value of cane yield relative to sucrose content increases for systems involving cogeneration (with a positive value for electricity) and lignocellulosic processing. The weighting changes from a ratio of about 10:1 for the production system where there is revenue from only production of sugar (as has been traditionally the case in sugar industries) up to 4:1, where ethanol and electricity are sold. This represents a substantial reweighting in selection in favor of yield, and reduces the heavy emphasis which has been traditionally applied to sucrose content.

### 7.4.3 Progress in Breeding for Energy Production

It has been suggested that much larger gains in sugarcane breeding programs in both biomass yields and industry profitability may be possible with a reorientation of selection criteria away from the very strong emphasis on sucrose content (e.g., Alexander, 1985; Botha, 2009). This arises for two main reasons. First, the presence of high-yielding individual clones which do not become commercial cultivars because of insufficient sucrose content. Usually such clones are discarded because their poor commercial value in a sugar-only production system because of the heavy weighting applied to sucrose content in assessing economic value. With a reweighting of sucrose versus cane yield (Table 7.3), it is possible that such clones would be ranked highly for economic value and, therefore, retained during selection. Second, changed selection criteria, as indicated above, should also favor clones more closely to the wild canes such as sugarcane × *S. spontaneum* hybrids (called F$_1$ clones) or sugarcane × F$_1$ crosses. In this way, the desirable traits of the wild clone ancestor, such as adaptation to stress environments and ratooning performance, may be more easily retained, in contrast to the situation where several more cycles of crossing to sugarcane parents would normally be required for attaining sufficient sucrose content in breeding programs targeting sugar production systems.

A number of sugarcane breeding programs have targeted use of related species (especially *S. spontaneum*, but also others) in hybridization with sugarcane for the purpose of achieving higher cane yields, particularly under adverse environmental conditions, and examples are described below. The objectives and varieties outputted from such programs have usually been targeted at sugar industries, but are also relevant for bioenergy as well. Some breeding programs have publicized development of high-yielding varieties suited for energy production systems, but actual commercial production for this purpose does not appear to have occurred yet.

Srivastava et al. (1994) compared performance in North Queensland, Australia, of a set of families with varying (one to five) generations removed from an *S. spontaneum* ancestor. In going from five generations removed from *S. spontaneum* to one generation, there was a marked decrease in commercial cane sugar content (from about 14% to 9%), an increase in fiber content (from about 13.5% to 17.5%) and a small increasing trend in cane yield (from about 72 t ha$^{-1}$ to 77.5 t ha$^{-1}$). Legendre and Burner (1995) reported in a similar study in Louisiana in which random clones from $F_1$, $BC_1$, $BC_2$, $BC_3$, and more advanced generations were evaluated. They found the early generations with a high component of *S. spontaneum* provided the highest biomass yields and dry matter percentage and concluded that clones from these generations have "great potential" for biomass production.

In Louisiana, Giamalva et al. (1984) reported high yields and high fiber content (28%) in the variety L 79-1002. This variety was released for potential commercial production in 2007, along with another two varieties with high fiber (16%) relative to commercial varieties, but with higher sugar content than L 79-1002 (Tew and Cobill, 2008). In Puerto Rico, Alexander (1985) reported on the variety US 67-22-2, a clone derived from sugarcane × *S. spontaneum* with yields well above comparative commercial sugarcane varieties in Puerto Rico. In Barbados, Martin-Gardiner and Rao (2001), Rao and Albert-Thenet (2005), and Kennedy (2005) reported on the development of early generation, interspecific, hybrid clones with high fiber content, specifically suited for electricity production. These clones had high yields, high fiber, good ratooning performance, and a wide range of brix levels, but low purity. Based on these results, it was suggested that such clones could be suited to production of ethanol (from juice) and electricity in Barbados.

In Mauritius, Ramdoyal and Badaloo (2007) evaluated $F_1$, $BC_1$, and $BC_2$ families (derived from crosses between commercial sugarcane × *S. spontaneum*), and observed highest vigor (along with high fiber and low pol, as expected) in the $F_1$ material. They suggested that the $F_1$ families offered the best potential for producing "fuel canes" for cogeneration, a high priority in Mauritius. They also indicated a plan to explore a wider range of *S. spontaneum* clones in developing further $F_1$ families and intercrossing $F_1$ hybrids, given the high variance found among families. In the Nansei islands in Japan, Terajima et al. (2007) reported interesting results from some sugarcane × *S. spontaneum* $F_1$ clones compared with commercial sugarcane varieties. The environments used are challenging for sugarcane production due to drought and poor soil. The results indicated a dramatic increase in yields per area of cane, sugar, brix, and fiber in the $F_1$ hybrids compared with commercial sugarcane varieties in ratoon crops. They concluded that these clones have potential to provide for simultaneous production of sugar and biomass ethanol in their environments.

Wang et al. (2008) conducted field trials in China and Australia to evaluate a common set of sugarcane × *S. spontaneum* families. They found some high biomass yielding clones from these families in both countries, with high fiber levels and high dry matter contents, and with high broad-sense heritabilities for these traits, suggesting good gains from selection for these traits could be readily made in small plots.

In the United States and Europe, one factor that will constrain widespread use of energy cane for biomass production is expected to be cold temperatures. However, it is widely recognized by sugarcane breeders that early generation hybrids between sugarcane and *S. spontaneum* express a higher level of cold tolerance than most sugarcane cultivars (e.g., Roach and Maynard, 1975; Symington, 1989). Sladden et al. (1991) reported on trials in Alabama comparing the sugarcane × *S. spontaneum* hybrid L79-1002 with Napier grass (*Pennisetum purpureum*), sugarcane, and sweet sorghum, found the energy cane was the most cold-tolerant and sugarcane the least. By contrast, however, Burner et al. (2009) reported on performance of several species including switchgrass (*Panicum virgatum*), sugarcane × *S. spontaneum* hybrids, and sugarcane × *Miscanthus* sp. (most likely all *Miscanthus sinensis*) in Arkansas, in environments that are considerably colder than existing sugarcane cultivars are adapted to. None of the sugarcane × *S. spontaneum* progeny tested regrew in the first ratoon crop after winter, indicating a lack of cold tolerance under the local conditions. However, as emphasized by the authors, only a small sample (four) of *S. spontaneum* parents was sampled. Given the diversity of *S. spontaneum*, it is possible that use of other clones may give different results.

There has been widespread interest among sugarcane breeders in use of the related genera *Erianthus* in production of high biomass sugarcane and energy cane. Many breeders consider that some *Erianthus* species and clones may provide high biomass yields, particularly in marginal environments, offering potential low-cost feedstock for future bioenergy production systems. However, published reports with adequate statistical analysis of yields are difficult to find. There are several reports of production of hybrids between *Saccharum* spp., especially *S. officinarum* and sugarcane cultivars and Erianthus arundinaceus (D'Hont et al., 1995; Piperidis et al., 2000; Cai et al., 2005; Nair et al., 2006; Lalitha and Premachandran, 2007); *Erianthus rockii* (Aitken et al., 2007), *Erianthus ravennae* (Janaki Ammal, 1941) and *Erianthus cilaris* (Kandasami, 1961). The main motivation for these efforts has been to support longer-term sugarcane improvement through introduction of new genetic diversity and/or genes contributing to desirable traits, and with the desire to introgress some favorable characteristics from *Erianthus* such as vigor, ability to handle stress environments (drought, water logging), and disease resistance. However, use of *Erianthus* for successful development of sugarcane cultivars has not been reported or occurred so far to our knowledge. Several factors have probably limited the contribution of *Erianthus* in sugarcane improvement programs to date: (i) production of *Saccharum* × *Erianthus* hybrids has been complicated by the difficulty in identifying true hybrids amongst populations of seedlings arising from crosses, (ii) Difficulties in producing fertile hybrids, or any hybrids at all, due to the apparently relatively large genetic distance between *Saccharum* and *Erianthus*, even larger than for other genera such as *Miscanthus* (Sobral et al., 1994; Alix et al., 1998; Cai et al., 2005). The ability to detect numerous specific *E. arundinaceus* markers allows for early and reliable identification of true hybrids during the breeding process (Piperidis et al., 2000; Cai et al., 2005). One successful example has been reported of fertile hybrids between *Saccharum* and *Erianthus*, and progeny produced from these hybrids and sugarcane, opening up opportunity for introgression of *Erianthus* genome components into sugarcane breeding programs (Cai et al., 2005). However, to date there have been few reports backed with statistically supported data demonstrating performance of *Saccharum* × *Erianthus* hybrids or their derivatives.

One point of caution needed in interpreting some results reported on performance for biomass is the likely impact of competition effects on yields and biomass in small plots. In most of the studies reported above, small single-row plots were used. Single-row plots are subject to considerable competition effects between adjacent genotypes (Jackson and McRae, 2001), with yields in some clones and genetic variances and estimates of heritability likely to be inflated in many cases. This may be particularly the case for some early generation hybrid clones which tend to have tall stalks.

## 7.5 Conclusion

Sugarcane and hybrids between sugarcane and its wild relatives provide leading candidates for bioenergy production, both for production of liquid biofuels and for generation of electricity (biopower). Sugarcane cultivars bred primarily for sucrose production already provide relatively low-cost feedstock for bioenergy in combination with raw sugar production, as illustrated by large-scale production of ethanol in Brazil and increasing cogeneration of electricity in many countries. Sugarcane has a number of potential advantages compared with other candidate bioenergy crops, including existing infrastructure for harvesting, transport, and processing, which is already in place for raw sugar production. Better use of the nonsucrose sugars and fiber component in cane to produce energy products is occurring and is expected to continue, as demand for renewable energy increases.

Appropriate directions for sugarcane breeding programs to take in response to emerging opportunities for production of bioenergy targeting future development of "energy cane" is an important issue for most sugarcane breeding programs. However, uncertainty about future renewable energy (electricity and biofuel) prices, sugar prices, and rates of future technological progress in developing low-cost lignocellulosic process technology creates uncertainty about optimal objectives and selection criteria in breeding programs. The major changes expected in sugarcane breeding programs targeting joint sugar plus energy production in future are: (i) allocation of a positive weighting toward fiber content (rather than slightly negative for targeting sugar production), but still not as high as that justified for sugar content, (ii) measurement and positive weighting given to total fermentable sugars, not just sucrose, and (iii) increased relative weighting to biomass yields generally compared with existing selection criteria for sugar production only, with corresponding reduced weighting to sucrose content. These changes may also be expected to encourage a greater use of wild germplasm, particularly *S. spontaneum* in sugarcane breeding programs, which may facilitate a broader adaptation and widespread commercially viable use of cane developed for energy production or joint sugar plus energy production, including into more cooler and drier regions than those currently used for sugarcane.

## Acknowledgments

Parts of work reported here were supported by the Sugar Research and Development Corporation of Australia. I thank several colleagues for useful comments and suggestions in reviewing this paper, particularly Graham Bonnett and George Piperidis. I also thank the numerous people in the Australian sugarcane industry who helped in providing information relating to modeling the economic impact of variation in different traits in sugarcane varieties.

## References

Aitken KS, Jackson PA, McIntyre CL. A combination of AFLP and SSR markers provides extensive map coverage and identification of homo(eo)logous linkage groups in sugarcane. Theor Appl Genet, 2005; 110: 789–801.

Aitken K, Li J, Wang L-P, Qing C, Fan Y-H, Jackson PA. Characterisation of intergeneric hybrids of *Erianthus rockii* and *Saccharum* using molecular markers. Genet Resour Crop Evol, 2007; 54: 1395–1405.

Albert-Thenet JR. Fuel cane for the production of electricity in Barbados. Proc. Barbados Society of Technologists in Agriculture, January 2003; Barbados.

Alexander AG. *The Energy Cane Alternative*, 1985; Elsevier, Amsterdam.

Alix K, Baurens FC, Paulet F, Glaszmann JC, D'Hont A. Isolation and characterisation of a satellite DNA family in the *Saccharum* complex. Genome, 1998; 41: 854–864.

Allsopp P, Samson P, Chandler K. Pest Management. In: Hogarth DM, Allsopp PG (eds) *Manual of Canegrowing*, 2000, pp. 291–338. Bureau of Sugar Experiment Stations, Brisbane.

Alwala S, Kimbeng C. Molecular genetic linkage mapping in Saccharum: strategies, resources and achievements. In: Henry R, Cole C (eds) *Genetics, Genomics and Breeding of Sugarcane*, 2010, pp. 69–96. CRC Press, Boca Raton, FL.

Andrietta MGS, Andrietta SR, Steckelberg C, Stupiello, ENA. Bioethanol—Brazil, 30 years of Proalcool. Int Sugar J, 2007; 109: 195–200.

Autrey LJC, Tonta JA. From sugar production to biomass utilization: the reform process to ensure the viability of the Mauritian sugarcane industry. Proc Int Soc Sugar Cane Technol, 2005; 25: 449–457.

Balat M, Balat H. Recent trends in global production and utilization of bio-ethanol fuel. Appl Energ, 2009; 86: 2273–2282.

Beale CV, Long SP. Can perennial C4 grasses attain high efficiencies of radiant energy conversion in cool climates? Plant Cell Environ, 1995; 16: 641–650.

Beeharry RP. Strategies for augmenting sugarcane biomass availability for power production in Mauritius. Biomass Bioenergy, 2001; 20: 421–429.

Berding N, Pendrigh RS. Breeding implications of diversifying end uses of sugarcane. Proc Aus Soc Sugar Cane Tech, 2009; 31: 24–38.

Berndes G. Bioenergy and water the implications of large-scale bioenergy production for water use and supply. Global Environ Chang, 2002; 12: 253–271.

Bonnett GD, Olivares-Villegas JJ, Berding N, Morgan T. Sugarcane sexual reproduction in a commercial environment: Research to underpin regulatory decisions for genetically modified sugarcane. Proc Aust Soc Sugar Cane Technol, 2010; 32: 1–9.

Botha FC. Energy yield and cost in a sugarcane biomass system. Proc Aust Soc Sugar Cane Technol, 2009; 31: 1–10.

Botha FC. Future prospects. In: Henry R, Cole C (eds) *Genetics, Genomics and Breeding of Sugarcane*, 2010, pp. 249–264. CRC Press, Boca Raton, FL.

Botha FC, Sawyer BJB, Birch RG. Sucrose metabolism in the culm of transgenic sugarcane with reduced soluble acid invertase activity. Proc Int Soc Sugar Cane Technol, 2001; 24: 588–591.

Bower R, Birch RG. Transgenic sugarcane plants via microprojectile bombardment. Plant J, 1992; 2: 409–416.

Brandes EW. Origin, classification and characteristics of sugarcane. In: Artschwager E, Brandes EW (eds) *Sugarcane, Agricultural Handbook 122*, 1958, pp. 1–35. USDA, Washington DC.

Braunack MV, Wood AW, Dick RG, Gilmour JM. The extent of soil compaction in sugarcane soils and a technique to minimise it. Sugar Cane, 1993: 12–18.

Breaux RD, Miller JD. Seed handling, germination and seedling propagation. In: Heinz DJ (ed.) *Sugarcane Improvement Through Breeding*, 1987, pp. 385–407. Elsevier, Amsterdam.

Bremer G. Problems in breeding and cytology of sugarcane. Euphytica, 1961; 10: 59–78.

Brown AHD, Daniels J, Latter BDH. Quantitative genetics of sugarcane. II. Correlations analysis of continuous characters in relation to hybrid sugarcane breeding. Theor Appl Genet, 1969; 39: 1–10.

BSES. *Laboratory Manual for Australian Sugar Mills*, 1984; Volume 1. Bureau of Sugar Experiment Stations, Brisbane, Australia.

Burner DM, Tew TL, Harvey JJ, Belesky DP. Dry matter partitioning and quality of Miscanthus, Panicum and Saccharum genotypes in Arkansas, USA. Biomass Bioenerg, 2009; 13: 610–619.

Cai Q, Aitken K, Deng HH, Chen XW, Fu C, Jackson PA, McIntyre CL. Verification of the introgression of *Erianthus arundinaceus* germplasm into sugarcane using molecular markers. Plant Breeding, 2005; 124: 322–328.

Calcino D, Kingston G, Haysom M. Nutrition of the plant. In: Hogarth M, Allsopp P (eds) *Manual of Canegrowing*, 2000, pp. 153–195. Bureau of Sugar Experiment Stations, Indooroopilly, Australia.

Carruthers SP. Solid biofuels: fuel crops. In: Carruthers SP, Miller PA, Vaughan CMA (eds) *Crops for Industry and Energy*. 1994 *CAS Report 15,* pp. 168–180. Centre for Agricultural Strategy, University of Reading.

Casu RE, Carlos TH, Glaucia M S. Functional genomics: transcriptomics of sugarcane – current status and future prospects. In: Henry R, Cole C (eds) *Genetics, Genomics and Breeding of Sugarcane*, 2010, pp. 167–192. CRC Press, Boca Raton, FL.

Chapman LS, Larsen PL, Jackson J. Trash conservation increases cane yield in the Mackay district. Proc Aust Soc Sugar Cane Technol, 2001; 23: 176–184.

Chen JCP, Chou C-C. *Cane Sugar Handbook – a Manual for Cane Sugar Manufacturers and their Chemists. 12th Edition*, 1993; John Wiley and Sons Inc., New York.

Chen W-H, Huang Y-J, Shen IS. Utilization of Miscanthus germplasm in sugarcane breeding in Taiwan. Proc Int Soc Sugar Cane Technol, 1985; 19: 641–647.

Clark MA. Sugars and nonsugars in sugarcane. In: Chen, W-H, Chou C-C (eds) *Cane Sugar Handbook – A Manual For Cane Sugar Manufacturers and Their Chemists, 12th Edition*, 1993, pp. 21–39. John Wiley and Sons Inc., New York.

Condon AG, Richards RA, Rebetzke GJ, Farquhar GD. Breeding for high water use efficiency. J Exp Botany, 2004; 55: 2447–2460.

Cotterill PP, Dean CA. *Successful Tree Breeding with Index Selection*, 1990; CSIRO, Melbourne, Australia.

Crook TD, Pope GM, Staunton SP, Norris, CP. A survey of field CCS versus mill CCS. Proc Aust Soc Sugar Cane Technol, 1999; 21: 33–37.

Daniels J, Smith P, Paton N, Roach, B. The origin of the genus Saccharum. ISSCT Sugarcane Breeders Newsletter, 1975; 36: 24–39 (cited in Daniels and Roach, 1987).

Daniels J, Roach BT. Taxonomy and evolution. In: Heinz DJ (ed.) *Sugarcane Improvement Through Breeding*, 1987, pp. 7–85. Elsevier, Amsterdam.

Deng HH, Liao Z-Z, Li Q-W, Lao FY, Fu C, Chen XW, Zhang CM, Liu SM, Yang YH. Breeding and isozyme marker assisted selection of F2 hybrids from *Saccharum* spp. x *Erianthus arundinaceus*. Sugarcane and Canesugar, 2002; 1: 1–5.

Denmead OT, Mayocchi, CL, Dunin, FX. Does green cane harvesting conserve soil water? Proc Aust Soc Sugar Cane Technol, 1997; 19: 139–146.

De Vries SC, van de Ven GWJ, van Ittersum MK, Giller KE. Resource efficiency and environmental performance of nine major biofuel crops, processed by first-generation conversion techniques. Biomass Bioenerg, 2010; 34: 588–601.

D'Hont A, Garsmeur O, McIntyre L. Mapping, tagging and map-based cloning of simply inherited traits. In: Henry R, Cole C (eds) *Genetics, Genomics and Breeding of Sugarcane*, 2010, pp. 97–116. CRC Press, Boca Raton, FL.

D'Hont A., Grivet L, Feldmann P, Rao S, Berding N, Glaszmann JC. Characterisation of the double genome structure of modern sugarcane cultivars (*Saccharum* spp) by molecular cytogenetics. Mol Gen Genet, 1996; 250: 405–413.

D'Hont A, Ison D, Alix K, Roux C, Glaszmann JC. Determination of basic chromosome numbers in the genus *Saccharum* by physical mapping of ribosomal RNA genes. Genome, 1998; 41: 221–225.

D'Hont A, Paulet F, Glaszman JC. Oligoclonal interspecific origin of 'North Indian' and 'Chinese' sugarcanes. Chromosome Res, 2002; 10: 253–262.

D'Hont A, Rao PS, Feldmann P, Grivel L, Islam-Faridi N, Taylor P, Glaszmann JC. Identification and characterisation of sugarcane intergeneric hybrids, *Saccharum officinarum* x *Erianthus arundinaceus*, with molecular markers and DNA in situ hybridization. Theor Appl Genet, 1995; 91: 320–326.

Falconer DS, Mackay TFC. *Introduction to Quantitative Genetics*, 4th Edition, 1996, Longman, Edinburgh Gate, England.

Food and Agriculture Organisation. *The State of Food and Agriculture 2008. Biofuels: Prospects, Risks and Opportunities*, 2008; Food and Agriculture Organisation, Rome, Italy.

Gallo-Meagher M, Irvine JE. Herbicide-resistant transgenic sugarcane containing the bar gene. Crop Sci, 1996; 36: 1367–1374.

Gerbens-Leenes W, Hoekstra AY, van der Meer, TH. The water footprint of bioenergy. Proc Natl Acad Sci, 2009; 106: 10219–10223.

Giamalva MJ, Clark SJ, Stein JM. Sugarcane hybrids of biomass. Biomass, 1984; 6: 61–68.

Goldemberg, J. The Brazilian biofuels industry. Biotechnol Biofuels, 2008; 1: 6.

Goldemberg J, Coelho ST, Guardabassi P. The sustainability of ethanol production from sugarcane. Energ Policy, 2008; 36: 2086–2097.

Grivet L, Arruda P. Sugarcane genomics: depicting the complex genome of an important tropical crop. Curr Opin Plant Biol, 2001; 5: 122–127.

Guimareas CT, Sobral BWS. The Saccharum complex: Relation to other Andropogoneae. Plant Breeding Rev, 1998; 16: 269–288.

Hale AL, Verimis JC, Tew TL, Burner DM, Legendre BL, Dunckelman PH. 50 years of sugarcane germplasm enhancement – roadblocks, hurdles and success. *International Society of Sugar Cane Technologists 9th Breeding and Germplasm workshop*, 2010, p. 35. BSES, Cairns.

Hazel DL. The genetic basis for constructing selection indexes. Genetics, 1943; 28: 476–490.

He S-C, Xiao F-H, Zhang F-C, He L-L, Yang Q-H. Collection and description of basic germplasm of sugarcane (Saccharum complex) in China. Int Sugar J, 1999; 101: 84–93.

Heinz DJ. *Sugarcane Improvement Through Breeding*, 1987; Elsevier, Amsterdam.

Heinz DJ, Tew TL. Hybridisation procedures. In: Heinz DJ (ed.) *Sugarcane Improvement Through Breeding*, 1987, pp. 313–342. Elsevier, Amsterdam.

Henry R, Cole C. *Genetics, Genomics and Breeding of Sugarcane*, 2010; CRC Press, Boca Raton, FL.

Herridge DF, Peoples MB, Boddey RM. Global inputs of biological nitrogen fixation in agricultural systems. Plant Soil, 2008; 311: 1–18.

Himmel ME, Ding SY, Johnson DK, Adney WS, Nimlos MR, Brady JW, Foust TD. Biomass recalcitrance: engineering plants and enzymes for biofuels production. Science, 2007; 315: 804–807.

Hobson PA, Edye LA, Lavarack BP, Rainey T. Analysis of bagasse and trash utilization options. SRDC Technical Report 2/2006; Sugar Research and Development Corporation, Australia.

Hodkinson TR, Chase MW, Lledo DM, Salamin N, Renvoise SA. Phylogenetics of Miscanthus, Saccharum and related genera (Saccharinae, Andropogoneae, Poaceae) based on DNA sequences from ITS nuclear ribosomal DNA and plastid trnL intron and trn-F intergeneric spacers. J Plant Res, 2002; 115: 381–392.

Hogarth DM 1987; Genetics of sugarcane. In: Heinz DJ (ed.) *Sugarcane Improvement through Breeding, 1987*, pp. 255–271. Heinz, Elsevier.

Hu F-D, Jackson PA, Basford K. Developing optimal selection systems in sugarcane breeding programs. Proc Aust Soc Sugar Cane Tech, 2008; 114: 162–174.

IBGE. Instituto Brasileiro de Geografia e Estatistica. Censo Agropecuario, 2006; www .ibge.gov.br/home/estatistica/economia/aropecuaria/censoagro/2006/default.htm. Accessed on 5 Feb 2009.

Inman-Bamber NG, Smith DM. Water relations in sugarcane and response to water deficits. Field Crop Res, 2005; 92: 185–202.

Irvine JE. *Saccharum* species as horticultural classes. Theor Appl Genet, 1999; 98:186–194.

Ivin PC, Doyle CD. Some measurements of the effect of tops and trash on cane quality. Proc Aust Soc Sugar Cane Technol, 1989; 11: 1–7.

Jackson PA. Genetic relationships between attributes in sugarcane clones closely related to *Saccharum spontaneum*. Euphytica, 1994; 79: 101–108.

Jackson PA. Breeding for improved sugar content in sugarcane. Field Crops Res, 2005; 92: 277–290.

Jackson PA, Bonnett G, Chudleigh P, Hogarth DM, Wood AW. The relative importance of cane yield and traits affecting CCS in sugarcane varieties. Proc Aust Soc Sugarcane Tech, 2000; 22: 23–29.

Jackson PA, Bull JK, McRae TA. The role of family selection in sugarcane breeding programs and the effect of genotype x environment interactions. Proc Int Soc Sugar Cane Tech Conf XXII Congress, 1996; 17: 261–271.

Jackson PA, McRae TA. Selection of sugarcane clones in small plots: effects of plot size and selection criteria. Crop Science, 2001; 41: 315–322.

Janaki Ammal EK. Intergeneric hybrids of Saccharum. J Genet, 1941; 41: 217–253.

Jannoo N, Grivet G, Dookun A, D'Hont A, Glaszmann J-C. Linkage disequilibrium among modern sugarcane cultivars. Theor Appl Genet, 1999; 99: 1053–1060.

Joyce PA, McQualter RB, Bernad MJ, Smith GR. Engineering for resistance in to SCMV in sugarcane. Acta Hort, 1998; 461: 385–391.

Kandasami PA. Interspecific and intergeneric hybrids of Saccharum spontaneum L. I. Functioning of gametes. Cytologia, 1961; 26: 117–123.

Kennedy AJ. Breeding improved cultivars for the Caribbean by utilization of total biomass production. Proc Int Soc Sugar Cane Technol, 2005; 26: 491–498.

Kent GA. Modelling the extraction processes of milling trains. Proc Aust Soc Sugar Cane Technol, 1997; 19: 315–321.

Lakshmanan P, Geijskes RJ, Aitken KS, Grof CLP, Bonnett GD, Smith, GR. Sugarcane biotechnology: the challenges and opportunities. In Vitro Cell Dev Biol, 2005; 41: 345–363.

Lalitha R, Premachandran MN. Meiotic abnormalities in intergeneric hybrids between *Saccharum spontaneum* and *Erianthus arundinaceus* (Graminae). Cytologia, 2007; 72: 337–343.

Laser M, Haiming J, Jayawardhana K, Lynd LR. Coproduction of ethanol and power from switchgrass. Biofuel Bioprod Bior, 2009; 3: 195–219. doi:10.1002/bbb.

Leal MRLV. The potential of sugarcane as an energy source. Proc Int Soc Sugar Cane Technol, 2007; 26: 23–33.

Legendre BL, Burner DM. Biomass production of sugarcane cultivars and early-generation hybrids. Biomass Bioenerg, 1995; 8: 55–61.

Leibbrandt NB, Synman SJ. Stability of gene expression and agronomic performance of a transgenic herbicide resistant sugarcane line in South Africa. Crop Sci, 2003; 43: 671–678.

Lingle S, Tew T, Hale A, Cobill R. Composition of residue from sugarcane and related species. Agronomy Abstracts 549-2, 2008. Agronomy Society of America, Madison, WI.

Loureiro ME, Barbosa MHP, Lopes FJF, Silverio FO. Sugarcane breeding and selection for more efficient biomass conversion in cellulosic ethanol. In: Buckeridge MS, Goldman GH (eds) *Routes to Cellulosic Ethanol*, 2011, pp. 199–239. Springer, New York, Heidelberg, Dordrecht, London.

Ma H, Albert, HH, Paull R, Moore PH. Metabolic engineering of invertase activities in different subcellular compartments affects sucrose accumulation in sugarcane cells. Aust J Plant Physio, 2000; 27: 1021–1030.

Macedo IC, Leal MRLV, Hassuani SJ. Sugar cane residues for power generation in the sugar/ethanol mills of Brazil. Energy Sust Dev, 2001; 1: 77–82.

Macedo IC, Leal MRLV, da Silva JEAR. Assessment of greenhouse gas emissions in the production and use of fuel ethanol in Brazil, 2004; (http://www.unica.com.br/i_pages/files/pdf_ingles.pdf)

Machado GR. *Sugarcane Variety Notes – An International Directory*, 7th Revision, 2001; Piracicaba, Brazil.

Martin-Gardiner M, Rao PS. Potential for newly enhanced sugarcane germplasm for sugar production in the Caribbean. 27th West Indies Sugar Technologists Conference, Trinidad, April 2001 (CD).

Matsuoka S, Ferro J, Arruda, P. The Brazilian experience of sugarcane ethanol industry. In: Tomes D, Lakshmanan P, Songstad D (eds) *Biofuels – Global Impact on Renewable Energy, Production Agriculture and Technological Advancements*, 2011, pp. 157–172. Springer, New York, Dordrecht, Heidelburg, London.

McGuire PJ, Inderbitzen M, Rich B, Kent GA. The effect of whole crop harvesting on crop yield. Proc Aust Soc Sugar Cane Technol, 2011; 33 (CD).

McQualter RB, Dale JL, Harding JA, McMahon, JA, Smith GR. Production and evaluation of transgenic sugarcane containing a Fiji disease virus (FDV) genome segment S9-derived synthetic resistance gene. Aust J Agric Res, 2004; 55: 139–145.

McQualter RB, Fong CB, Meyer K, Van Dyk DE, O'Shea MG, Nicholas JW, Vitanen PV, Brumbley SM. Initial evaluation of sugarcane as a production platform for p-hydroxybenzoic acid. Plant Biotechnol J, 2005; 3: 29–41.

Miller JD. Combining ability and yield component analysis in a five-parent diallel cross in sugarcane. Crop Sci, 1977; 17: 545–547.

Miller KF, Ingram GD, Murry JD. Exhaustion characteristics of Australian molasses. Proc Aust Soc Sugar Cane Technol, 1998; 20: 506–513.

Milligan SB, Gravois KA, Bischoff KP, Martin FA. Crop effects on genetic relationships between sugarcane traits. Crop Sci, 1990; 30: 927–931.

Moller D, Broadfoot R, Bell S, Bakir H. Whole crop processed at Broadwater mill: Impacts on process operations. Proc Aust Soc Sugar Cane Technol, 2010; 32: 559–572.

Monteith JL. Climate and the efficiency of crop production in Britain. Philos Trans R Soc Lond, 1977; 281: 277–294.

Monteith JL. Reassessment of maximum growth rates for C3 and C4 crops. Exp Agr, 1978; 14: 1–5.

Moore PH. Temporal and spatial regulation of sucrose accumulation in the sugarcane stem. Aust J Plant Physiol, 1995; 22: 661–679.

Moore PH. Integration of sucrose accumulation processes across hierarchical scales: towards developing an understanding of the gene-to-crop continuum. Field Crops Res, 2005; 92: 119–135.

Moore PH, Nuss KJ. Flowering and flower synchronisation. In: Heinz DJ (ed.) *Sugarcane Improvement Through Breeding*, 1987, pp. 273–311. Heinz, Elsevier.

Mukherjee SK. Origin and distribution of *Saccharum*. Bot Gaz, 1957; 119: 55–61.

Nair NV, Selvi A, Sreenivasan TV, Pushpalatha KN, Sheji M. Characterisation of intergeneric hybrids of *Saccharum* using molecular markers. Genet Resour Crop Evol, 2006; 53: 163–169.

Oliverio JL, Carmo VB, Gurgel MA. The DSM – Dedini sustainable mill: a new concept in designing complete sugarcane mills. Proc Int Soc Sugar Cane Technol, 2010; 27: (CD ROM).

Panjie RR, Babu CN. Studies in Saccharum spontaneum. Distribution and geographic association of chromosome numbers. Cytologia, 1960; 25: 152–172.

Papini-Terzi FS, Rocha FR, Vêncio RZN, Felix JM, Branco DS, Waclawovsky AJ, Bem LEVD, Lembke CG, Costa MDL, Nishiyama MY, Vicentini R, Vincentz MGA, Ulian EC, Menossi M, Souza GM. Sugarcane genes associated with sucrose content. BMC Genomics, 2009; 10: 120. doi:10.1186/1471-2164-10-120.

Pastina MM, Pinto LR, Olivereira KM, de Souza AP, Garcia AAF. Molecular mapping of complex traits. In: Henry R, Cole C (eds) *Genetics, Genomics and Breeding of Sugarcane*, 2010, pp. 117–148. CRC Press, Boca Raton, FL.

Pimental D. Patzek T. Ethanol production: energy and economic issues related to U.S. and Brazilian sugarcane. Nat Resour Res, 2007; 16: 235–242.

Piperidis G, Christopher MJ, Carroll BJ, Berding N, D'Hont A. Molecular contribution to selection of intergenic hybrids between sugarcane and the wild species *Erianthis arundinaceus*. Genome, 2000; 43: 1033–1037.

Piperidis G, D'Hont A. Chromosome composition analysis of various Saccharum interspecific hybrids by genomic in situ hybridisation (GISH). Int Soc Sugar Cane Technol Congress, 2001; 11: 565.

Raboin L-M, Pauquet J, Butterfield MA, D'Hont A, Glaszmann J-C. Analysis of genome-wide linkage disequilibrium in the highly polyploidy sugarcane. Theor Appl Genet, 2008; 116: 701–714.

Rae AL, Grof CPL, Casu RE, Bonnett GD. Sucrose accumulation in the sugarcane stem: pathways and control points for transport and compartmentation. Field Crops Res, 2005; 92: 159–168.

Ramdoyal K, Badaloo MGH. An evaluation of interspecific families of different nobilised groups in contrasting environments for breeding novel sugarcane clones for biomass. Proc Int Soc Sugar Cane Technol, 2007; 26: 632–643.

Rao PS, Albert-Thenet JR. Fuel cane biomass potential for year round energy production. Proc Int Soc Sugar Cane Technol, 2005; 25: 537.

Rao PS, Kennedy A. Genetic improvement of sugarcane for sugar, fibre, and biomass. National Agriculture Conference, 2004; Ministry of Agriculture and Rural Development, Barbados.

Rein P. *Cane Sugar Engineering*, 2006; Bartens, Berlin.

Renouf MA, Wegener MK, Nielsen LK. An environmental life cycle assessment comparing Australian sugarcane with US corn and UK sugar beet as producers of sugars for fermentation. Biomass Bioenerg, 2008; 32: 1144–1155.

Ricaud C, Egan BT, Gillaspie AG, Hughes CG. *Diseases of sugarcane—Major Diseases*, 1989; Elsevier, Amsterdam.

Roach B. Origin and improvement of the genetic base of sugarcane. Proc Aust Soc Sugarcane Technol, 1989; 11: 34-48.

Roach BT, Maynard EO. Assessment of cold tolerance in sugarcane hybrids derived from traditional and new germplasm. Sugarcane Breed Newsl, 1975; 35: 43–49.

Robertson FA, Thorburn PJ. Trash management – consequences for soil carbon and nitrogen. Proc Aust Soc Sugar Cane Technol, 2000; 22: 225–229.

Robertson MJ, Muchow RC. Future research challenges for efficient crop water use in sugarcane production. Proc Aust Soc Sugar Cane Technol, 1994; 16: 193–200.

Robertson MJ, Muchow RC, Wood AW, Campbell JA. Accumulation of reducing sugars by sugarcane: effects of crop age, nitrogen supply and cultivar. Field Crops Res, 1996a; 49: 39–50.

Robertson MJ, Wood AW, Muchow RC. Growth of sugarcane under high input tropical conditions. I. Radiation use, biomass accumulation and partitioning. Field Crops Res, 1996b; 48: 11–25.

Rott P, Bailey RA, Comstock JC, Croft BJ, Saumtilly S. *A Guide to Sugarcane Diseases*, 2000; CIRAD, Montpellier.

Sainz MB, Dale J. Towards cellulosic ethanol from sugarcane bagasse. Proc Aust Soc Sugar Cane Tech, 2009; 31: 18–23.

Singels A, Donaldson RA, Smit MA. Improving biomass production and partitioning in sugarcane: theory and practice. Field Crops Res, 2005; 92: 291–303.

Skinner JC, Hogarth DM, Wu KK. Selection methods, criteria, and indices. In: Heinz DJ (ed) *Sugarcane Improvement through Breeding*, 1987, pp. 255–271. Elsevier.

Sladden SE, Bransby DI, Aiken GE, Prine GM. Biomass yield and composition, and winter survival of tall grasses in Alabama. Biomass Bioenerg, 1991; 1: 123–127.

Smith HF. A discriminant function for plant selection. A Eug, 1936; 7: 240–250.

Sobral BWS, Braga DPV, Lahood ES, Keim P. Phylogenetic analysis of chloroplast restriction enzyme site mutations in the *Saccharinae Griseb* subtribe of the *Andropogoneae Dumort* tribe. Theor Appl Genet, 1994; 87: 843–853.

Spangler R, Zaitchik B, Russo E, Kellog E. Andropogoneae evolution and generic limits in Sorghum (Poaceae) using ndhf sequences. System Bot, 1999; 24: 267–281.

Srivastava BL, Cooper M, Mullins RT. Quantitative analysis of the effect of selection history on sugar yield adaptation of sugarcane clones. Theor Appl Genet, 1994; 87: 627–640.

Stevenson GC. *Genetics and Breeding of Sugar Cane*, 1965; Longmans, London.

Swinford JM, Boevey TMC. The effects of soil compaction due to infield transport on ratoon cane yields and soil physical characteristics. Proc Sth Afr Sugar Tech Ass, 1984; 58: 198–203.

Symington WM. Commercial potential of Macknade nobilizationsfor yield, sugar content and stress tolerance. Proc Aust Soc Sugar Cane Tech, 1989; 11: 48–53.

Tanner CB, Sinclair TR. Efficient water use in crop production: research or re-search? In: Taylor HM, Jordan WR, Sinclair, TR (eds) *Limitations to Efficient Water Use in Crop Production*, 1983, pp. 1–27. ASA, Madison, WI.

Terajima Y, Matsuoka M, Irei S, Sakaigaichi T, Fukuhara S, Ujihara K, Ohara S, Sugimoto A. Breeding for high-biomass sugarcane and its utilization in Japan. Proc Int Soc Sugar Cane Tech, 2007; 26: 759–763.

Tew TL. World sugarcane variety census – year 2000. Sugar Cane International March/April 2003, pp. 12–18.

Tew TL, Cobill RM. Genetic improvement of sugarcane (*Saccharum* spp.) as an energy crop. In: Vermerris W (ed.) *Genetic Improvement of Bioenergy Crops*, 2008, pp. 249–272. Springer, New York.

Thorburn PJ, Horan HL, Biggs JS. The impact of trash management on sugarcane production and nitrogen management: a simulation study. Aust Soc Sugar Cane Technol, 2004; 26: (CD ROM).

Vermerris W. Composition and biosynthesis of lignocellulosic biomass. In: *Genetic Improvement of Bioenergy Crops*, 2008, pp. 89–129. Springer, NY.

Waclawovsky AJ, Sato PM, Lembke CG, Moore PH, Souza, GM. Sugarcane for bioenergy production: an assessment of yield and regulation of sucrose content. Plant Biotech. J, 2010; 8: 263–276.

Walker DIT. Breeding for disease resistance. In: Heinz DJ (ed.) *Sugarcane Improvement through Breeding*, 1987, pp. 7–85. Elsevier, Amsterdam

Walter, A, Ensinas, AV. Combined production of second generation biofuels and electricity from sugarcane residues, Energy, 2010; 35: 874–879.

Wang L-P, Jackson PA, Lu X, Fan Y-H, Foreman JW, Chen X-K, Deng, H-H, Fu C, Ma L, Aitken KS. Evaluation of sugarcane x *S. spontaneum* progeny for biomass composition and yield components. Crop Sci, 2008; 48: 951–961.

Watt D, Butterfield M, Huckett B. Proteomics and metabolomics. In: Henry H, Cole C (eds) *Genetics, Genomics and Breeding of Sugarcane*, 2010, pp. 193–228. CRC Press, Boca Raton, FL.

Wei X, Jackson PA, Hermann S, Kilian A, Heller-Uszynska K, Deomano E. Simultaneously accounting for population structure, genotype by environment interaction and spatial variation in marker-trait associations in sugarcane. Genome, 2010; 53: 973–981.

Wei X-M, Jackson PA, McIntyre CL, Aitken KS, Croft B. Associations between DNA markers and resistance to diseases in sugarcane and effects of population substructure. Theor Appl Genet, 2006; 114: 155–164.

Wei X-M, Jackson PA, Stringer J, Cox M. Relative economic genetic value (rEGV) – an improved selection index to replace net merit grade (NMG) in the Australian sugarcane variety improvement program. Proc Aust Soc Sugar Cane Tech, 2008; 114:174–181.

Welbaum GE, Meinzer FC. Compartmentation of solutes and water in developing sugarcane stalk tissue. Plant Physiol, 1990; 93: 1147–1153.

Wilcox T, Garside A, Braunack M. The sugarcane cropping system. In: Mac H, Peter A (eds) *Manual of Cane Growing*, 2000, pp. 127–140. Bureau of Sugar Experiment Stations, Indooroopilly, Australia.

Wood AW. Management of crop residues following green harvesting of sugarcane in north Queensland, Soil Till Res, 1991; 20: 69–85.

Wu KK, Burnquist W, Sorrels ME, Tew TL, Moore PH, Tanksley SD. The detection and estimation of linkage in polyploids using single dose restriction fragments. Theor Appl Genet, 1992; 83: 294–300.

Chapter 8
# Breeding Maize for Lignocellulosic Biofuel Production

Natalia de Leon, Shawn M. Kaeppler, and Joe G. Lauer

*Department of Agronomy, University of Wisconsin, 1575 Linden Drive, Madison, WI 53706, USA*

## 8.1 Introduction

Bioenergy derived from lignocellulosic feedstocks includes ethanol via fermentation, electricity via combustion, biodiesel via extraction, and various forms of fuel derived by extraction of precursors followed by chemical synthesis and purification. Maize has an early role in the emerging lignocellulosic biofuel industry because it is widely grown across the United States, providing a readily available source of feedstock.

## 8.2 General Attributes of Maize as a Biofuel Crop

The aerial, nongrain portion of a maize plant is called stover, and includes cobs, leaves and leaf sheaths, stalks, and tassels (Figure 8.1). The current value of maize grain is approximately $200 ton$^{-1}$, which is 3–6 times the projected value of lignocellulosic feedstocks in a viable bioeconomy. Farmers can realize added benefit from their crop by harvesting and selling maize cobs and stover, in addition to the grain, with no other inputs beyond those required for producing a high-yielding grain crop. Options to gather cobs and stover are currently in development and might include specialized combines that separate and gather both grain and cobs or stover, or baling discarded stover using currently available machinery (Shinners et al., 2009).

The lignocellulosic components which are most likely to be harvested include cobs, cobs plus husks, and leaves which remain following harvest with an ear-snapper head. A proportion of the aerial part of the plant is harvested by a specialized combine with a cutter head and built-in stover chopper, or discarded stover gathered with a baler following collection of grain (Shinners et al., 2009; Shinners et al., 2011). Cobs represent approximately 10–15% of the total stover and are energy dense, but are prone to spontaneous ignition when stored on large piles.

---

*Bioenergy Feedstocks: Breeding and Genetics*, First Edition. Edited by Malay C. Saha, Hem S. Bhandari, and Joseph H. Bouton.
© 2013 John Wiley & Sons, Inc. Published 2013 by John Wiley & Sons, Inc.

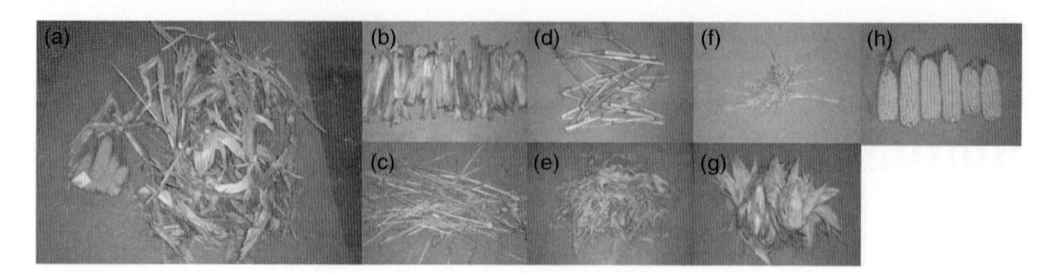

**Figure 8.1.** Maize plant parts and whole plant components at grain physiological maturity stage. (a) Whole plant; (b) leaf sheath; (c) midribs; (d) stalks; (e) leaf blades; (f) tassel; (g) husks; (h) cobs/ears. *(For color details, see color plate section.)*

Plant remnants that are discarded following harvest by a standard combine with an ear-snapper head include cobs, husks, and leaves. These can be harvested using commonly available baling equipment and can be readily stored. This material represents 45–50% of the total stover, is of relatively high quality, and has the benefit of requiring little specialized harvest equipment. A greater proportion of stover can be collected using specialized combines that cut the stalk at a specified height above the ground, and include a chopper to process large plant material. Using this approach, 30–95% of the stover could be collected; however, producers would possibly need to make substantial investments in new equipment.

The amount and type of stover that should be harvested for biofuel production are partially determined by practical considerations such as availability of necessary implements. There is also concern that removing too much stover from fields will detrimentally affect soil health and carbon sequestration. The concentration of soil organic carbon and structural stability are important considerations for environmentally sustainable stover removal, and factors such as minimizing soil erosion and soil nutrient depletion are also considerations (Mann et al., 2002; Wilhelm et al., 2007). Studies have demonstrated that removing more than approximately 25% of the available stover can have significant consequences on the long-term sustainability of the soil. The extent of the impact, however, depends on the specific soil characteristics such as slope and erosion susceptibility, as well as management practices such as adoption of no-till (Graham et al., 2007; Blanco-Canqui and Lal, 2009). Selectively harvesting specific maize plant parts such as cobs or the above-ear portion of the plant is likely to provide a higher-quality feedstock (Hansey et al., 2010) while maintaining soil sustainability, including soil structure and composition (Johnson et al., 2010).

If grain is the primary product of a maize crop and stover a secondary product, the relationship between grain and stover yield is important in predicting future yield trends. Average grain yield has increased 745% from 1930 to the present (USDA-NASS, 2011). Lorenz et al. (2010) reported that the average harvest index has remained relatively constant over the past 50 years indicating that selection for increased grain yield per hectare has resulted in a linear increase in stover yield per unit of land for materials adapted to the U.S. Midwest area. The evaluation of a set of approximately 50 diverse hybrids adapted to the Midwest US showed a positive relationship between harvest index and total biomass production (Figure 8.2). These results indicate that it is reasonable to expect that stover yield will increase at the same rate as grain yield into the future. However, research on experimental and commercial hybrids also suggests that there is substantial variability for harvest index, and that there is the possibility,

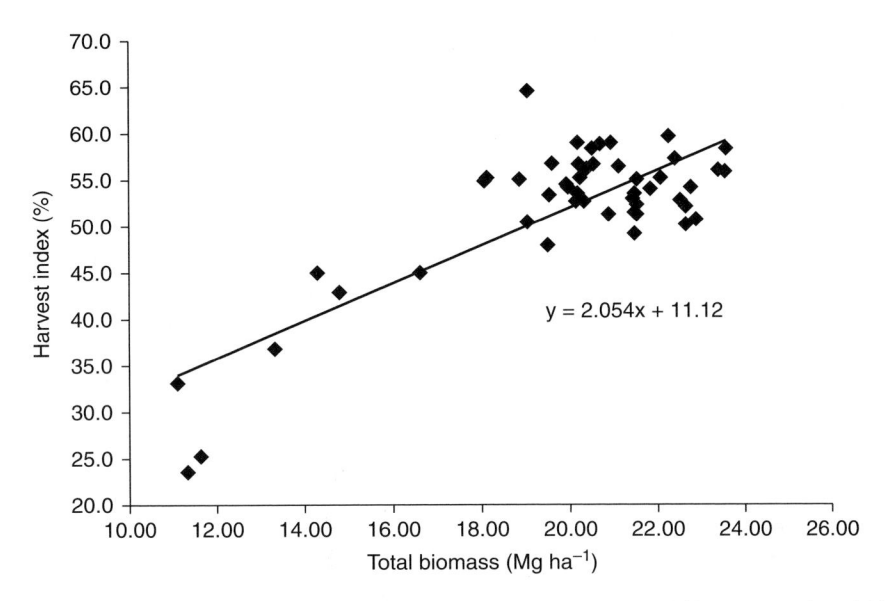

**Figure 8.2.** Relationship between harvest index (ratio of grain over total biomass) and total biomass for a set of 50 cultivars selected to be diverse for grain yield, forage yield, and forage quality. Total biomass includes stover (with cobs) plus grain. Materials were evaluated at Madison and Arlington, WI during 2005 and 2006 (modified from Lorenz et al., 2009b)

even among currently available commercial hybrids, to choose varieties with elite grain yield and high stover yield (Lorenz et al., 2009b)

## 8.3 Potential Uses of Maize Stover for Bioenergy

The current vision of the lignocellulosic biofuel industry has substantial focus on ethanol production (Perlack et al., 2005). Ethanol production is likely to occur by a multistep process including physical, chemical, and or enzymatic deconstruction of biomass, followed by fermentation. Methods of deconstruction currently with high visibility include ammonia fiber expansion (AFEX), dilute acid hydrolysis, liquid hot water, aqueous ammonia, and ionic liquids. Comparison of various methods using switchgrass (Kim et al., 2011) hints at interactions between method, genotype, and harvest time. This study can reasonably be extended to grasses in general, and is consistent with the notion that it will be possible to optimize biomass quality for specific processes by breeding and production practices. Following pretreatment is the fermentation step. Fermentation organisms vary in species and sugar preference (pentoses, hexoses, or both). A goal is development of a consolidated bioprocessing organism that could simultaneously produce deconstruction enzymes and utilize pentose and hexose carbon sources, but such an organism is not yet commercially available. Given a pretreatment step that makes carbohydrates readily available for fermentation, the additional quality traits that are likely to be important are reduction in inhibitory compounds, and optimizing the hexose/pentose proportion to best suit the fermentation organism.

In addition to ethanol production, lignocellulosic biomass may be used to produce energy by other approaches. With the exception of biomass-based processes that operate at high temperatures, such as fast pyrolysis, most biological conversion processes have increased efficiency when less recalcitrant biomass sources are utilized. One viable approach is combustion which is highly efficient at capturing released energy, produces energy indiscriminately from many molecular compounds, and may be the most efficient way to utilize highly recalcitrant plant components or compounds. In fact, recalcitrant compounds such as lignin are already being planned for utilization in ethanol plants to generate energy by burning, including electricity to power the plant and heat to incubate the fermentation organisms. Next-generation biofuels may include compounds other than ethanol for example methanol and butanol. Examples of gas fuels include biogas and syngas. The latter can in turn be transformed into a liquid. In addition to liquid and gas fuels, solid biofuels including dried manure which and charcoal are produced from plant-derived carbon-containing substrates. Specific compositional characteristics of the raw material (feedstock) utilized will directly affect the efficiency of conversion. Certain processes involving, for example, fermentation benefit from feedstocks that possess high carbohydrate concentration as well as lower cell wall recalcitrance (i.e., higher microbial accessibility to the carbohydrates). On the other hand, although processes such as pyrolysis or gasification still require high concentration of carbohydrates, they are less sensitive to levels of cell wall recalcitrance. Examples of such processes, resulting biofuel types and most desirable biomass compositional characteristics for them, are described in Table 8.1.

Exciting opportunities exist to increase the energy density of plant biomass. Lignin has 1.7 times and lipids have 2 times the energy density per weight of cellulose (Ohlrogge et al., 2009). Therefore, strategies based on transgenics and/or natural variation that would increase the proportion of energy-dense compounds have the potential to dramatically improve the economics of producing biofuels. Production of more or novel energy-dense molecules in stable organs such as stalks during the period in which a plant would normally senesce has the dual advantages of maximizing photosynthate accumulation during the full growing season and mitigating the loss of biomass that currently accompanies late-season harvest and biomass storage (Ohlrogge et al., 2009). In this context, metabolic engineering holds substantial promise in engineering novel varieties specifically suited for biofuel production.

## 8.4 Breeding Maize for Biofuels

Maize has served mankind over the past 10,000 years in many capacities and production systems. In addition to being a source of food and feed, maize is now expected to also be an important source of feedstock for energy production. Substantial genetic and phenotypic variation exists in breeding populations and germplasm accessions to rapidly repurpose maize as an important contributor to the developing bioeconomy. The remainder of this chapter will provide further details related to breeding and development of maize varieties for biofuel production.

### 8.4.1 Selection Criteria

Relative performance of varieties for biofuels is most directly calculated as a function of stover yield per unit of land and the quality of that stover relative to specific conversion processes. In addition, various sustainability criteria are also likely to be of relevance such as carbon sequestration, nitrogen efficiency, drought tolerance, and ecosystem services. Below, we will discuss each of these parameters in more detail.

**Table 8.1.** Description of the primary types of second-generation biofuels and related characteristics sought from biomass sources utilized for the production of such biofuel products. This is not intended to be an exhaustive list of all existing biofuel types, but just a highlight of the most commonly found ones (mostly adapted from Luque et al., 2008 and additional sources indicated below). Chemical conversion processes of oil crops and vegetables to biodiesel as well as fructose and glucose to 2,5-dimethylfuran (DMF) are not included in this table since, up to date, they typically do not directly use green biomass as raw material.

| Conversion | Biofuel | Technologies | Raw material | Ideal Biomass Composition | Advantages | Challenges |
|---|---|---|---|---|---|---|
| Biological | Bioethanol | Hydrolysis (for complex sugars) and fermentation | Starch, sugar, and biomass | High carbohydrate concentration <br><br> Low recalcitrance | Higher octane values than gasoline <br><br> Can be mixed with gasoline <br><br> Reductions on carbon dioxide emissions compared to petrol | Hydrophobic Lower energy density than gasoline (~30%) <br><br> More corrosive than gasoline |
| Biological | Biobutanol | Saccharification/ fermentation | Sugar crops and biomass | High carbohydrate concentration <br><br> Low recalcitrance | Lower energy content than gasoline but higher than bioethanol <br><br> Can be used directly in a gasoline engine <br><br> It is less corrosive and less soluble in water than ethanol | Potential toxicity issues |

(continued)

Table 8.1. (Continued)

| Conversion | Biofuel | Technologies | Raw material | Ideal Biomass Composition | Advantages | Challenges |
|---|---|---|---|---|---|---|
| Biological | Biogas (mostly methane, carbon dioxide)[a] | Microbial digestion under anaerobic conditions | Organic materials | High carbohydrate concentration  Low recalcitrance | Utilizes diversity of organic materials  Potential use in transportation and electricity | Relatively high generation and emission of $CO_2$ and $N_2O$ in the process |
| Thermochemical | Syngas or biosynthetic natural gas (mostly combination of carbon monoxide and hydrogen)[b] | Gasification at high temperatures or fast pyrolysis followed by alternating flashes of steam and air | Biomass | High carbohydrate concentration  Less sensitive to recalcitrance levels | Can be directly burned in combustion and it can be transformed to a synthetic petroleum  Can use the whole biomass including lignin | Issues of storability due to its gas state at ambient temperatures  produces large quantities of tars  Half the energy content of natural gas |
| Thermochemical | Biomethanol[b] | Can be produced via gasification of biomass to syngas followed by catalytic synthesis | From biomass via syngas | High carbohydrate concentration  Less sensitive to recalcitrance levels | Easier to recover from syngas compared to ethanol | Highly toxic and flammable  Low volumetric energy density |
| Thermochemical | Synthetic fuels[b] | Several different processes | From biomass via syngas | High carbohydrate concentration  Less sensitive to recalcitrance levels | Cleaner than traditional fuels  Can be mostly used in diesel engines | Process is still relatively expensive |

[a]Khalid et al., 2011 and Lastella et al., 2002.
[b]Munasinghe and Khanal, 2010.

## 8.4.2 Stover Yield

Stover yield per unit of land is determined by many traits that have been under selection relative to grain yield including abiotic and biotic stress tolerance, density tolerance, stability across environments, seedling vigor, and standability. Uniquely, harvest index, resource partitioning, and maturity are likely to be weighted differently in dual-purpose and dedicated biofuel varieties relative to dedicated grain varieties. In addition, hybrids with greater biomass may be prone to stalk and root lodging, so some traits already selected in grain hybrids may become of greater consequence.

Abiotic stresses such as drought and heat are likely to remain important in breeding for improved feedstock crops. Among the most important pathogens, fungal diseases such as Gray Leaf Spot (caused by *Cercospora zeae-maydis*), Northern Leaf Blight (caused by *Exserohilum turcicum*), and Common Rust (caused by *Puccinia sorghi*) as well as root rots (usually caused by a complex of several soil-borne organisms including *Pythium*, *Fusarium*, *Helminthosporium*, among others) are likely to remain important, primarily for grass crops. Ideotypes that favor tall and more voluminous plants are likely to be more common for biofuel cultivars than their primarily grain-producing counterparts. Selection for resistance to pests such as European Corn Borer (caused by *Ostrinia nubilalis*), Fall Armyworm (*Spodoptera frugiperda*), the Western Corn Rootworm (*Diabrotica virgifera*), and the Northern Corn Rootworm (*Diabrotica barberi*), which are among the most common pests in the Midwest of the United States, will be important primarily due to the severe impact they can have on standability.

Due to concerns with the potential impact of stover removal on soil sustainability, intercropping and the utilization of living mulch systems is likely to have a greater role in the production of biofuel maize crops than purely grain-producing systems. In addition to providing soil protection after maize harvest, these cropping systems have the potential to provide more effective weed control early in the season. As with maize varieties grown for grain, biofuel types will be favored by fast canopy closure that provides more aggressive competition to weeds early in the growing season and utilization of living mulches such as legume cover crops will be advantageous for that purpose. The identification of ideal crops to be utilized in such cropping systems and the potential competition for water and other resources from both crops is currently an area of active research (Zemenchik et al., 2000)

Stover yield is determined by the overall size of the plant which is a function of the number of leaves, the length of the internodes, size of organs (e.g., leaf width and length and stalk diameter), and the density of the tissues for a given planting density. In addition, partitioning of resources to grain and stover – harvest index – may also be amenable to selection when considering maize varieties for biofuel production. In a preliminary study in 2010, we surveyed a set of maize inbreds and testcrosses of those inbreds and determined the impact of component traits on stover yield. Plant height and internode length were all positively correlated with stover yield on the inbred, per se evaluation, when flowering time was accounted for as a covariate (Figure 8.3). Substantial variation also exists for stalk diameter, but this variation is manifested to a greater extent in low-density planting.

Later-flowering time is positively correlated with overall grain and stover yield as early flowering genotypes have fewer nodes and leaves (Salvi et al., 2007). Earlier flowering also reduces the length of time that plants can accumulate carbohydrates during the available growing season. Maturity will have to be restricted if a variety is to be used as a dual-purpose crop as the grain must still mature and dry down, both to facilitate harvest as well as minimize additional cost of drying, as this represents a negative factor in calculating energy balance. There has been some interest in the potential of using day-length sensitive varieties that do

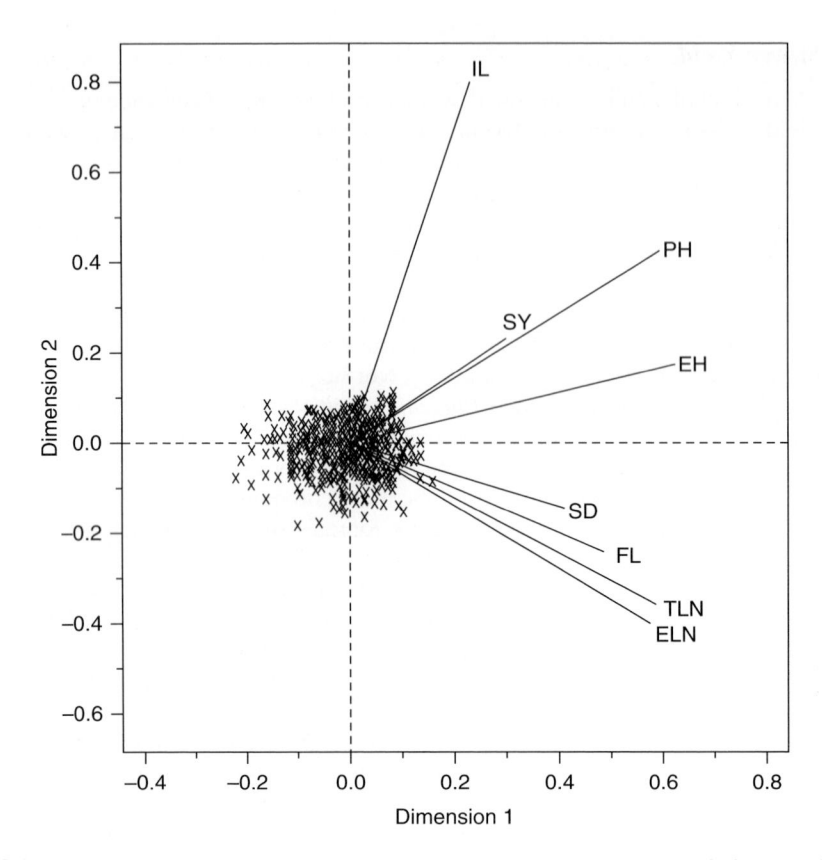

**Figure 8.3.** Biplot analysis of 534 inbreds from the Wisconsin Diverse Association Panel (Hansey et al., 2010) evaluated at one location in South Central Wisconsin with two replications in 2010. PH, plant height; SY, stover yield; IL, internode length; EH, ear height; SD, stalk diameter; FL, flowering time; TLN, total leaf number; ELN, position of the leaf sustaining the ear; X, position of each of the 534 genotypes. Dimension 1 explains 49% of the variation, dimension 2 explains an additional 21% of the variation.

not flower in certain regions, or flower very late, as dedicated biofuel crops. The concept is that plants which do not flower can accumulate biomass until the very end of the season, maximizing use of available sunlight. Available data suggest that while varieties that do not flower in a given environment do accumulate more stover biomass, they accumulate less total biomass than those that produce grain. A study in Wisconsin, including 64 hybrids from the germplasm enhancement of maize (GEM) program, was conducted at three locations during 2009 and 2010. The goal was to evaluate the biomass production of Midwest-adapted and commercial hybrids compared with more Southern-adapted materials and exotic types. The results from this study indicated that total biomass production was not substantially different between the two groups (Table 8.2).

In another experiment to assess the viability of utilizing photoperiod-sensitive (nonflowering) germplasm, we evaluated a series of 25 hybrids in more than 10 locations throughout the state of Wisconsin from 2006 to 2008. Of the 25 hybrids, 10 were adapted to the Wisconsin environments, whereas the other 15 were from tropical origin. The silage yield for the adapted

**Table 8.2.** Biomass yield of 65 hybrids evaluated for stover yield in 2009 and 2010. The hybrids represented a wide range of maturities. Ears were stripped and yield is expressed on a dry matter basis.

| Hybrids | Biomass Yield (Mg ha$^{-1}$) | | |
|---|---|---|---|
| | Average | Minimum | Maximum |
| Midwest adapted | 7.76 | 6.82 | 7.26 |
| Southern adapted | 8.36 | 9.19 | 9.19 |
| | | LSD (0.05) = 0.16 | |

material was either equal or superior to the tropical set (Table 8.3). In addition, even if the total aerial biomass of a nonflowering and a conventional hybrid were equivalent, the disproportionately higher value of the grain still makes use of nonflowering hybrids sensible only in situations where the importance of stover yield per acre supersedes income per acre.

### 8.4.3 Maximum Biomass Yield and the Effects of Time and Latitude

Most US private companies and state universities conduct grain performance trials and some have silage trials. Relatively few maize hybrid trials evaluate biomass production. Therefore, the best estimate for describing biomass yield increases may lie with silage trials. It is important to highlight, however, that silage trial results likely underestimates the maximum potential biomass yield, which is determined nearer to physiological maturity. This is because maize for silage is harvested at an immature stage when grain is still filling. Most maize silage trials are harvested between 50% and 25% kernel milk, therefore an additional 5–10% grain yield would be expected by the end of the season (Afuakwa and Crookston, 1984).

The highest recorded silage yield occurred at Arlington, Wisconsin during 1996 when a hybrid produced a replicated mean of 30.3 Mg ha$^{-1}$. Among highest producing locations in Wisconsin are the southern locations of Lancaster, Arlington, and Madison. Forage yield decreases 1.82 Mg ha$^{-1}$ ($R^2 = 0.68$) for each degree increase in latitude in Wisconsin (Table 8.4). On the other hand the rate of change for maize silage yield over time is increasing.

**Table 8.3.** A set of 25 hybrids evaluated for silage (grain and stover combined) yield and moisture in a combination of 10 in-farm and at the research station locations. Adapted hybrids produced grain in these environments, but the tropical hybrids flower late therefore the majority of the biomass is stover.

| | | Forage Yield Mg ha$^{-1}$ | Forage Moisture g kg$^{-1}$ |
|---|---|---|---|
| 2006 | Adapted | 13.4 | 650 |
| | Tropical | 13.4 | 730 |
| | LSD (0.10) | NS | 30 |
| 2007 | Adapted | 15.0 | 550 |
| | Tropical | 12.4 | 730 |
| | LSD (0.10) | 1.8 | 30 |
| 2008 | Adapted | 16.1 | 540 |
| | Tropical | 12.6 | 760 |
| | LSD (0.10) | 1.3 | 20 |
| | Average hybrid | 17.9 | 640 |

**Table 8.4.** Forage yields of commercial hybrids in the University of Wisconsin corn silage performance trials.

| Location | Latitude Deg. Min. Sec. | Mean Mg ha$^{-1}$ | Max |
|---|---|---|---|
| Antigo | 45° 8′ 8.0″ N | 14.8 | 17.5 |
| Arlington | 43° 17′ 48.2″ N | 21.2 | 30.2 |
| Ashland | 46° 35′ 47.8″ N | 14.2 | 21.6 |
| Chippewa Falls | 44° 57′ 2.3″ N | 16.7 | 22.3 |
| Coleman | 45° 2′ 11.1″ N | 17.6 | 23.2 |
| Fond du Lac | 43° 43′ 35.2″ N | 18.7 | 28.2 |
| Galesville | 44° 4′ 45.0″ N | 20.3 | 28.9 |
| Hancock | 44° 7′ 0.98″ N | 16.3 | 20.6 |
| Lancaster | 42° 49′ 34.7″ N | 19.4 | 28.6 |
| Madison | 43° 4′ 36.9″ N | 21.8 | 27.1 |
| Marshfield | 44° 38′ 28.9″ N | 15.9 | 23.5 |
| Rhinelander | 45° 38′ 13.5″ N | 16.1 | 23.6 |
| Spooner | 44° 29′ 51.9″ N | 15.4 | 28.1 |
| Valders | 44° 2′ 15.7″ N | 16.3 | 25.8 |

Between 1930 and 1995, maize silage yield has increased 0.22 Mg ha$^{-1}$ year$^{-1}$ in Wisconsin (USDA-NASS, 2010). Between 1996 and 2009, Wisconsin silage yield has increased 0.47 Mg ha$^{-1}$ year$^{-1}$. Moreover, the rate of increase varies spatially across the United States (Figure 8.4). Since 1990, most of the silage yield increase has occurred in Wisconsin, Minnesota, South Dakota, and North Dakota. Whereas the rest of the country has seen little change in silage yield.

As previously mentioned, harvest index is a trait that has the potential for manipulation in the context of dual-purpose hybrids (Lorenz et al., 2010). If average grain yields are doubled by a target date such as 2030, then it is reasonable to expect stover yields to concomitantly double. However, within current hybrids of similar maturity and grain-yield potential, there is substantial variation in stover yield (Lorenz et al., 2009b). In general, there has been selection in temperate grain hybrids for moderate height to minimize lodging and to facilitate mechanical harvest.

Traits such as increased lateral branch formation, leaf number, and the exploitation of regrowth capacity of some axillary meristems in grasses are potential alternatives to be utilized by breeders and geneticists (Sakamoto and Matsuola, 2004; Sakamoto et al., 2005; Torney et al., 2007; Hansey and de Leon, 2011). Single genes such as the dominant *Leafy1* (*Lfy1*) provides extra nodes and leaves above the ear, lower ear positioning, and highly lignified stalks in maize (Shaver, 1983) which are traits that can have potential beneficial impact for biomass production. Field evaluation of *Lfy1* hybrids across different environments showed that, on average, these materials surpassed wild-type hybrids in terms of total biomass yields, but had lower harvest index compared to their wild-type counterparts (Dwyer et al., 1998). The concentration of carbohydrate on the portion of the stover above the main ear was superior for *Lfy1* hybrids compared to the wild type, primarily during the periods of anthesis and grain-filling stage.

Increasing the proliferation of lateral branches can also alter the production of biomass in maize. Mutants such as *grassy tiller1* (*gt1*) and *teosinte branched1* (*tb1*) have been associated with the activation of lateral meristems and reduced apical dominance in maize (Doebley et al., 1997; Colasanti, 2001). The presence of abundant long lateral branches and potential increased aboveground biomass production have been seen in plants homozygous for these mutations,

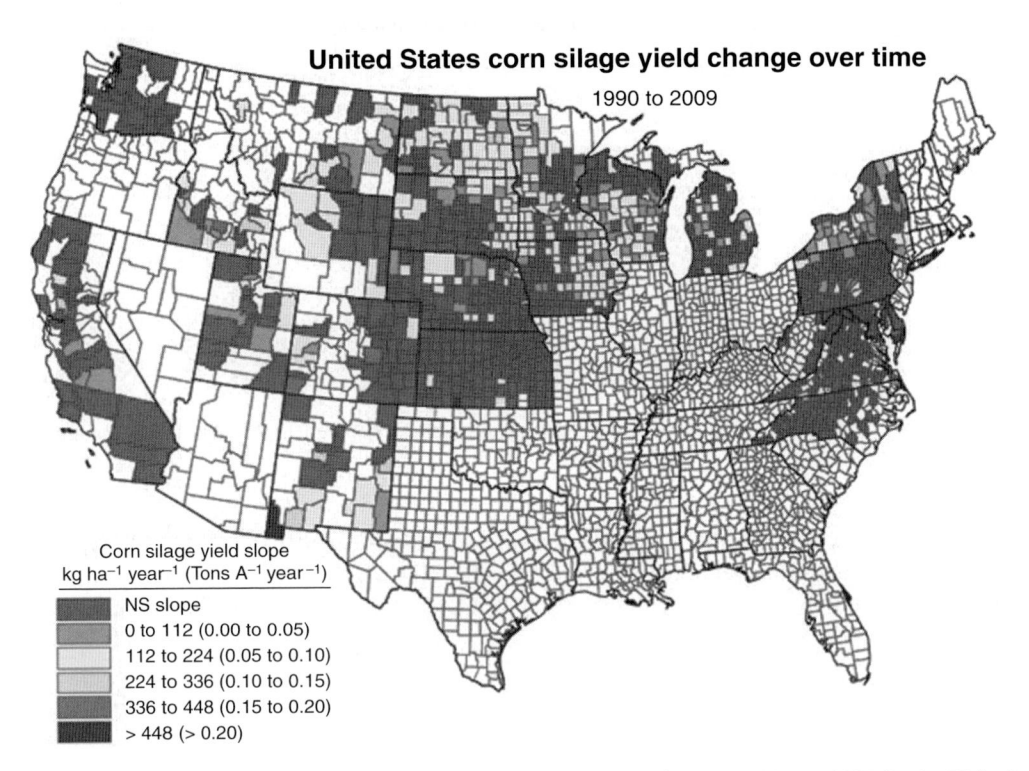

**Figure 8.4.** County average corn silage yield increase (kg ha$^{-1}$) from 1990 to 2009 in the United States. *Source:* USDA-NASS, 2010.

primarily when grown under low planting density conditions (Doebley et al., 1995; Hansey and de Leon, 2011). Figure 8.5 provides a pictorial demonstration of the biomass production potential of tillering relative to single-stalked genotypes.

Another important morphological characteristic differentiating maize from other sources of biofeedstock is the occurrence of a strong sink (ear) that continuously imports carbon and nitrogen assimilates primarily during the grain-filling period. This remobilization process is expected to affect stover composition differently depending on genetic background (Coors et al., 1997; Hirel et al., 2005). About 70% of the nitrogen removed by maize for silage ends up in the grain (Randall and Vetsch, 2003). Grain production is, therefore, energetically more costly than green biomass. As discussed before, this will lead to the proposition that nonflowering ideotypes, such as photoperiod-sensitive materials which will not flower in more temperate locations, for example, could have an important role for the production of feedstock from maize, if stover and grain yield become equally valuable.

### 8.4.4 Stover Quality

The composition of stover, hereafter referred to as stover quality, can dramatically influence the production of bioenergy from a specific lot of material. Quality parameters include the concentration of simple and complex carbohydrates, the proportion of pentoses and hexoses,

**Figure 8.5.** Transversal section of a *gt1* maize plant (left) compared to its wild-type counterpart (right). (*For color details, see color plate section.*)

the concentration of minerals, the concentration and composition of lignin and the types of lignocellulose bonds to other components of the cell wall such as cellulose and hemicellulose, and the concentration and types of inhibitory compounds. The importance of these parameters is highly dependent on the process by which bioenergy is produced from the stover. For example, a process that involves chemical, physical, and/or enzymatic pretreatment followed by fermentation with a yeast strain that utilizes only hexose sugars might be primarily affected by the concentration and availability of that type of carbohydrate. Alternatively, if stover is burned to produce electricity, a completely different parameter, such as the concentration of silicon, might be the most important quality factor. The availability of diverse feedstock conversion technologies including acid hydrolysis, enzymatic hydrolysis, thermochemical methods such as syngas as well as direct combustion demand different feedstock ideotypes (Hamelinck et al., 2005). Increased biomass production is a desirable characteristic underlying all these systems; however, the ideal biomass structure and composition are expected to be system dependent.

The possibility of increasing the polysaccharide to lignin ratio has been proposed as a potential alternative to increase feedstock quality for fermentation-based processes. The compositional evaluation of a large number of samples is necessary for experiments aiming to study genetic variability and realize feedstock improvements. Standard methods for biochemical conversion of cellulosic biomass to ethanol such as simultaneous saccharification and fermentation (SSF) processes are labor intensive and not viable for assessing large numbers of samples. Other assays, however, have been developed that have greater throughput and are more suited for larger sample numbers (Dien et al., 2006; Isci et al., 2008). These evaluations typically mimic the steps of cellulosic ethanol production which involve some type of physical size reduction and pretreatment of the plant biomass followed by enzymatic hydrolysis of cell wall polysaccharides and fermentation of simple sugars released. The measurable result involves either released of glucose and pentose or ethanol production.

Alternatively, researchers might prefer to determine the composition of a specific sample. That would involve determination of structural carbohydrate and lignin concentration and composition. Several protocols have been proposed that generally utilize small tissue amounts and are very high throughput (Decker et al., 2009; Santoro et al., 2010). The information collected from these methods can be used for relative comparisons of diverse samples or to predict relative performance for attainable ethanol yields. The National Renewable Energy Laboratory (NREL; Golden, Colorado, USA) developed a near-infrared spectroscopy (NIRS) calibration that allows for the predictions of constituent concentrations specifically for maize stover (Hames et al., 2003). However, the observed variation in structural composition has not yet been linked to attainable ethanol yield via pretreatment and SSF in maize.

Forage quality protocols primarily used in ruminant nutrition have been shown to be valuable for the evaluation of feedstock conversion potential (Lorenz et al., 2009a). The throughput of all forage protocols have been substantially automated (Vogel et al., 1999) and through the utilization of NIRS technology, a large number of samples can now be processed in a very timely manner.

Breeding has primarily focused on increasing total dry matter digestibility. This, however, has resulted in relatively minor alterations on the cell wall lignification. Major reductions in lignin concentrations have rendered plants with poor agronomic performance and not necessarily higher conversion efficiencies. Recent advances in our understanding of the composition of cell walls, lignin composition, and the specific characteristics of the cross-linking of lignin with other components in the cell wall of specific tissue types are providing potential new direct targets for selection and breeding (Grieder et al., 2012; Jung et al., 2012).

In addition to the genetic improvement of cultivars for the production of high-quality, low-cost biomass, exploring alternative harvesting techniques provides another venue to enhance the ethanol yield per monetary unit spent on feedstock production as well as processing. Selectively removing the most digestible parts of the plant has the potential to maximize the biomass conversion efficiency while allowing for more plant residues to persist on the field, therefore, maintaining soil structure and sustainability. In maize, recent studies have indicated that the stalk represents the highest proportion (approximately 50%) of the dry matter produced by an average plant at grain physiological maturity but also the most recalcitrant. Leaves, conversely, represent between 21% and 30% of the total dry biomass and are more digestible (e.g., Garlock et al., 2009; Pordesimo et al., 2005; Tolera and Sundstol, 1999; Hansey et al., 2010).

### *8.4.5 Sustainability Parameters*

Sustainability of crop production is important for all crops and systems, but parameters such as energy balance and carbon sequestration are especially important in the context of replacing energy from fossil fuels with bioenergy.

A primary challenge with maize, and annual grasses in general, is that annual species requiring nitrogen fertilization fare poorly relative to perennials in terms of energy use, carbon sequestration, and mineral nutrient balance (Fazio and Monti, 2011). If harvested stover is a value added or by-product of a grain production system, then this imbalance is amortized or ameliorated by considering it as an investment that has already been made. Perennializing maize production systems using cover crops (Zemenchik et al., 2000), some of which may also fix nitrogen, can improve the sustainability of the maize production system. Hybrids may require selection for their ability to compete in living mulch systems, although little information is available regarding specific genotype by environment interactions in these systems. Overall, utilization of maize stover for the biofuel industry in dual-purpose production systems will contribute in the near term to providing needed feedstock. In the long term, reasonable arguments can be made for replacement of maize acreage, especially in highly erodable and fragile land, with dedicated perennial biofuel crops.

## 8.4.6 Breeding Methods

In the United States, maize is almost exclusively grown as a row crop and primarily as hybrids derived from the cross of two inbred lines from different heterotic groups. This heterotic group structure has substantial influence on how germplasm is utilized in most maize breeding programs. Heterotic group assignments have been historically based on pedigree information and/or combining ability, primarily as it relates to grain yield. The same heterotic patterns have been maintained in the improvement of maize for silage production.

Silage production is currently the second largest use of maize in the United States following grain production (USDA, 2011). Maize for silage is ideally harvested at around 65% moisture and includes both grain and stover. The grain portion provides a readily available source of easily digestible sugar and, therefore, plays a pivotal role in determining the overall quality of the forage. As such, until the early 1990s, forage maize breeding in the United States emphasized the increase in grain productivity in order to increase the energy content of the overall forage (Allen et al., 2003). However, it is now widely accepted that variability in digestibility of the neutral detergent fiber (NDFD represents the digestibility of the total cell wall concentration determined by Van Soest detergent analysis) is an important determinant on the potential energy contribution from the fiber fraction. On the other hand, harvesting of maize as a source of biomass for biofuel production is expected to occur primarily at grain physiological maturity to allow the collection of grain for food and feed and the remaining available biomass for the production of biofuel. The later harvesting time compared to silage permits reduced moisture content of the stover as well as full maturation of the grain to be commercialized as a co-product. Since the focus of breeding of maize for biofuel production is on biomass yield and composition and not just grain yield, the appropriateness of current heterotic groups in breeding maize for biofuel is still an open question.

Breeding strategies to improve biomass composition for biofuel conversion will depend on the specific process adopted by biorefineries processing these complex carbohydrates. The most dominant method currently proposed for conversion of lignocellulosic biomass to ethanol comprises physical size reduction and thermochemical pretreatment followed by a so-called SSF protocol. The purpose of pretreatment is to increase accessibility to complex carbohydrate components such as cellulose and hemicelluloses. The process of hydrolysis and fermentation of those complex sugars typically follow such pretreatment. Although the overall process is well accepted, details related to key steps in this process still remain unknown. The lack of a clear procedure to process biomass limits the breeder's ability to make long term plan decisions of particular compositional components to specifically focus his/her breeding efforts. Cultivars with greater biomass yield, carbohydrate concentration, and reduced recalcitrance would be better fit for use as sources of feedstocks for energy production and could substantially advance the sustainability of this industry. On the other hand, if other processing methods that do not involve hydrolysis and fermentation such as fast pyrolysis and gasification become prevalent in the industry, a different set of optimal compositional characteristics will be desired in the raw material used for the process (Table 8.1).

More variability has been found for forage yield than for polysaccharide concentration and degradability, therefore breeding efforts to increase biomass yield emerge as the most efficient strategy to increase overall liquid fuel yield per unit of land in production (Kirkpatrick et al., 2006; Lorenz et al., 2009b; Grieder et al., 2012).

Methods for improving dual-purpose crops where both grain and biomass are relevant, such as maize, is more complex than most other dedicated bioenergy crops because the effect that changes to the stover portion of the plant will have on grain yield and quality needs to be considered in the context of their relative economic importance. It is important to emphasize however, that variability for both traits is plentiful (Lorenz et al., 2009b) and there is evidence that grain and stover yield and quality can be improved simultaneously. To simultaneously improve grain and stover yield and quality, a useful strategy might involve some type of independent culling strategy, where grain yield is set at a minimum acceptable level and a combination of biomass yield and different compositional characteristics are then used in an index for a second step in the selection process. The relative weight of the compositional traits will likely be a function of the correlation of each specific measured compositional trait and the final product (liters of liquid fuel). On the other hand, relative weight of biomass yield versus grain yield will be a function of the relative economic value of these two components in the market.

Recent studies highlighted the potential utilization of silage varieties as germplasm source for feedstock production (Lorenz et al., 2009b; Barriere et al., 2009; Grieder et al., 2012). Factors differentiating the top-performing silage varieties from the top-performing feedstock varieties are likely to include physiological process affecting sugar remobilization during the last two weeks of grain filling in maize.

## 8.5 Single Genes and Transgenes

In the context of ethanol production, the most substantial roadblock to maximal utilization of carbohydrates is separation of the cellulose matrix from lignin. Substantial endogenous variation for cell wall recalcitrance has been observed in several crop species (Lundvall et al., 1994; Casler and Vogel, 1999; Rooney et al., 2007; Lorenz et al., 2009b; Bhandari et al., 2010). Such range in digestibility has been primarily associated with variation in total lignin concentration as well as the cross linking of hemicelluloses to lignin by ferulate molecules (Grabber et al., 2008). Breeding for improved forage varieties has relied on such variation and has successfully identified superior genotypes in maize (Frey et al., 2004; Gustafson et al., 2010) as well as other species (Jung et al., 1997; McLaughlin and Kszos, 2005; Anderson et al., 2008).

Large mutations conferring lesions in the lignin pathway, such as the *brown midrib* (*bm*) have been discovered and their effect on the composition of the cell wall has been well characterized in several crop species including maize and sorghum (*Sorghum bicolor* L.), (Barriere and Argillier, 1993; Saballos et al., 2008; Sattler et al., 2010). In maize, four *bm* mutations were identified in the late 1960s, with a fifth one recently discovered (Sattler et al., 2010). These mutations present a characteristic reddish coloration on their midribs and sometimes also stalks, the cause of which is still not completely understood. Two of the four *bm* mutations in maize, the *bm1* and *bm3* genes, have been identified and the other two, *bm2* and *bm4*, have been genetically mapped to bins 1.11 and 9.07 of the maize genome, respectively.

Maize mutant *bm1* affects the activity of the cinnamyl alcohol dehydrogenase (CAD) which is the last enzyme in the lignin monolignol pathway. CAD function involves catalyzing the reduction of cinnamyl aldehydes (coniferyl, coumaryl, and sinapyl aldehydes) to their analogous cinnamyl alcohols prior to the incorporation of these monomers into the lignin polymer. Subsequently, significant reduction on all, H, G, and S subunits are commonly observed in plants carrying the *bm1* mutation. Such reductions are partially restored by the incorporation of coniferyl and sinapyl aldehydes into the lignin polymer. CAD enzymatic activity has been shown to be reduced in both below- and aboveground plant organs, but not completely turned off. This is an indication that the *bm1* is not a null mutation of CAD, but rather an alteration in the transcriptional machinery located outside of the coding region of this gene (Halpin et al., 1998; Marita et al., 2003). The *bm1* locus has been mapped to the orthologous ZmCAD2 gene in maize; however, a specific mutation associated with this gene has not yet been identified. Studies of the CAD2 family across several species such as tobacco, Arabidopsis, rice, among others, have demonstrated its essential function for the synthesis of monolignol (reviewed by Sattler et al., 2010).

Maize mutant *bm3* is the most widely used mutation commercially. This mutant encodes a caffeic O-methyltransferases (COMTs) and it is a member of the conserved O-methyltransferase family. The function of genes from this family has been identified in dicots and monocots. COMT affects the penultimate step of monolignol biosynthesis by transferring a methyl group from S-adenosyl-methionine to the 5-hydroxyl group of 5-hydroxy-coniferyl substrates to form sinapyl products. This substitution renders substantial reduction on the syringyl lignin and slight reduction of p-hydroxyphenyl-monomeric and guaiacyl monomeric composition of lignin of *bm3* plants. Additionally, elevated levels of a novel lignin monomer, 5-hydroxy guaiacyl, resulting from the reduction of COMT activity has been observed which subsequently generate accumulation of 5-hydroxy coniferyl alcohol in these mutant plants (reviewed by Sattler et al., 2010).

Although maize mutant *bm2* has not been cloned yet, similar phenotypic expression of the homolog of the sorghum gene has suggested that this mutant typically presents unaltered levels of the H-subunit, significant reductions on the G-subunit, and either unaltered or increased presence of the S-subunit compared to nonmutant types. Contrasting with *bm1* and *bm3* mutant plants, plants carrying the *bm2* mutation show no observable accumulation of unusual subunit types of the lignin polymer (Chabbert et al., 1994b; Vermerris and Boon, 2001; Marita et al., 2003; Sattler et al., 2010). Subtle changes in lignin composition have been observed in *bm4* mutant plants compared to their wild-type counterparts. As in the case of *bm2*, cloning of the *bm4* mutant has not yet been accomplished (Marita et al., 2003; Sattler et al., 2010).

Cell walls of plants carrying different *bm* mutations, especially the *bm3* and to some extent *bm1*, typically have improved digestibility for ruminants (reviewed by Barriere and Argillier, 1993) and improved ethanol potential from lignocellulosic biomass (Lorenz et al., 2009b). Similar examples can be found in alfalfa where the downregulation of key enzymes in the lignin pathway has produced alfalfa (*Medicago sativa*) plants with increased conversion efficiencies (Jackson et al., 2008).

Significant variation on the expression of the *bm3* gene has been documented (Miller et al., 1983; Lee and Brewbaker, 1984; Gentinetta et al., 1990; Brenner et al., 2010) in maize. Additionally, associated reductions on whole plant as well as grain yields (Barriere and Argillier, 1993; Lorenz et al., 2009b; Sattler et al., 2010) and detrimental effects on plant protection traits (Lauer and Coors, 1997; reviewed by Pedersen et al., 2005) are commonly observed in many genetic backgrounds when *bm3* is introduced. Modern *bm* maize hybrids have clearly improved overall phenotypic and agronomic characteristics and, therefore, provide more consistent productivity performance compared to older types; nevertheless, overall performance is generally inferior to non-*bm* counterparts (personal observation).

Most recently, the identification of the *seedling ferulate ester* (*sfe*) maize mutant represents another potential source of improved digestibility as this mutation reduces cross-linking of hemicelluloses and precursors of lignin, therefore, increasing accessibility of microorganisms to the cell wall (Jung and Phillips, 2010). An initial evaluation of this mutation revealed that the decreased presence of ferulate esters in seedlings produced adult plants with reduced cross-linking of ferulate and ether which consequently increased the digestibility of *sfe* mutant plants while only producing a slight decrease in cell wall content. Interestingly, in contrast with observations made with *bm* mutants, primarily, *bm3*, *sfe* does not seem to adversely affect yield and agronomic performance of the mutant plants compared to their wild-type counterparts (Jung and Phillips, 2010).

## 8.6 Future Outlook

Maize stover is a bountiful agricultural residue in the United States and a potentially very important source of feedstock for the production of liquid fuels and other sources of bioenergy (Perlack et al., 2005). Among the most promising aspects of using maize stover as a feedstock is the ability of commercializing the stover as a source of biomass in addition to grain, which already is a high-value commodity. Given the economic importance of maize grain as a commodity worldwide, the ability for these plants to produce larger amounts of biomass with high conversion efficiency potential will have to occur while still maintaining the desirable grain productivity, at least in high maize production areas such as the U.S. Maize Belt. Other models, such as nonflowering maize varieties and biomass ideotypes are likely to have a place in areas of the United States and the world where grain commercialization is limiting.

The most feasible short- to medium-term goals for improving maize for biomass production are likely to involve increasing maize biomass production by modifying plant architecture, density tolerance, resistance to abiotic stresses, and nutrient acquisition. In addition to the economic importance and its potential to be immediately useful for the production of biomass for biofuel production, maize is an excellent monocot model system for other dedicated bioenergy grass crops such as switchgrass and Miscanthus.

## References

Afuakwa JJ, Crookston RK. Using the kernel milk line to visually monitor grain maturity in maize. Crop Sci, 1984; 24: 687–691.

Allen MS, Coors JG, and Roth GW. In: Buxton DR, Muck RE, Harrison JH (eds) *Silage Science and Technology*, 2003, pp. 547–608. ASA–CSSA–SSSA, Madison, WI.

Anderson WF, Casler MD, Baldwin BS. Improvement of perennial forage species as feedstock for bioenergy. In: Vermerris W (ed.) *Genetic Improvement of Bioenergy Crops*, 2008; 2: 347–376. Springer Science+Business Media, NY.

Barriere Yves, Argillier O. Brown-midrib genes of maize review. Agronomie, 1993; 13: 865–876.

Barriere, Yves, Mechin V, Riboulet C, Guillaumie S, Thomas J, Bosio M, Fabre F, Goffner D, Pichon M, Lapierre C, Martinant J-P. Genetic and genomic approaches for improving biofuel production from maize. Euphytica, 2009; 170: 183–202.

Bhandari HS, Saha MC, Bouton JH. Genetic variation in lowland switchgrass (Panicum virgatum L.). In: Huyghe C (ed.) *Sustainable use of Genetic Diversity in Forage and Turf Breeding*, 2010; 2: 67–71. Springer Science+Business Media, NY.

Blanco-Canqui H, Lal R. Corn stover removal for expanded uses reduces soil fertility and structural stability. Soil Sci Soc Am J, 2009; 73: 418–426.

Brenner EA, Zein I, Chen Y, Andersen JR, Wenzel G, Ouzunova M, Eder J, Darnhofer B, Frei U, Barrière Y, Lübberstedt T. Polymorphisms in O-methyltransferase genes are associated with stover cell wall digestibility in European maize (*Zea mays* L.). BMC Plant Biol, 2010; 10: 27.

Casler MD, Kenneth PV. Accomplishments and impact from breeding for increased forage nutritional value. Crop Sci, 1999; 39: 12–20.

Chabbert B, Tollier MT, Monties B, Barriere Y, Argillier O. Biological variability in lignification of maize: expression of the *brown midrib bm2* mutation. J Sci Food Agr, 1994b; 64: 455–460.

Colasanti J. Some observations on the *grassy tillers (gt1)* mutant. MNL, 2001; 75: 2–3.

Coors JG, Albrecht KA, Bures EJ. Ear-fill effects on yield and quality of silage corn. Crop Sci, 1997; 37: 243–247.

Decker SR, Brunecky R, Tucker MP, Himmel ME, Selig MJ. High-throughput screening techniques for biomass conversion. Bioenergy Res, 2009; 2: 179–192.

Dien BS, Jung HJG, Vogel KP, Casler MD, Lamb JFS, Iten L, Mitchell RB, Sarath G. Chemical composition and response to dilute acid pretreatment and enzymatic saccharification of alfalfa, reed canary grass, and switchgrass. Biomass Bioenerg, 2006; 30: 880–891.

Doebley J, Stec A, Gustus C. *teosinte branched1* and the origin of maize: evidence for epistasis and the evolution of dominance. Genetics, 1995; 141: 333–346.

Doebley J, Stec A, Hubbard L. The evolution of apical dominance in maize. Nature, 1997; 386: 485–488.

Dwyer LM, Stewart DW, Glenn F. Silage yields of leafy and normal hybrids. In: *53rd Proceedings of Annual Corn and Sorghum Research Conference Chicago, IL*, 1998, pp. 193–216. American Seed Trade Association, Washington, DC.

Fazio S, Monti A. Life cycle assessment of different bioenergy production systems including perennial and annual crops. Biomass Bioenerg 2011; 35: 4868–4878.

Frey, TJ, Coors JG, Shaver RD, Lauer JG, Eilert DT, Flannery PJ. Selection for silage quality in the Wisconsin quality synthetic and related maize populations. Crop Sci, 2004; 44: 1200–1208.

Garlock RJ, Chundawat SPS, Balan V, Dale BE. Optimizing harvest of corn stover fractions based on overall sugar yields following ammonia fiber expansion pretreatment and enzymatic hydrolysis. Biotechnol Biofuels, 2009; 2: 29.

Gentinetta E, Bertolini M, Rossi I, Lorenzoni C, Motto M. Effect of brown midrib-3 mutant on forage quality and yield in maize. J Genet Breed, 1990; 44: 21–26.

Grabber JH, Mertens DR, Kim H, Funk C, Lu F, John R. Cell wall fermentation kinetics are impacted more by lignin content and ferulate cross-linking than by lignin composition. J Sci Food Agric, 2008; 89: 122–129.

Graham RL, Nelson R, Sheehan J, Perlack RD, Wright LL. Current and potential U.S. corn stover supplies. Agron J, 2007; 99: 1–11.

Grieder C, Dhillon BS, Schipprack W, Melchinger AE. Breeding maize as biogas substrate in Central Europe: I. Quantitative-genetic parameters for testcross performance. Theor Appl Genet, 2012; 124: 971–980.

Gustafson T, Coors JG, de Leon N. Evaluation of S2-topcross selection for maize (Zea mays L.) silage yield and quality in the Wisconsin quality synthetic population. Crop Sci, 2010; 50: 1795–1804.

Halpin C, Holt K, Chojecki J, Oliver D, Chabbert B, Monties B, Edwards K, Barakate A, Foxon GA. *Brown-midrib* maize (*bm1*) – a mutation affecting the cinnamyl alcohol dehydrogenase gene. Plant J, 1998; 14: 545–553.

Hamelinck CN, van Hooijdonk G, Faaij APC. Ethanol from lignocellulosic biomass: techno-economic performance in short-, middle- and long-term. Biomass Bioenerg, 2005; 28: 384–410.

Hames BR, Thomas SR, Amy D, Sluiter, Roth CJ, Templeton DW. Rapid biomass analysis—New tools for compositional analysis of corn stover feedstocks and process intermediates from ethanol production. Appl Biochem Biotechnol, 2003; 105: 5–16.

Hansey CN, de Leon N. Biomass yield and cell wall composition of corn (*Zea mays* L.) with alternative morphologies planted at variable densities. Crop Sci, 2011; 51: 1005–1015.

Hansey CN, Lorenz AJ, de Leon N. Cell wall composition and ruminant digestibility of various maize tissues across development. Bioenerg Res, 2010; 3: 28–37.

Hirel B, Martin A, Terce-Laforgue T, Gonzalez-Moro MB, Estavillo JM. Physiology of maize I: A comprehensive and integrated view of nitrogen metabolism in a C4 plant. Physiol Plantarum, 2005; 124: 167–177.

Isci A, Murphy PT, Anex RP, Moore KJ. A rapid simultaneous saccharification and fermentation (SSF) technique to determine ethanol yields. Bioenerg Res, 2008; 1: 163–169.

Jackson LA, Shadle GL, Zhou R, Nakashima J, Chen F, Dixon RA. Improving saccharification efficiency of alfalfa stems through modification of the terminal stages of monolignol biosynthesis. Bioenerg Res, 2008; 1: 180–192.

Johnson, JMF, Wilhelm WW, Karlen DL, Archer DW, Wienhold B, Lightle DT, Laird D, Baker J, Ochsner TE, Novak JM, Halvorson AD, Arriaga F, Barbour N. Nutrient removal as a function of corn stover cutting height and cob harvest. Bioenerg Res, 2010; 3: 342–352.

Jung H-J, Phillips RL. Putative *Seedling Ferulate Ester* (*sfe*) maize mutant: morphology, biomass yield, and stover cell wall composition and rumen degradability. Crop Sci, 2010; 50: 403–418.

Jung H-J, Samac DA, Sarath G. Modifying crops to increase cell wall digestibility. Plant Sci, 2012; 185–186: 65–77.

Jung H-J, Sheaffer CC, Barnes DK, Halgerson JL. Forage quality variation in the U.S. alfalfa core collection. Crop Sci, 1997; 37: 1361–1366.

Kim Y, Mosier NS, Ladisch MR, Pallapolu VR, Lee YY, Garlock R, Balan V, Bale BE, Danohoe BS, Vinzant TB, Elander RT, Falls M, Sierra R, Holtzapple MT, Shi J, Ebrik MA, Redmond T, Yang B, Wyman CE, Warner RE. Comparative studies of enzymatic digestibility of switchgrass varieties and harvests processed by leading pretreatment technologies. Bioresource Technol, 2011; 102: 11089–11096.

Khalid A, Arshad M, Anjum M, Mohmood T, Dawson L. The anaerobic digestion of solid organic waste. Waste Manage, 2011; 31:1737–1744.

Kirkpatrick KM, Lamkey KR, Paul Scott M, Moore KJ, Haney LJ, Coors JG, Lorenz AJ. Identification and characterization of maize varieties with beneficial traits for biobased industries. International Plant Breeding Symposium, 2006, Mexico City, Mexico. p. 127. (http://www.cimmyt.org/english/docs/proceedings/IPBS06-Abstracts.pdf)

Lastella GC, Testa, Comacchia G, Notomicola M, Voltasio F, Sharma VK. Anaerobic digestion of semi-solid organic waste: biogas production and its purification. Energ Convers Manage, 2002; 43:63–75.

Lauer JG, Coors JG. Brown Midrib Corn. Field Crop Res, 1997; 28: 31–11.

Lee MH, Brewbaker JL. Effects of brown midrib-3 on yields and yield components of maize. Crop Sci, 1984; 24: 105–108.

Lorenz AJ, Anex RP, Isci A, Coors JG, de Leon N, Weimer PJ. Forage quality and composition measurements as predictors of ethanol yield from maize stover. Biotechnol Biofuels, 2009b; 2: 5.

Lorenz AJ, Coors JG, de Leon N, Wolfrum EJ, Hames BR, Sluiter AD, Weimer PJ. Characterization, genetic variation, and combining ability of maize traits relevant to the production of cellulosic ethanol. Crop Sci, 2009a; 49: 85–98.

Lorenz AJ, Gustafson T, de Leon N, Coors JG. Breeding maize for a bioeconomy: A literature survey examining harvest index and stover yield and their relationship to grain yield. Crop Sci, 2010; 50: 1–12.

Lundvall JP, Buxton DR, Hallauer AR, George JR. Forage quality variation among maize inbreds: *In vitro* digestibility and cell wall components. Crop Sci, 1994; 34: 1672–1678.

Luque R, Herrero-Davila L, Campelo JM, Clark JH, Hidalgo JM, Luna D, Marinas JM, Romero AA. Biofuels: a technological perspective. Energy Environ Sci, 2008; 1:542–564.

Mann L, Tolbert V, Cushman J. Potential environmental effects of corn (*Zea mays* L.) stover removal with emphasis on soil organic matter and erosion. Agr Ecosyst Environ, 2002; 89: 149–166.

Marita JM, Vermerris W, Ralph J, Hatfield RD. Variations in the cell wall composition of maize brown midrib mutants, J Agric Food Chem, 2003; 51: 1313–1321.

McLaughlin SB, Kszos LA. Development of switchgrass (*Panicum virgatum*) as a bioenergy feedstock in the United States. Biomass Bioenerg, 2005; 28: 515–535.

Miller JE, Geadelmann JL, Marten GC. Effect of the brown midrib-allele on maize silage quality and yield. Crop Sci, 1983; 23: 493–496.

Munasinghe PC, Khanal SK. Biomass-derived syngas fermentation into biofuels: opportunities and challenges. Bioresource Tech, 2010; 101:5013–5022.

Ohlrogge J, Allen D, Berguson B, DellaPenna D, Shachar-Hill Y, Stymne S. Driving on biomass. Science, 2009; 324: 1019–1020.

Pedersen JF, Vogel KP, Funnell DL. Impact of reduced lignin on plant fitness. Crop Sci, 2005; 45: 812–819.

Perlack RD, Wright LL, Turhollow AF, Graham RL, Stokes BJ, Erbach DC. Biomass as feedstock for a bioenergy and bioproduct industry: The technical feasibility of a billion-ton annual supply. DOE & USDA, 2005.

Pordesimo LO, Hames BR, Sokhansanj S, Edens WC. Variation in corn stover composition and energy content with crop maturity. Biomass Bioenerg, 2005; 28: 366–374.

Randall G, Vetsch J. Assessing Soil N Availability Using the Illinois Nitrogen Soil Test for Corn After Soybeans, 2003. http://sroc.cfans.umn.edu/prod/groups/cfans/@pub/@cfans/@sroc/@research/documents/asset/cfans_asset_165290.pdf (accessed on 21 December 2012).

Rooney WL, Blumenthal J, Bean B, Mullet JE. Designing sorghum as a dedicated bioenergy feedstock. Biofuel Bioprod Bior, 2007; 1: 147–157.

Saballos A, Vermerris W, Rivera L, Ejeta G. Allelic association, chemical characterization and saccharification properties of *brown midrib* mutants of sorghum (*Sorghum bicolor* (L.)Moench). Bioenerg Res, 2008; 1: 193–204.

Sakamoto T, Matsuola M. Generating high-yielding varieties by genetic manipulation of plant architecture. Curr Opin Biotech, 2004; 15: 144–147.

Sakamoto T, Morinaka Y, Ohnishi T, Sunohara H, Fujioka S, Ueguchi-Tanaka M, Mizutani M, Sakata K, Takatsuto S, Yoshida S, Tanaka H, Kitano H, Matsuoka M. Erect leaves caused by brassinosteroid deficiency increase biomass production and grain yield in rice. Nat Biotechnol, 2005; 24: 105–109.

Salvi S, Sponza G, Morgant M, Tomes D, Niu X, Fengler KA, Meeley R, Ananiev EV, Svitashev S, Bruggemann E, Li B, Hainey CF, Radovic S, Zaina G, Rafalski J-A, Tingey SV, Miao G-H, Phillips RL, Tuberosa R. Conserved noncoding genomic sequences associated with a flowering-time quantitative trait locus in maize. PNAS, 2007; 104: 11376–11381.

Santoro N, Cantu SL, Tornqvist C-E, Falbel TG, Bolivar JL, Patterson SE, Pauly M, Walton JD. A high-throughput platform for screening milligram quantities of plant biomass for lignocellulose digestibility. Bioenerg Res, 2010; 3: 93–102.

Sattler SE, Funnell-Harris DL, Pedersen JF. Brown midrib mutations and their importance to the utilization of maize, sorghum, and pearl millet lignocellulosic tissues. Plant Sci, 2010; 178: 229–238.

Shaver DL. Genetics and breeding of maize with extra leaves above the ear. In: *38th Proceedings of Annual Corn and Sorghum Research Conference, Chicago, IL*, 1983, pp. 161–180. American Seed Trade Association, Washington DC.

Shinners K, Boettcher GC, Hoffman DS, Munk JT, Weimer PJ. Single pass harvest of corn grain and stover: Performance of three harvester configurations. T ASABE, 2009; 52: 51–60.

Shinners KJ, Wepner AD, Muck RE, Weimer PJ. Aerobic and anaerobic storage of single-pass, chopped corn stover. Bioenerg Res, 2011; 4: 61–75.

Tolera A, Sundstol F. Morphological fractions of maize stover harvested at different stages of grain maturity and nutritive value of different fractions of the stover. Anim Feed Sci Technol, 1999; 81: 1–16.

Torney F, Moeller L, Scarpa A, Wang K. Genetic engineering approaches to improve bioethanol production form maize. Curr Op Biotech, 2007; 18: 193–199.

USDA – United States Department of Agriculture. National Agricultural Statistics Service. Crop Production 2010 Summary, 2011. http://usda01.library.cornell.edu/usda/nass/CropProdSu//2010s/2011/CropProdSu-01-12-2011_revision.pdf (accessed on 21 December 2012).

USDA-NASS. National Agricultural Statistics Service [Online], 2010. USDA-NASS, Washington, DC. Available at www.nass.usda.gov/index.asp (accessed on 11 June 2010; verified 11 June 2010)

USDA-NASS. National Agricultural Statistics Service [Online], 2011. USDA-NASS, Washington, DC. Available at www.nass.usda.gov/index.asp (accessed on 20 August 2012; verified 20 August 2012).

Van Soest PJ. Nutritional Ecology of the Ruminant Cornell University Press. 1994.

Vermerris W, Boon JJ. Tissue-specific patterns of lignification are disturbed in the *brown midrib2* mutant of maize (*Zea mays* L.), J Agric Food Chem, 2001; 49: 721–728.

Vogel KP, Pedersen JF, Masterson SD, Toy JJ. Evaluation of a filter bag system for NDF, ADF, and IVDMD forage analysis. Crop Sci, 1999; 39: 276–279.

Wilhelm WW, Johnson JMF, Karlen DL, Lightle DT. Corn stover to sustain soil organic carbon further constrains biomass supply. Agron J, 2007; 99: 1665–1667.

Zemenchik RA, Albrecht KA, Boerboom CM, Lauer JG. Corn production with Kura clover as a living mulch. Agron J, 2000; 92: 698–705.

**Plate 2.6.** Hybrid vigor in lowland switchgrass. The large central plant is the progeny of a cross and is contrasted with the relatively diminutive flanking plants that are the progeny of self-pollination of one of the hybrid's parents.

**Plate 4.1.** *Miscanthus × giganteus* "Illinois clone" cultivated at Sapporo, Japan.

---

*Bioenergy Feedstocks: Breeding and Genetics*, First Edition. Edited by Malay C. Saha, Hem S. Bhandari, and Joseph H. Bouton.
© 2013 John Wiley & Sons, Inc. Published 2013 by John Wiley & Sons, Inc.

**Plate 4.2.** Major *Miscanthus* species: (a) *M. sinensis*, (b) *M. sacchariflorus,* (c) *M. sinensis* var. *condensatus*, and (d) *M. floridulus* in wild population in Japan.

**Plate 5.1.** The four main types of *Miscanthus* used in breeding for biomass are (a) *M. sacchariflorus* (shown is Japanese type, tetraploid), (b) *M. sinensis* (typical for all Asia, diploid), (c) *M. sacchariflorus* (shown is Chinese type, diploid), (d) *M. floridulus* (shown is Taiwanese type, diploid) growing in the field in Aberystwyth, Wales, UK. Dr. Lin Huang is 1.6 m tall.

**Plate 5.3.** The diversity trial at Aberystwyth. Each of 244 *M. sinensis*, *M. sacchariflorus*, and hybrid genotypes is replicated in four randomized blocks.

**Plate 5.4.** Variation in spring emergence from the overwintering rhizomes in four selected genotypes at Aberystwyth (photos taken on 18 April, 2009). These exemplar plants are growing in the same phenotype diversity trial at Aberystwyth as shown in Figure 5.3. Japanese genotypes (a) and (b) are variants of *M. sacchariflorus* and are later emerging than (c) and (d), which are variants of *M. sinensis*. In (c) there is evidence of low-temperature chlorosis (yellowing), while (d), a selection made in Sweden, has both early canopy closure and dark green leaves.

**Plate 6.1.** Panicles of five primary race divisions of sorghum beginning in the lower left clockwise bicolor, kafir, durra, caudatum, and guinea.

**Plate 6.3.** The effect of the sorghum conversion program on adaptation in more temperate climates (College Station, Texas). Genotype 1 is T × 406 which is the donor parent for dwarfing and maturity insensitivity genes. Genotype 2 is the converted version of IS5670C. Genotype 3 is the unconverted IS5670.

**Plate 8.1.** Maize plant parts and whole plant components at grain physiological maturity stage. (a) Whole plant; (b) leaf sheath; (c) midribs; (d) stalks; (e) leaf blades; (f) tassel; (g) husks; (h) cobs/ears.

**Plate 8.5.** Transversal section of a *gt1* maize plant (left) compared to its wild-type counterpart (right).

**Plate 9.1.** Prairie cordgrass at Urbana, IL, August 2012.

**Plate 9.2.** Big bluestem at Urbana, IL, August 2012.

**Plate 9.3.** Little bluestem at Urbana, IL, August 2012.

**Plate 9.4.** Eastern gamagrass at Urbana, IL, August 2012.

**Plate 12.1.** Effect of *Sebacina vermifera* on early growth of switchgrass (after 6 weeks of cocultivation). One-month-old rooted switchgrass clones of NF/GA-993 were cocultivated with two different strains (MAFF 305828 and MAFF 305830) of *S. vermifera*.

Chapter 9
# Underutilized Grasses

Arvid Boe[1], Tim Springer[2], D.K. Lee[3], A. Lane Rayburn[3], and J. Gonzalez-Hernandez[1]

[1] South Dakota State University, Brookings, SD 57007, USA
[2] Southern Plains Range Research Station, Woodward, OK 73801, USA
[3] University of Illinois, Urbana, IL 61801, USA

## 9.1 Introduction

Perennial warm-season grasses have been recognized for having several properties, such as high rates of net photosynthesis, energy and labor savings, and reduced soil and nutrient losses, which make them better suited for biofuel production than many annual crops (U.S. DOE, 2006). Two species of perennial grass that have been used as energy crop models are switchgrass (*Panicum virgatum* L.) and Miscanthus (*Miscanthus* × *giganteus*). Although both species are deemed suitable for biomass production on marginal land, neither is well adapted to land that is marginal for conventional crops due to poorly drained saline soils or coarse-textured droughty soils. For example, about $15 \times 10^6$ ha of poorly drained land are considered to have severe limitations for cropping (Land Capability Classes IIIw, IVw, Vw, and VIw), yet they are being farmed. In addition, 560,000 ha in the CRP would fit into that same category. Together, those two land uses account for about 50% of the total area of land within those capability classes in the United States (Table 9.1). The other two major land use categories, pasture and rangeland, are composed of perennial vegetation not amenable to conversion to perennial grass biomass feedstock production.

The species covered here, that is, prairie cordgrass (*Spartina pectinata* Link), big bluestem (*Andropogon gerardii* Vitman), little bluestem (*Schizachyrium scoparium* (Michx.) Nash), sand bluestem (*Andropogon hallii* Hack.), and eastern gamagrass (*Tripsacum dactyloides* (L.)) have potential to expand the area of marginal land on which production of perennial grass lignocellulosic feedstocks can be profitable, sustainable, and environmentally beneficial. These species are ideally adapted to occupy and be productive in agricultural and environmental niches where switchgrass and miscanthus are not adapted. Such niches tend to be on lower and upper landscape positions where soil moisture levels, either excessive or deficient, limit the number of species that can persist and be productive. Also, the use of multiple species enhances biodiversity, stabilizes yields across spatial and temporal variation, and increases environmental benefits on heterogeneous landscapes (Gonzalez-Hernandez et al., 2009).

---

*Bioenergy Feedstocks: Breeding and Genetics*, First Edition. Edited by Malay C. Saha, Hem S. Bhandari, and Joseph H. Bouton.
© 2013 John Wiley & Sons, Inc. Published 2013 by John Wiley & Sons, Inc.

Table 9.1. Land cover/use (hectares) on non-Federal rural land by Land Capability Class (LCC) and subclass (w, poor drainage) in 2007, excluding forest land and other rural land.

| LCC | Crop | CRP | Pasture | Rangeland |
| --- | --- | --- | --- | --- |
| IIIw | 11,061,320 | 340,038 | 2,959,821 | 1,346,463 |
| IVw | 2,339,442 | 112,185 | 1,591,245 | 1,407,780 |
| Vw | 1,024,731 | 58,426 | 1,420,902 | 2,100,897 |
| VIw | 548,208 | 50,382 | 519,170 | 1,812,699 |

Adapted from U.S. Department of Agriculture, 2009.

## 9.2 Prairie Cordgrass

### 9.2.1 Importance

Prairie cordgrass (Figure 9.1) is native to all of North America north of Mexico up to 60°N except for the Gulf States and AZ, CA, and NV in the United States (Barkworth et al., 2007). It is dominant in low prairie where soils are too wet (Land Capability Classes IVw and Vw) for switchgrass, maize (*Zea mays* L.) and other grain, forage, and biofuel crops. Weaver (1954, 1960) described a prairie cordgrass-dominated community that covered hundreds of hectares along rivers and tributaries in the tallgrass prairie. It also frequently occurs in open dry prairie, on high ground along railroad rights-of-way (Mobberley, 1956), and in roadside drainage channels, especially on alluvial soils, in the northern Great Plains and Midwest. Several other native North American species of *Spartina*, most notably saltmeadow cordgrass (*S. patens* (Aiton) Muhl.) and smooth cordgrass (*S. alterniflora* Loisel.)), are important for stabilizing saline marshes and sandy soils along the Atlantic and Gulf Coasts (e.g., Fine and Thomassie, 2000; Skaradek, 2006; Owsley, 2009). Cultivars of these species (Table 9.2) are vegetatively propagated due to poor seed production. Above- and belowground biomass production has been studied in smooth cordgrass (Gross et al., 1991) and genetic variation for morphometric traits was described for saltmeadow cordgrass (Silander and Antonovics, 1979).

Prairie cordgrass is tolerant of moderate salinity and is valued for wetland revegetation of irregularly flooded salt marshes (Montemayor et al., 2008), streambank stabilization, wildlife habitat, and forage (Jensen, 2006). Although highly adapted to high soil moisture conditions, prairie cordgrass outproduced switchgrass on well-drained prime land (Boe and Lee, 2007) and produced comparable biomass yields to switchgrass on dry marginal land (Boe et al., 2009). Although trials containing populations of prairie cordgrass from a wide latitudinal gradient have not been conducted, Boe and Lee (2007) found differences for biomass production among seven populations of prairie cordgrass from South Dakota.

In a 7-year study in England, prairie cordgrass from CT in the United States averaged 13 Mg ha$^{-1}$ without fertilizer. However, in two of the years, it produced >18 Mg ha$^{-1}$ (Potter et al., 1995). On wet marginal land (Argiaquolls) in SD, biomass production of prairie cordgrass was three times that of switchgrass (Boe, unpublished data, 2011). Skinner et al. (2009) evaluated "Red River" prairie cordgrass, switchgrass, big bluestem, Indiangrass, and eastern gamagrass in riparian areas in the Northeast. Red River had low mortality, the greatest vegetative spread, and relatively high biomass yield compared with big bluestem and Indiangrass. Its biomass yield, compared with switchgrass and eastern gamagrass, varied among locations. Prairie cordgrass, because of its early growth in the spring and high tiller density throughout the growing season, also showed promise for biomass production in a short growing season in Canada (Madakadze et al., 1998).

**Figure 9.1.** Prairie cordgrass at Urbana, IL, August 2012. (*For color details, see color plate section.*)

The ability to accumulate large amounts of belowground biomass has been observed in *Spartina*. Gross et al. (1991) reported >80 Mg ha$^{-1}$ of biomass in the top 15 cm of soil in salt marshes dominated by smooth cordgrass. Prairie cordgrass also has potential for carbon storage due to its large proaxes and extensive rhizome and deep root systems (Weaver and Fitzpatrick, 1932; Boe et al., 2009). The biomass of rhizomes and proaxes in the upper 15 cm of soil exceeded 20 Mg ha$^{-1}$ in stands of "Red River" (Boe et al., 2009). Prairie

**Table 9.2.** Variation among five natural populations of prairie cordgrass and "Red River" for biomass and morphological traits in a spaced-plant nursery in Illinois. SD-109 is from SD, ND-101 is from ND, and IL-99, IL-102, and IL-105 are from IL.

| Trait | Population ||||||
|---|---|---|---|---|---|---|
| | SD-109 | ND-101 | Red River | IL-99 | IL-102 | IL-105 |
| Mg DM ha$^{-1}$ | 15 | 9 | 9 | 21 | 25 | 22 |
| Tillers m$^{-2}$ | 151 | 400 | 308 | 167 | 130 | 217 |
| g tiller$^{-1}$ | 9.9 | 2.3 | 3.1 | 13.3 | 19.9 | 10.5 |
| Phytomers tiller$^{-1}$ | 8.3 | 5.9 | 6.5 | 9.1 | 9.2 | 8.7 |
| Height (m) | 1.9 | 1.3 | 1.2 | 2.1 | 2.1 | 2.1 |

cordgrass is the highest-producing perennial grass in temporary wetland areas in much of North America. It could produce >20 Mg ha$^{-1}$ in the northern Great Plains and Great Lakes regions (Table 9.2). Prairie cordgrass resumes growth in the spring about 3 weeks before switchgrass from cataphyll-protected tillers that emerge in autumn. It also is more tolerant of late spring frosts than switchgrass and miscanthus (Boe, unpublished data, 2011). It will persist and be productive in wet saline soils that switchgrass and miscanthus cannot tolerate. Since, up to now, very little collection, evaluation, and breeding work has been done on this species, the potential for genetic improvement by conventional plant breeding methods is high.

## 9.2.2 Genetic Variation and Breeding Methods

*Germplasm Resources*

Two selected classes of germplasms, "Red River" and "Atkins", of prairie cordgrass have been developed (Table 9.3). "Red River Germplasm" was developed from natural populations in MN, ND, and SD. Both seed and vegetative propagules are commercially available (Jensen,

**Table 9.3.** Cultivar (C) and selected germplasm (S) releases for prairie cordgrass, saltmeadow cordgrass, and smooth cordgrass.

| Release Name | Type | Collection Area | Primary Agency | Year of Release |
|---|---|---|---|---|
| *Prairie Cordgrass* | | | | |
| Atkins germplasm | S | NE | Manhattan PMC[a], KS | 1998 |
| Red River natural germplasm | S | MN, ND, SD | Bismarck PMC, ND | 1998 |
| *Saltmeadow Cordgrass*[b] | | | | |
| Avalon | C | NJ | Cape May PMC, NJ | 1986 |
| Flageo | C | NC | Jimmy Carter PMC, GA | 1990 |
| Sharp | C | LA | Brooksville PMC, FL | 1994 |
| Gulf Coast | C | LA | Golden Meadow PMC, LA | 2003 |
| *Smooth Cordgrass*[b] | | | | |
| Vermilion | C | LA | Golden PMC, LA | 1989 |
| Bayshore | C | MD | Cape May PMC, NJ; MDAES | 1992 |

[a]USDA-NRCS Plant Materials Center.
[b]Descriptions of some releases of saltmeadow cordgrass and smooth cordgrass can be found in Alderson and Sharp (1994).

Table 9.4. Prairie cordgrass germplasm collections that are currently under evaluation for biomass and other agronomic traits at Urbana, IL.

| State | Ploidy | Number of collections |
|---|---|---|
| Connecticut | 4x | 2 |
| Iowa | 4x/8x | 12 |
| Illinois | 4x/6x | 43 |
| Indiana | 4x | 1 |
| Kansas | 4x/8x | 10 |
| Maine | 4x | 4 |
| Massachusetts | 4x | 1 |
| Minnesota | 8x | 8 |
| Missouri | 4x | 6 |
| Nebraska | 8x | 5 |
| N. Dakota | 8x | 2 |
| New Jersey | 4x | 1 |
| Oklahoma | 8x/4x | 4 |
| S. Dakota | 8x | 9 |
| Wisconsin | 4x | 5 |
| New York | 4x | 1[a] |
| New Hampshire | 4x | 1[a] |

[a] USDA-NRCS collection.

2006). "Atkins Germplasm" was developed from a natural population in NE. Vegetative material is available, but no seed is produced (R. Wynia, personal communication, 2011). Several cultivars of saltmeadow cordgrass and smooth cordgrass (Alderson and Sharp, 1994) have been released (Table 9.2) and are used for stabilization of highly sensitive (e.g., sand dunes) environments. Since interspecific hybridization is common in *Spartina*, the value of these materials for breeding purposes should also be evaluated.

Recently, D.K. Lee collected seed from natural populations in the eastern and Midwestern USA and J. Gonzalez-Hernandez collected seed and vegetative propagules from 87 different sites in the Great Plains (Table 9.4). Many of those accessions are being evaluated in SD and IL. Earlier, seed collections from seven populations from eastern SD were evaluated with three switchgrass cultivars for 5 years (Boe and Lee, 2007; Boe et al., 2009). Based on that evaluation, among- and within-population selection for biomass, disease resistance, and lodging resistance resulted in an improved germplasm. This germplasm is currently being compared with Red River and switchgrass on wet marginal land in the northern Great Plains, and a cultivar release is anticipated within 3 years. Collection and evaluation of prairie cordgrass is also ongoing by USDA-NRCS in the northeastern USA.

*Sexual Reproduction and Breeding Systems*

Protogyny is the mode of pollination in smooth cordgrass. Stigmas emerge about 3–5 days before anthers on individual inflorescences. Large differences for seed set occurred between self-pollinated (23%) and cross-pollinated (52%) panicles (Fang et al., 2004). Comparatively little is known about the mode of pollination in prairie cordgrass. Protogyny is observed in populations from the northern Great Plains and Midwest. Ongoing seed set studies on prairie cordgrass have indicated higher seed set than that observed for smooth cordgrass. Panicles collected from several different populations over several years indicated seed set of 70–80% across a wide range of temporal and spatial variations (Boe and Lee, unpublished data, 2011).

## Chromosome Numbers and Morphology

Prairie cordgrass is a multiple cytotype species resulting from intraspecific polyploidy which appear geographically distributed (Marchant, 1968). Populations in the eastern USA are tetraploid with 2n = 4x = 40 chromosomes. The largest chromosomes are about 1 μm long. The somatic nuclear DNA content of the tetraploid is ≈1.6 pg with a 1C content of ≈399 Megabases (Mb) (Sumin et al,. 2010). The western type has 2n = 8x = 80 chromosomes, whose morphologies are similar to the eastern type. The western type appears to be an auto-octaploid. The origin of the western type is hypothesized to be the result of an unreduced egg gamete (40 chromosomes) fertilized with an unreduced pollen grain (40 chromosomes) of the eastern tetraploid. This mechanism is a well-established model for the production of polyploid plants (Chen and Ni, 2006). In addition, ploidy evolution is well documented in the genus *Spartina* with ploidy induction still playing a pivotal role in the adaption and invasiveness of *Spartina* ssp. (Ainouche et al., 2004, 2009). The somatic genome size of the octaploid is ≈3 pg with a 1C DNA content of ≈374 Mb (Sumin et al., 2010). Reduction in 1C genome size has been associated with increasing ploidy level. According to Bennett (1987) and Leitch and Bennett (2004), genome downsizing is a widespread phenomenon that occurs in many polyploidy species. The range for the octaploid cytotype is from roughly MO and westward. Both tetraploids and octaploids appear to have normal meiosis with 20 and 40 bivalents in metaphase I, respectively (Reeder, 1977; Marchant, 1968). Both cytotypes have good fertility and seed set.

In 2010, a new cytotype was discovered growing among natural tetraploid populations in IL (Sumin et al., 2010). This IL cytotype has 6x = 60 chromosomes that are similar in morphology to both the tetraploid and octaploid cytotypes. The IL cytotype is hypothesized to be the result of an unreduced gamete (40 chromosomes) combining with a reduced gamete (20 chromosomes) to result in a 60 chromosome auto-hexaploid. Chromosome pairing and reproductive parameters have yet to be determined for the hexaploid cytotype. The somatic genome size of the hexaploid cytotype is ≈2.3 pg with a 1C genome size of ≈381 Mb (Sumin et al., 2010). The intermediate 1C genome size of the hexaploid as compared to the diploid and octaploid support the hypothesis that the 1C genome size is decreasing as ploidy level increases in prairie cordgrass. Although seemingly competitive with and in some cases appearing invasive to tetraploid populations, at present, the hexaploid cytotype has yet to be found outside this one area in IL (Sumin et al., unpublished data, 2011). Interestingly, no diploid progenitor species of prairie cordgrass, or in fact any other *Spartina* ssp. has been identified (Ainouche et al., 2009).

## Molecular Genetics and Genomics

An AFLP analysis within a germplasm collection from 87 different populations in the Great Plains (IA, KS, ND, NE, and SD) demonstrated high levels of genetic diversity and polymorphism. Out of a total of 1890 DNA fragments, 81% were polymorphic. Most of the genetic variations (83.5%) occurred within populations with no distinct clustering associated with ecoregions. Genetic diversity did not increase with increased geographic distance (Dwire, 2010).

An investigation of the transcriptome of three lines of prairie cordgrass indicated a close relationship between genome sequences of prairie cordgrass and sorghum, based on the number of contigs (12,600) annotated to the sorghum gene sequences compared to the number annotated to maize (10,775). More than 550,000 sequence reads were obtained resulting in the formation of 26,300 contigs and the identification of 841 SSR regions (Gedye et al., 2010). Additional SSR markers have been developed from genomic sequences and a tentative linkage map has been developed. Ongoing transcriptome analysis in rhizomes, leaves, and roots will facilitate the development of additional genomic tools for this species (Gonzalez-Hernandez, unpublished data, 2011).

*Growth Habit and Propagation*

Prairie cordgrass is strongly rhizomatous. Rhizomes start to form in early summer (Mueller, 1941), can reach lengths of 40 cm (Weaver and Fitzpatrick, 1932), and are mostly confined to the upper 20 cm of the soil (Weaver and Fitzpatrick, 1932; Boe et al., 2009). The number of rhizomes produced per proaxis varied with soil conditions, but was usually two to three for "Red River". Seedling transplants placed on 1 m centers spread rapidly by rhizomes making identification of individual plants difficult after the second growing season. Variation for growth habit occurs among and within populations (Boe et al., 2009). Two extreme growth habits (i.e., phalanx and guerilla) (Harper, 1985) can be identified, but intermediate types are prevalent (Boe et al., 2009). Procedures for successful vegetative propagation were described by Jensen (2006). Rhizomes of prairie cordgrass can lie dormant in the soil under drought conditions for several years (Weaver and Albertson, 1944). The morphologies of the proaxis and the aerial portions of vegetative and reproductive tillers of prairie cordgrass were described by Boe and Lee (2007) and Boe et al. (2009).

*Resistance to Abiotic Stresses*

Although salt tolerant mechanisms have not been studied, prairie cordgrass is recognized as having a level of tolerance higher than other tall grasses (Tober et al., 2007; Montemayor et al., 2008). Reduction in seed germination percentage and germination rate of prairie cordgrass was lower than switchgrass under saline conditions. Prairie cordgrass maintained a constant level of salt concentration in the plant by excreting salt through glands onto the leaf surfaces, which is the same mechanism of the salt marsh species smooth cordgrass (Warren et al., 1985; Hendricks and Bushnell, 2008). "Red River" showed greater tolerance to flooding and saturated soils than big bluestem, switchgrass, Indiangrass, and eastern gamagrass in riparian areas in the Northeast (Skinner et al., 2009).

*Potential Pest Problems*

Little is known about the number or severity of diseases on prairie cordgrass. Mankin (1969) listed six species of fungi found on prairie cordgrass in SD. Prairie cordgrass populations from southeastern SD were more resistant to ash-cordgrass rust (*Puccinia sparganioides* Ellis & Tracy) than populations from east-central ND (Boe and Stein, unpublished data, 2011). "Red River" was more susceptible to ash-cordgrass rust than indigenous populations in trials in IL (Lee, unpublished data, 2011). Careful attention is being paid to identification of diseases and their potential impact on biomass and seed production in new collections of natural populations (Gonzalez-Hernandez et al., 2009).

A 40% reduction in biomass production was observed in natural stands of prairie cordgrass infested by *Ischnodemus falicus* (say) (Hemiptera: Lygaeidae) in KS (Johnson and Knapp, 1996). We have observed similar symptoms of stunting, yellowing, premature drying of leaves, and reduced biomass in the northern Great Plains and Midwest. The insects overwinter as adults at the bases of tillers and feeds, on newly emerged leaves in the spring. We have observed >20 individuals leaf$^{-1}$ of *I. falicus* feeding on young blades in April. Tillers in heavily infested stands are severely stunted with little internode development throughout the growing season. Losses in seed production in natural stands due to insect predation were estimated at 50% in six Midwestern states. Larvae of a moth, *Aethes spartinana* (Barnes and McDunnough), were responsible for most of the damage. The larvae generally fed on a series of consecutive spikelets, with high infestations (>50% insect damage) concentrated in the center of spikes. Because larvae are concealed by moving into adjacent spikelets and later tunneling into stems, they may be difficult to control using insecticides. Breeding and seed production efforts may be severely limited without efforts to manage *A. spartinana* (Prasifka et al., 2012).

*Selection Methods in Cultivar Development and Genetic Improvement Challenges*

Since no cultivars of prairie cordgrass are available, the first generation of new cultivars for bioenergy production will likely come from selection among and within natural populations. Superior populations may be increased and released as selected germplasms or cultivars. In some circumstances, there may be demand for selected and source-identified classes of releases, as they may be preferred over cultivars in regions that require the seed planted to be from a natural population in close proximity to the planting site. In the case of source-identified releases, neither selection for biomass or other traits nor testing of the germplasm in trials with other populations have been done.

Biomass yield in most populations of grasses is of low-to-moderate heritability (Casler et al., 1996). Therefore, for the second generation of new cultivars, family selection or selection of parents based on performance of their half-sib progenies would be expected to give greater progress than mass selection. Half-sib families will be grown in transplanted spaced- or single-row plots, or small seeded plots. For spaced plants and for transplanted rows, selecting individual plants within individual families is an option (Vogel and Pedersen, 1993). It is also likely that some breeding programs will utilize marker-assisted selection (Gonzalez-Hernandez et al., 2009).

Biomass production varied among populations from widely separated geographic regions, but yield levels of adapted germplasms were high in their areas of origin (Table 9.2). Therefore, no serious challenges to genetic improvement for biomass are evident at this time. However, potentially serious biomass-impacting disease and insect species have been identified and careful attention to the identification of potential genetic sources of disease and insect resistance is needed.

### 9.2.3 Future Goals

The immediate goal is a set of new cultivars or other types of releases of prairie cordgrass that are collectively adapted to wet marginal land in the northern Great Plains, Midwest, and Northeast. These cultivars should be able to be established from seed, be disease resistant, and be tolerant of moderate soil salinity. They should produce >20 Mg ha$^{-1}$ in habitats where they are the best adapted species available. Selection for morphological components of biomass yield, such as tiller mass, should be investigated for potential for increasing biomass from evaluations of spaced plants. These new releases should also protect soil and water, provide wildlife habitat, store carbon in roots, crowns, and rhizomes, and provide recreational opportunities (e.g., hunting). These releases will complement other perennial and annual grain, forage, and bioenergy crops in a multiple species approach to biomass production, resource conservation, and sustainable agriculture on heterogeneous landscapes in North America.

Longer range goals include: (1) more extensive germplasm collections across the geographic range of prairie cordgrass in North America; (2) expanded efforts in molecular genetics and cytogenetics to increase knowledge of genetic variation among and within populations and for development of molecular markers to assist in selection for useful traits; and (3) development of a suite of sustainable agronomic practices. At this time, seed production levels are low, stand establishment from seed is unpredictable, and little is known about seed longevity in different types of storage environments. Agronomic experiments conducted over 2 or more years at multiple locations to generate that critical knowledge base for switchgrass are needed for prairie cordgrass.

## 9.3 Bluestems

### 9.3.1 Importance

Big bluestem (Figure 9.2) and little bluestem (Figure 9.3) are indigenous to all of USA with the exception of CA, ID, NV, OR, and WA (Hitchcock, 1951; Barkworth et al., 2007). The distribution of sand bluestem is restricted to sandy soils in the Central Plains, Midwest, and Southwest (Sims and Risser, 2000; Barkworth et al., 2007). Big and little bluestem are dominants of the tallgrass prairie, where big bluestem is prevalent on mesic slopes and lowlands, and little bluestem dominates dry slopes on uplands. Little bluestem is also dominant on upland slopes in the mixed-grass prairie, whereas big bluestem is more restricted to bases of slopes and valleys where soil moisture is more plentiful. Together, the tallgrass and mixed-grass prairies occupy more than $100 \times 10^6$ ha, which is nearly 40% of the total area of natural grassland of the contiguous USA (Sims and Risser, 2000). The tallgrass and mixed-grass prairie regions are considered to be highly promising areas for biofuel industry (e.g., U.S. DOE, 2006). All three species are important forages for domestic livestock and wild ungulates in natural and planted grasslands, and they are also widely planted for revegetation, conservation, wildlife habitat, and beautification purposes (Boe et al., 2004).

There is a paucity of data from long-term biomass trials for bluestems. The most extensive trial was conducted by Jacobson et al. (1986) and recently summarized by Tober et al. (2008). Seven cultivars of big bluestem were evaluated for 5 years at five locations in ND, SD, and MN. The highest mean cultivar–location yield was 6.7 Mg ha$^{-1}$ for "Sunnyview" (Boe et al., 1999) at Upham, ND (48.6°N 100.7°W). The highest yield for any cultivar during any year was 8.9 Mg ha$^{-1}$ for Sunnyview at Upham during the first year after establishment (Tober et al., 2008). Out of the 42 cultivar–location combinations, 14 had yields >4.5 Mg ha$^{-1}$. In general, cultivars developed from natural populations south of 43°N produced more biomass than those from north of 43°N. In mid-August, maturity ranged from early head for Kaw to mature seed for Bison. Eight cultivars of switchgrass evaluated in that study produced about 25% more biomass than big bluestem. The little bluestem yield was only about 50% of switchgrass and 65% of big bluestem (Jacobson et al., 1986).

In the Northeast, the highest-yielding big bluestem cultivar, "Niagara", produced 7.6 Mg ha$^{-1}$ while the highest-yielding switchgrass cultivar, "Carthage", produced 10.1 Mg ha$^{-1}$ (Jung et al., 1990). In a 3-year study in IA, Kaw big bluestem produced more biomass (6.8 Mg ha$^{-1}$ vs. 6.3 Mg ha$^{-1}$) than "Blackwell" switchgrass in 1 year and comparable biomass in the other 2 years (Hall et al., 1982). In NE, strains of big bluestem from southern NE, KS, and OK

**Figure 9.2.** Big bluestem at Urbana, IL, August 2012. (*For color details, see color plate section.*)

outproduced strains from northern NE, ND, and IA. However, big bluestem × sand bluestem hybrids (Newell and Peters, 1961) outyielded the southern strains by 15% (Newell, 1968). In Canada, "Dacotah" switchgrass outproduced "Bison" big bluestem at two locations, and the yields of Bison and Dacotah exceeded those of Badlands little bluestem (Jefferson et al., 2002). "Nebraska 28" switchgrass produced 33% more biomass (5.6 Mg ha$^{-1}$) than Bison in SD (Lee et al., 2009).

**Figure 9.3.** Little bluestem at Urbana, IL, August 2012. (*For color details, see color plate section.*)

Studies in native grasslands (e.g., Redmann, 1975) showed biomass production of little bluestem, the dominant species on sloping landscape positions with coarse-textured soils and low organic matter, nitrogen, and phosphorus concentrations, was at least 70% of the biomass production of other dominant species of grasses on better soils on lower landscape positions. Cornelius (1946) evaluated little bluestem, switchgrass, and big bluestem for 5 years on eroded land of low fertility at Manhattan, KS. Harvested in early October, little bluestem produced

7.7 Mg ha$^{-1}$, followed by big bluestem (6.3 Mg ha$^{-1}$), and switchgrass (5.7 Mg ha$^{-1}$). However, there is a lack of long-term studies on dry marginal land, where the bluestems are better adapted than switchgrass. Studies in semi-arid environments and on droughty soils are expected to ensure that the bluestems will be competitive with switchgrass in monocultures or mixtures for long-term biomass production. Consequently, they will be important components of dedicated biomass production systems on dry and/or eroded marginal land.

Another important component of biomass feedstock production is the concentration and yield of holocellulose and ash concentration. In Canada, cool-season grasses had higher biomass yields and higher ash concentrations than warm-season grasses, but warm-season grasses had higher concentrations of holocellulose. Switchgrass, big bluestem, and little bluestem had similar holocellulose concentrations. However, little bluestem had lower ash concentrations than switchgrass or big bluestem at several site-years (Jefferson et al., 2004).

### 9.3.2 Genetic Variation and Breeding Methods

*Germplasm Resources*

Boe et al. (2004) summarized the common garden evaluations from the 1940s to the 1960s to describe genetic variation in the indigenous bluestems in the Great Plains. It was observed that the southern populations were more productive, utilized more of the growing season, and were more disease resistant than northern populations (e.g., Cornelius, 1947). However, winter hardiness and persistence are issues with the southern types if planted more than 500 km north of their area of origin, and disease and general nonadaptation is a problem for northern types at southern locations (Moser and Vogel, 1995). Germplasms for big, sand, and little bluestem released from 1950 to 2001 can be found in Boe et al. (2004). Releases not included there, either from inadvertent omission or release dates post-publication, are presented in Table 9.5. Gulf bluestem (*Schizachyrium maritimum* (Chapm.) Nash) and dune bluestem (*Schizachyrium littorale* (Nash) E.P. Bicknell) are two other bluestems which have close relationship to little bluestem and have ability to root from stems when covered with soil or sand. Their potential long-range value is for understanding the genetics of this valuable growth habit in the genus *Schizachyrium*. These species are important for stabilizing sandy areas along the Great Lakes and Atlantic and Gulf Coasts.

Three major germplasm collection efforts have recently been conducted in the United States for big bluestem. Dr. Kenneth Vogel, USDA-ARS, Lincoln, NE, collected from remnant prairies from southern MN and SD east through IL and south through NE and northern KS and east through IA and northern MO. Composite populations for Plant Hardiness Zones 4 and 5 have been developed through two cycles of random mating and reciprocal recurrent phenotypic selection (Vogel and Pedersen, 1993). In addition, selected half-sib families are currently being evaluated as source populations for potential cultivar release. Dr. Michael Casler, USDA-ARS, Madison, WI, collected big bluestem from 34 prairies in WI. Germplasm releases and registrations from these populations are planned. Selection among half-sib families is ongoing for biomass production and other agronomic traits. Dr. Paul Salon, USDA-NRCS, Big Flats PMC, NY, collected about 80 accessions of big bluestem in the Northeast. Biomass production and conservation potential have been evaluated at two locations in NY. In Manitoba, Phan and Smith (2000) collected plants from 14 natural populations of little bluestem from locations that varied for soil texture, soil moisture, exposure, and slope.

Recent collections have also been made throughout the Central Plains for little bluestem and sand bluestem. Fifteen accessions of sand bluestem were collected by T. Springer from southern

**Table 9.5.** Cultivar (C) and selected germplasm (S) releases for big bluestem, sand bluestem, and little bluestem. A comprehensive list of releases of cultivars and other germplasms prior to 2000 can be found in Boe et al. (2004), with descriptions for some in Alderson and Sharp (1994).

| Release Name | Type | Collection Area | Primary Releasing Agency | Year of Release |
|---|---|---|---|---|
| *Big Bluestem* | | | | |
| OZ-70 germplasm | S | AK, IL, MO, OK | Elsberry PMC[a], MO | 2003 |
| Bonanza | C | NE | ARS, Lincoln, NE | 2004 |
| Goldmine | C | KS | ARS, Lincoln, NE | 2004 |
| Prairie view Indiana germplasm | S | IN | Rose Lake PMC, MI | 2005 |
| Refuge germplasm | S | AR | Elsberry PMC, MO | 2006 |
| Hampton germplasm | S | MO | Booneville PMC, AR | 2007 |
| *Sand Bluestem* | | | | |
| Cottle County germplasm | S | TX | E. "Bud" Smith PMC, TX | 2002 |
| Chet | C | Central and southern Great Plains | USDA-ARS, Woodward, OK | 2004 |
| *Little Bluestem* | | | | |
| OK select germplasm | S | OK | E. "Bud" Smith PMC, TX | 2002 |
| Prairie view Indiana germplasm | S | IN | Rose Lake PMC, MI | 2005 |
| Ozark germplasm | S | IL, MO | Elsberry PMC, MO | 2010 |

[a] PMC=USDA-NRCS Plant Materials Center.

mixed-grass prairie sites in CO, NM, OK, and TX. Controlled crosses of these accessions are currently being made in the greenhouse using a complete diallel mating design (Hallauer and Miranda, 1981). Selection for superior biomass and seed production traits is in progress.

*Sexual Reproduction and Breeding Systems*

The breeding system for big, sand, and little bluestem is putatively xenogamy (Anderson and Aldous, 1938; Law and Anderson, 1940; Norrmann et al., 1997). Selfing generally reduced vigor in most, but not all $S_1$ lines of big bluestem (Law and Anderson, 1940). However, vigor of $S_1$ lines of little bluestem was comparable to that of progenies produced from cross-pollination (Anderson and Aldous, 1938). Andromonoecy is the most common form of sex expression in big bluestem (Norrmann et al., 1997; Boe et al., 2004), but the presence of bisexual flowers and formation of caryopses in pedicellate spikelets in SD (Boe et al., 1983; 1989), KS (Birkholz, 1970), MN (McKone et al., 1998), and OK (Springer, 1991) indicated that hermaphroditic plants occurred across a wide range of environments. In SD, seed set was higher in sessile (72%) than in pedicellate (61%) spikelets of hermaphroditic genotypes, and caryopses from sessile spikelets were 50% heavier than those from pedicellate spikelets (Boe et al., 1989).

Low seed set has been determined to be a significant factor for reducing seed production in big bluestem (Cornelius, 1950), even when other components of seed yield are high (Boe and Ross, 1983). Reasons for low seed set include failure of anthers to exert and release pollen (Cornelius, 1950) and predation of ovaries and developing caryopses by the bluestem seed midge (*Stenodiplosis wattsi* Gagné). Reduction in seed yield to the midge exceeded 40% in NE (Carter et al., 1988). We observed greater effects of genotype and location for inflorescences genet$^{-1}$ and spikelets inflorescence$^{-1}$ than for seed set.

*Chromosome Numbers*

The base chromosome number for Andropogoneae is x = 10 (Gould and Shaw, 1983). Big bluestem is predominantly hexaploid (e.g., Gould, 1956; Riley and Vogel, 1982; Norrmann et al., 1997). However, plants with 90 chromosomes occur in varying frequencies in populations across North America (Keeler, 1990; Norrmann et al., 1997). Meiosis is regular (i.e., 30 II) in hexaploids but irregular in enneaploids (Norrmann et al., 1997). Enneaploid plants were taller and produced more biomass than hexaploid plants, but hexaploids produced seeds with higher germination (Keeler and Davis, 1999). Boe et al. (1980) reported aneuploidy in several plants from a population in northeastern SD. Little bluestem is mostly 2n = 40 (Dewald and Jalal, 1974). However, Gould (1956) reported a range of 2n = 20 to 2n = 100. Sand bluestem chromosome counts range from 2n = 60 to 2n = 100 (Gould, 1956). Cultivars of big and sand bluestem have 60 chromosomes (Riley and Vogel, 1982).

*Molecular Genetics and Genomics*

Gustafson et al. (1999) found extensive RAPD variation within big bluestem populations. Gustafson et al. (2004) also quantified genetic variation among and within remnant and restored populations of big bluestem. They found no association between size of remnant population and genetic variation. Restored populations, although genetically different than remnant populations were as genetically diverse as remnant populations, and a greater genetic similarity was found between restored populations than between restored populations and a remnant from close proximity. Although propagation is primarily vegetative in natural populations of prairie grasses, Keeler et al. (2002), using protein electrophoresis showed that more than 30 different genotypes of big bluestem occurred in 100 m$^2$ areas.

A recent study of 34 different populations of big bluestem from natural prairies in WI found that 96% of the variation for AFLP markers was within populations. Significant variation among eco-regions and collection sites was detected, but accounted for less than 1% and 3% of the total variation, respectively (Price and Casler, 2010). Looking ahead to achieving compatibility between preservation of the genetic integrity of local remnants and commercial goals for expanded areas of native grasses including bluestems, Gustafson et al. (2005) reported cases where local sources were more vigorous and pest resistant than nonlocal sources. They also found that local and nonlocal populations in close proximity, so that pollen exchange was likely still maintained their genetic identities after 20 or so years. Little bluestem also showed much greater within-population genetic variation than among-population variation (Huff et al., 1998; Fu et al., 2004); however, geographic distance was a significant component of the total variation.

*Growth Habit and Propagation*

Big, sand, and little bluestem are long-lived perennials that are normally propagated commercially by seed, except for genotypes that are vegetatively propagated for ornamental purposes. Big bluestem and sand bluestem are rhizomatous, although the internodes of the rhizomes are longer for sand bluestem (usually >2 cm) than for big bluestem (Barkworth et al., 2007). Vegetative propagation for clonal evaluations or formation of polycross nurseries is easily accomplished with 2-year-old plants that will normally have >50 tillers. Little bluestem is composed of three varieties. Creeping bluestem (*S. scoparium* var. *stoloniferum* (Nash) Wipff) is strongly rhizomatous; whereas the other two varieties are usually cespitose. Two-year-old spaced plants of cespitose little bluestem can have up to 350 tillers, and are easily propagated for clonal evaluations and replicated polycross nurseries using single- or multiple-tillered ramets (Boe and Bortnem, 2009).

*Resistance to Abiotic Stresses*

Since the collective ranges of the three species span from about 25°N to 50°N in North America, there is considerable genetic variation for tolerance of temperature extremes, moisture extremes, and variability in soil physical and chemical characteristics (Boe et al., 2004). For example, big and little bluestem are indigenous to regions of the mixed-grass prairie where average annual precipitation is <350 mm, in regions of the tallgrass prairie where it exceeds 1000 mm, and in the eastern USA where it exceeds 1100 mm. Temperature extremes in these grasslands can range from −50°C in the northwest regions of the mixed-grass prairie to >45°C in the southeast region of the tallgrass prairie (Holechek et al., 2001). Where topographic position influenced plant distributions, little bluestem was found on tops of dunes and north-facing slopes, whereas sand bluestem tended to occupy south-facing slopes (Schacht et al., 2000). Of the perennial warm-season grasses currently being considered for biofuel crops, the bluestems are the most drought tolerant (Weaver, 1954). Their adaptation to droughty soils gives them potential for biomass production on landscape positions in semi-arid and subhumid/humid regions where switchgrass and miscanthus would likely not persist.

*Potential Pest Problems*

Nearly 100 disease organisms have been isolated from the genus *Andropogon* and about 50 from the genus *Schizachyrium* in the United States, most of which were fungi that collectively can infect all developmental stages (Farr et al., 1989). However, their economic impact is not known. Ascochyta leaf blight (*Ascochyta brachypodi* (Sydow) Sprague and A.G. Johnson) appeared to be the most important disease on big and little bluestem in the northeastern USA. Susceptibility differed among cultivars of big bluestem, with Niagara, a cultivar from the Northeast having greater resistance than Kaw and Pawnee, which are from the Great Plains (Zeiders, 1982). Kernel smut (*Sphaecolotheca occidentalis* (Seym.) Clint.) occurred on big bluestem in Iowa (Snetselaar and Tiffany, 1991). Cultivars of bluestem from the central and southern Great Plains would be more resistant to disease than those from the northern Great Plains. Krupinsky and Tober (1990) found no differences between big bluestem cultivars from KS, NE, SD, and ND for damage from *Phyllosticta andropogonivora* Sprague & Roderson, which causes leaf spot. "Cimarron", a composite population of little bluestem composed of strains from KS, CO, NM, TX, and OK was more resistant than "Aldous", which was selected from a population of little bluestem near Manhattan, KS.

The bluestem seed midge is a threat to seed production of the bluestems in the Great Plains (K. P. Vogel and R. Mitchell, personal communications, 2011). An insecticide has been labeled for control of this midge. However, more research is needed to determine the best management practices for control of this insect. *Conioscinella nuda* (Adams) could affect seed production of bluestem. This fly has been reared from inflorescences of big and little bluestem in the Southern Great Plains (Springer et al., 1989) and big bluestem in the Northern Great Plains (Boe and McDaniel, 1990). Springer et al. (1989) observed the larvae feeding on the ovary of the sessile spikelets of big bluestem.

*Other Traits of Interest*

Traits of interest, in addition to biomass and seed yield, are feedstock quality, pest resistance, and seedling vigor. Concurrent selection for forage quality and yield resulted in new high-biomass-producing cultivars with improved animal performance (Vogel et al., 2006a, 2006b); thus, developing new cultivars that are superior for both forage and biomass purposes has been successful. Unlike switchgrass and many other forage grasses, big, sand, and little bluestem produce floriferous branches from axillary meristems of aerial phytomers (Boe et al., 2000; Boe and Bortnem, 2009). The response of those meristems in different environments is a significant source of phenotypic plasticity (de Kroon et al., 2005) in biomass production. Genetic variation for frequency of phytomers with aerial branches and biomass production from axillary branches exists in bluestems (Boe and Bortnem, 2009).

*Selection Methods in Cultivar Development and Genetic Improvement Challenges*

Vogel (2000) concluded that there was ample additive genetic variation among and within populations of big and sand bluestem for progress from selection. For little bluestem, considerable genetic variation remains *in situ* in natural populations. Most regions of the original tallgrass prairie once dominated by big bluestem have been converted to conventional cropping, and consequently only remnants remain. Badlands Ecotype little bluestem is an example of a recent product from a coordinated germplasm collection effort by the USDA-NRCS. Vogel (2000) and Boe et al. (2004) provided detailed histories of breeding of indigenous bluestems in North America through 2000. Cultivar and other types of releases subsequent to the publication of those chapters are presented here in Tables 9.2 and 9.3. Prior to 2000, 65% of the releases were cultivars. However, of the 15 releases since 2000, 3 were cultivars, 8 were selected-type releases, and 4 were source-identified releases (Belt and Englert, 2008). All of the selected and source-identified releases since 1996 were by USDA-NRCS Plant Materials Centers (Boe et al., 2004; Table 9.5).

The two most recently released cultivars of big bluestem, Bonanza and Goldmine, were selected for forage yield and digestibility through three generations of restricted, recurrent selection (Burton, 1974; Vogel and Pedersen, 1993). Bonanza and Goldmine had higher yields than their parent cultivars (Vogel et al., 2006a, 2006b). Collectively, the selected class of releases of big, sand, and little bluestem represent genetic diversity in remnants of the northeastern (IN and MI), southeastern (AR, LA, and MO), and southern (OK and TX) tallgrass prairie (Kucera 1992) (Table 9.5). The OZ-70 Germplasm was selected for late maturity, forage yield and quality, and rust resistance (Bruckerhoff, 2004). The Prairie View Indiana Germplasm is a composite of 20 collections from Indiana (Leif and Burgdorf 2005b). The Refuge Germplasm was selected for short columnar stature and resistance to lodging (Bruckerhoff and Cordsiemon, 2009). The Hampton Germplasm of big bluestem (King, 2008) demonstrated superiority for vigor, disease resistance, forage production, and forage quality. The Cottle County Germplasm of sand bluestem is recommended for sandy soils in TX and OK (Houck, 2005a). The three select releases of little bluestem encompass a collective area of adaptation from TX, OK, AR, and LA (OK Select Germplasm (Houck, 2005b)) north through the Ozark region (Ozark

Germplasm (Cordsiemon and Kaiser, 2010)) to Indiana (Prairie View Indiana Germplasm (Leif and Burgdorf, 2005a)) (Table 9.5).

The source-identified germplasms of big bluestem released since 2000 were collected from natural stands in NC (Suther Germplasm (Skaradek 2008a)) and the southern Lower Peninsula of Michigan (Anonymous, 2012). The recently released source-identified germplasms of little bluestem (Suther Germplasm (Skaradek 2008b)) were collected from remnant prairies in NC and the southern Lower Peninsula of MI (Southlow Michigan Germplasm (Durling et al., 2007)).

In addition to the releases presented in Boe et al. (2004) and here in Table 9.5, there are 19 additional available accessions (8 of which are previously mentioned cultivars) of big bluestem, 32 available accessions (6 of which are previously mentioned cultivars) of little bluestem, and 5 accessions (4 of which are previously mentioned cultivars) of sand bluestem in the USDA National Plant Germplasm System. Within a year or so, 120 more accessions of big bluestem from collections made from border-to-border along the eastern edge of SD will become available (M. Harrison-Dunn, personal communication, 2011). Presently, the most common breeding methods being employed for improving biomass production in the bluestems are restricted, recurrent phenotypic selection (Burton, 1974), among-family selection, and among- and within-family selection (Vogel and Pedersen, 1993) (M.D. Casler, P. Salon, and K. P. Vogel, personal communications, 2011), with the ultimate goal of development of new high-producing synthetic cultivars consisting of genotypes with good general combining ability. Normally, each cycle of selection takes at least 4 years (Vogel, 2000).

The difficulties associated with capitalizing on nonadditive genetic variation in the development of hybrids in cross-pollinated perennial grasses with small florets were pointed out by Vogel (2000). The potential for development of hybrids in the bluestems is unknown. Currently, cultivars and recent germplasm collections from natural stands are the principal base populations for most genetic improvement programs focused on increasing biomass production. Undoubtedly, the select and source-identified releases from the USDA-NRCS Plant Materials Centers will also be important sources of genetic variation for biomass breeding programs.

Since biomass yield potential in native warm-season grasses, including bluestems, is strongly related to latitudes of origin of the base natural populations (e.g., Cornelius, 1947; Newell, 1968; Tober et al., 2008), breeding programs located in USDA Plant Hardiness Zones 4 and 5 in the central USA (i.e., USDA-ARS at Lincoln, NE and Madison, WI for big bluestem, and Brookings, SD for little bluestem) are placing emphasis on persistence within high-yielding populations from southern origins. In the northern Great Plains, Sunnyview big bluestem, which has its origin in southeastern SD (42.7°N 96.7°W), was the highest-yielding cultivar (Tober et al., 2008). It had an excellent stand after 10 years at Upham, ND (48.6°N 100.7°W). The latitudinal distance between Elk Point and Upham is about 700 km. The cultivar Kaw, origin Manhattan, KS (39.2°N 96.6°W) and cultivars with origins in NE, also persisted for 5 years at Upham; however, they did not produce as much biomass as Sunnyview. Phenological development in early August at Upham was mid-anthesis for Sunnyview and vegetative (pre-stem elongation) for Kaw. "Aldous" little bluestem, which is also from Manhattan, sustained severe winter injury at Upham (Jacobson et al., 1986). These results indicated that populations of big bluestem from SD, NE, and KS can be highly productive and persistent at locations greater than 500 km north of their origins. It remains to be seen if germplasms with origins within Zones 6 to 8 are presently, or can be naturally selected to be, persistent in Zones 4 and 5; or if hybridization between genotypes from different zones is more effective for improving long-term stability of biomass production.

Evaluation of breeding populations in space-planted nurseries is a common practice in forage and biomass improvement programs for perennial grasses. However, performance of a genotype in a spaced-plant environment is not a reliable predictor of the performance of its half-sib progeny in rows or swards. Therefore, the most rapid progress from selection will require new designs or selection techniques in spaced-plant nurseries or selection among half-sib families in rows or preferably small swards. Missaoui et al. (2005) successfully identified half-sib families that produced high amounts of biomass in rows and swards by selecting high-yielding spaced plants and suggested that further evaluation of the high-yielding half-sib families in the wide plant-spacing honeycomb design would enable the breeder to utilize both among- and within-family selection.

Vogel (2000) pointed out that grass breeders have made progress for disease tolerance in native grasses, evident in the superiority of cultivars compared with native collections from the same region. Since so little attention has been paid to diseases of the bluestems, particularly the rusts, screening for disease resistance should be an objective of breeding programs. Mankin (1969) listed three species of rust, two species of smut, and two leaf spot diseases collected from big bluestem at Brookings, SD. If area planted with bluestems expands in response to a new bioenergy industry based on perennial warm-season grasses, the incidence and impact of diseases on biomass production, feedstock quality, and stand longevity could be expected to increase. Therefore, identification to species of the disease organisms involved and elucidation of their life histories, including determination of alternate hosts, is needed.

Increased problems due to diseases and insects would be expected, with increases in area planted. Therefore, field surveys, insect collections, and elucidation of life histories, including identification of potential biological controls, are needed to determine the insect fauna associated with the wild bluestems and, in particular, those species that may have an economic impact. Switchgrass breeding programs in South Dakota and at the Noble Foundation in Oklahoma are screening individual plants for moth and midge infestation as part of their selection protocol. As we learn more about the insects that feed on the bluestems, this procedure will likely be used for them, as well.

Little effort has been made in the development of molecular markers for use in marker-assisted selection or in genetic engineering of the bluestems. However, the protocols for plant regeneration in big and little bluestem callus cultures of inflorescence origin were developed by Chen et al. (1977) and Songstad et al. (1986), respectively. The *in vitro* approaches and chromosomal stability of tissue-culture-derived populations were described in Chen and Boe (1988). More recently, plant regeneration from immature inflorescences via somatic embryogenesis was successful for creeping bluestem (*Schizachyrium scoparium* (Michx.) Nash var. *stoloniferum* (Nash) J. Wipff) (Chakravarty et al., 2001), and mature caryopses served well as explants for callus induction for transformation and regeneration of big and little bluestem plantlets (Li et al., 2009).

Because of the extensive amount of genetic variation among and within populations, at least the first generation of new cultivars selected for high biomass production will be accomplished by using conventional plant breeding approaches (Vogel, 2000).

### 9.3.3 Future Goals

Since biomass production for the bluestems has been consistently low in trials from the Great Plains to the Northeast, increasing biomass production is the primary future goal. However, most of the trials have not been conducted on dry marginal land of low soil fertility where bluestems could be expected to perform well. Therefore, to complete the suite of available cultivars of native warm-season grasses that will collectively provide adapted populations for sculpturing (Jacobson et al., 1994) heterogeneous landscapes so as to increase biodiversity, stabilize biomass production, and enhance ecological goods and services from land areas dedicated to the production of herbaceous biomass feedstocks, high-yielding cultivars of the bluestems are needed for dry marginal land. These new cultivars of bluestems will complement new cultivars of switchgrass, prairie cordgrass, miscanthus, and other species in a multiple species, ecologically minded approach to biofuel production.

The most rapid and most efficient path to improving biomass production in the bluestems in regions where biomass production is restricted by the length of the growing season, and where prevailing winter conditions limit survival to putatively adapted types, will involve developing new persistent cultivars that utilize more of the growing season than populations that are indigenous to the region.

## 9.4 Eastern Gamagrass

### 9.4.1 *Importance*

Eastern gamagrass (Figure 9.4) is indigenous from MA to NE (40°N latitude) southward through KS, OK, and TX into northeastern Mexico and eastward to the Atlantic and Gulf coasts (Hitchcock, 1951). It is highly variable with strains adapted to prairies, coastal plains, semiarid regions, deep sandy soils, rocky outcrops, river banks, and openings in forested areas. Eastern gamagrass is valued for grazing, stored forage, soil amelioration, and conservation. The usage of eastern gamagrass has increased during the past decades because of renewed interest in its use for pasture and soil conservation.

Eastern gamagrass initiates growth earlier in the spring than most other warm-season grasses; consequently, it can substitute for rangeland when range grasses are at a critical growth stage. Many factors affect eastern gamagrass biomass production, i.e., moisture, nitrogen fertilization,

**Figure 9.4.** Eastern gamagrass at Urbana, IL, August 2012. (*For color details, see color plate section.*)

harvest frequency, plant density, and interactions among these factors (Moyer and Sweeney, 1995; Brejda et al., 1996; Springer et al., 2003). However, moisture, by far, is the most limiting factor in the production of eastern gamagrass. Eastern gamagrass grows well where annual rainfall exceeds 800 mm, and acceptable growth is possible in areas where annual rainfall is as low as 400 mm.

Biomass yields of eastern gamagrass vary with management and area of production. Biomass production trials have been conducted in several states in the eastern USA. Levels of N fertilizer and harvest frequency varied, but in general yields ranged from 7 Mg ha$^{-1}$ to 16 Mg ha$^{-1}$ For most locations with average annual precipitation of approximately 650 mm, dry matter yields range from 5 to greater than 10 Mg ha$^{-1}$. At Woodward, OK, annual dry matter yields ranged from 15 to 20 Mg ha$^{-1}$ when the combined precipitation and irrigation totaled 38 mm week$^{-1}$ from April 1 to August 15 and when 280 kg N ha$^{-1}$ was applied in three split applications on April 1, June 1, and July 15.

Nitrogen fertilization and harvest frequency are important factors in the production of biomass from eastern gamagrass (Brejda et al., 1997). Nitrogen fertilization rates vary from 60 kg ha$^{-1}$ in a single broadcast application to over 300 kg ha$^{-1}$ broadcast in equal split applications (Springer and Dewald, 2004). Three or four harvests are possible during the growing season using a 4-week harvest interval and two or three harvests are possible using a 6-week harvest interval (Brejda et al., 1996). Total biomass yields were higher for the 6-week than the 4-week harvest interval (C. Dewald, unpublished data, 1974). Springer et al. (2003) reported the effects of plant density on biomass yield. They found that in young stands, higher plant densities (10.7 plants m$^{-2}$) generally produced higher dry matter yields compared with 3-year-old or older stands.

The total cost associated with conversion of eastern gamagrass biomass into bioenergy (million Btu) ranged from approximately $4.50 to $9.50 GJ$^{-1}$ (Nelson et al., 1994), which was "competitive with current prices and future projections for other fuel sources such as natural gas and fuel oil". Eastern gamagrass is also a good candidate for conversion to ethanol. Weimer et al. (2005) estimated the yield of ethanol from four warm-season grasses, including eastern gamagrass, and found a linear relationship between *in vitro* ruminal gas production and ethanol production from saccharification and fermentation similar to that of switchgrass. Weimer and Springer (2007) found considerable variation among genotypes, locations, and harvest dates and suggested that eastern gamagrass might be better suited for bioenergy production than for grazing.

### 9.4.2 Genetic Variation and Breeding Methods

*Growth Habit and Propagation*

Eastern gamagrass is a long-lived perennial grass with a caespitose growth habit. It produces many leaves that are up to 2 cm wide and 70 cm long. It displays a variety of growth forms from decumbent to erect. Erect growth forms can reach heights of 3.5 m where nutrients and moisture are not limiting. Its stem base is characteristic of a proaxis and consists of tillers, single shoots, compound shoots, and reproductive shoots in various stages of development (Dewald and Louthan, 1979). Vegetative propagation is a proven method for the establishment of eastern gamagrass; however, vegetative establishment is very expensive and labor intensive and vegetative propagules are not readily available. The success of transplanting eastern gamagrass was enhanced by using compound shoots over single shoots because compound shoots have more reserves (Dewald and Sims, 1981). Single and compound shoots transplanted during the winter dormant period will survive over 90% of the time.

Sowing seeds is the easiest method for the establishment of eastern gamagrass; but seed costs are high, and stand establishment can be difficult due to several biotic and abiotic factors. The intact spikelet of eastern gamagrass consists of a caryopsis enclosed by the indurate fruit case making it impossible to visually determine seed purity. Seed purity can be improved by the removal of empty fruit cases (Ahring and Frank, 1968). The indurate fruit case also provides a mechanical form of seed dormancy (Springer et al., 2001) that can mostly be broken by cold, moist stratification at 5°–10°C for 4–8 weeks (Ahring and Frank, 1968; Anderson, 1985; Springer et al., 2001); nevertheless, considerable variation has been found to exist among eastern gamagrass seed lots for germination and stratification durations (Ahring and Frank, 1968). Seeds of some eastern gamagrass accessions may also exhibit physiological seed dormancy. Several chemical agents (potassium nitrate, sodium hypochlorite, gibberellic acid, carbon dioxide) have been used in attempts to stimulate germination of intact spikelets; however, none have been entirely successful (Ahring and Frank, 1968; Anderson, 1985; Tian et al., 2003; Finneseth, 2010). Winter-sown eastern gamagrass intact spikelets have resulted in stands that were equal to or greater than stands established in the spring with stratified seeds (Mueller et al., 2000).

An alternative to sowing intact spikelets in the winter is to sow caryopses in the spring. Anderson (1985) and Springer et al. (2001) reported that germination was enhanced by the removing of the indurate fruit case. Tian et al. (2002) reported complete germination of caryopses by scarifying the pericarp region over the embryo. The disadvantage to this approach is that the specialized equipment needed to dehull seed without significantly damaging the caryopsis has not been developed.

*Biomass Yield and Distribution*

Experiments were conducted at the USDA-ARS Southern Plains Range Research Station, Woodward, OK, to evaluate the effects of harvest frequency and clipping height on biomass yield and nutritive value of eastern gamagrass (Dewald, unpublished data, 1975). The highest tonnage was achieved when clipped at 60-day intervals. But the highest TDN per acre was at 45-day intervals (Dewald, unpublished data, 1990).

The management goals have been driven by the frequency at which eastern gamagrass was harvested and to the levels to which it was fertilized. Forage yields for eastern gamagrass vary across the eastern USA (from between 6.6 and 7.5 Mg ha$^{-1}$ in AR, FL, IL, KS, NY, OK, and TN to >11.5 Mg ha$^{-1}$ in MO and MS) in relationship to variation among locations for management practices, such as fertilizer rate, harvest frequency, and stubble height (Faix et al., 1980; Kalmbacher et al., 1990; Graves et al., 1997; Aiken and Springer, 1998). Eastern gamagrass forage is distributed fairly uniformly across the growing season. In general, 34% of its growth occurs by June 13, 50% by July 7, 77% by July 18, 85% by August 11, 94% by August 28, and 99% by September 20. Dewald speculated that maximum production was possible from one cutting on or about September 15. This corresponds to approximately 2190 cumulated growing degree days at Woodward, OK using the "optimum day method" of calculation for growing degree days (Barger, 1969).

*Resistance to Abiotic Stresses*

Eastern gamagrass is adapted to various edaphic conditions. It often occurs along stream banks and on moist prairie sites and can survive short-term flooding. Under water-logged conditions, aerenchyma cells in the roots provide a pathway for $O_2$ to reach actively growing roots. Consequently, eastern gamagrass is proficient at penetrating clay hard-pan layers during saturated soil conditions (Clark et al., 1998). Higher $O_2$ levels in the roots may also reduce the toxic effects of Mn and other potentially toxic elements through oxidation–reduction. Eastern gamagrass tolerates acid soils (pH 4.1–4.2) with Al concentrations of 64–77% of the cation exchange capacity. It can tolerate Al concentrations that were lethal to many other plant species (Foy, 1997).

Colonization of eastern gamagrass roots by arbuscular mycorrhiza and free-living N-fixing bacteria may also help with water relations during drought and flooding and nutrient acquisition and uptake (Brejda et al., 1994). Clark et al. (1998) found two species of arbuscular mycorrhiza colonizing eastern gamagrass roots. Brejda et al. (1994) reported that eastern gamagrass roots provided a suitable environment for the growth of several free-living bacterial species, many of which were capable of nitrogenase activity. Nitrogen fixed by bacteria could supplement the N needs of eastern gamagrass. Conditions favorable for associative N fixation include: reduced soil nitrogen, moist soils, an available energy source, and a low $O_2$ environment (Klucas, 1991).

Under moderate drought, eastern gamagrass' deep root system allows it to obtain water and nutrients from lower in the soil profile; thus, maintaining good biomass yield and persistence during moderate drought conditions. After severe drought events, eastern gamagrass stands should not be defoliated to allow time for the plants to regenerate. Coyne and Bradford (1985) found genotypes of eastern gamagrass to vary for drought traits. They found one genotype had nearly optimal stomatal function for controlling water loss without limiting its photosynthetic activity, while another genotype had greater water use efficiency with reduced photosynthetic performance. Coyne and Bradford (1985) hypothesized that greater drought tolerance could be attained in eastern gamagrass by combining the traits of these two genotypes into a single accession.

*Potential Pest Problems and Resistance*

Very little information is available for eastern-gamagrass-associated insects and plant pathogens. Insects known to infest eastern gamagrass include: aphids (*Rhopalosiphum maidis* Fitch, *R. padi* L., and *Schizaphis graminum* Rondani; Seifers et al., 1993; Piper et al., 1996), leafhoppers (*Dalbulus* spp. and *Cocrassana riepmai*; Nault et al., 1983; Blocker and Larsen, 1991), leafminers (Agromyzidae), corn ear worm (*Heliothis zea* Boddie), maize billbug (*Sphenophorus maidis* Chittenden; Maas et al., 2003), southern cornstalk borer (Diatraea crambidoides Grote; Krizek et al., 2003; Maas and Springer, 2005), and the southwestern corn borer (*Diatraea grandiosella* Dyar; Dewald, unpublished data, 1978). Maize dwarf mosaic virus (MDMV; Thornberry et al., 1966; Seifers et al., 1993; Piper et al., 1996) and sugarcane mosaic virus strain MDMV-B (SCMV-MDMV-B; Seifers et al., 1993; Piper et al., 1996) have been isolated from breeding nurseries and native populations of eastern gamagrass. Both viruses are known to be vectored by aphids; however, the aphid species that transmitted these viruses to eastern gamagrass are unknown. Piper et al. (1996) speculated that severe infections of viruses could reduce forage and seed yields.

Bacterial and fungal diseases that potentially could infect eastern gamagrass are bacterial leaf spot (*Xanthomonas* sp.), common corn rust (*Puccinia sorghi* Schwein.), southern corn rust (*Puccinia polysora* Underw.), tropical corn rust (*Physopella pallescens* (Arth.) Cummins & Ramacher), *Physopella zeae* (Mains) Cummins & Ramacher, and anthracnose leaf blight (*Colletotrichum graminicola* (Ces.) G.W. Wils (Seymour and Miller, 1974; Shanmuganathan, 1974; Schieber, 1975; Handley et al., 1990)). Under the right weather conditions each of these diseases is capable of affecting the forage production of eastern gamagrass. Virtually nothing is known on genetic resistance to these or other insects or plant pathogens of eastern gamagrass. With more widespread use of eastern gamagrass, plant diseases and insects common to corn may also affect eastern gamagrass forage and seed production. Substantial research is needed to isolate, identify, and study the disease and insect pests of eastern gamagrass.

*Other Traits of Interest*

Eastern gamagrass can become iron deficient (chlorotic, light green to yellow with white growth) when grown on calcareous (high pH) soils. Developing cultivars with tolerance to high pH soils would extend its range of adaptation into the southern High Plains of North America. Germplasm with tolerance to high pH soils need to be collected, evaluated, and used to breed the trait into superior germplasm. The transfer of traits in eastern gamagrass, although not impossible, can be hindered by its breeding system that will be outlined below.

*Selection Methods in Cultivar Development and Genetic Improvement Challenges*

Eastern gamagrass consists of diploid ($2n = 2x = 36$), triploid ($2n = 3x = 54$) (Dewald et al., 1992; Dewald and Kindiger, 1996), tetraploid ($2n = 4x = 72$)(Salon and Pardee, 1996), pentaploid ($2n = 5x = 90$), and hexaploid ($2n = 6x = 108$) cytotypes (Farquharson, 1954; de Wet et al., 1982), where diploids reproduce exclusively by sexual reproduction and polyploid cytotypes reproduce by apomixis (Brown and Emery, 1958; Burson et al., 1990; Sherman et al., 1991; Leblanc et al., 1995), with the exception of SG4X-1, an induced tetraploid that reproduces sexually (Salon and Pardee, 1996; Salon and Earle, 1998). The main apomictic mechanism is diplospory of the antennaria type (Burson et al., 1990; Leblanc et al., 1995). Though, very low incidences of the taraxacum type of diplospory have been reported (Leblanc et al., 1995).

Polyploids originated from sexual diploids through the production of unreduced gametes, with unreduced megasporocytes being more common than unreduced microsporocytes (Harlan and de Wet, 1975). Diploid *Tripsacum* gave rise to tetraploid *Tripsacum* through the intermediate triploid cytotype followed by fertilization of unreduced female gametes which resulted in an increase in ploidy level (Dewald and Kindiger, 1994; Kindiger and Dewald, 1994, 1997). Fertile triploids reproduce as facultative apomicts and provide a bridge for the exchange of traits between sexual diploids and apomictic cytotypes (Dewald et al., 1992). The diploid forms of *Tripsacum* are all morphologically distinct and allopatric in their distribution, whereas polyploid forms are not always easily distinguishable on either a morphological or geographic basis. Thus, numerous intermediate cytotypes can exist in natural populations and contribute to the heterogeneity of polyploid species (Leblanc et al., 1995).

Table 9.6 includes recent releases not included in Springer and Dewald (2004). The diploid cultivars "Pete" (Fine et al., 1990) and "Iuka IV" were derived from bulk populations of diploid germplasm. Pete was derived from a bulk population of 70 accessions collected from KS and OK. Seed was advanced through two generations of natural selection under combine harvesting resulting in increased uniformity of maturity. Iuka IV was developed from a bulk population of 22 accessions collected from AK, KS, OK, and TX. Seed was advanced through four generations with only the first seed harvest of each generation being used for the next generation in an effort to select plant material with reduced dormancy.

The diploid cultivars "Martin" and "St. Lucie" (Table 9.6) were derived from single plant collections from wild populations. Martin was selected for its blue-green foliage and dark purple culms. St. Lucie was selected for its blue foliage and dark red culms. Except for the

**Table 9.6.** Cultivar releases for eastern gamagrass. A comprehensive list of releases of cultivars and other germplasms prior to 2000 can be found in Springer and Dewald (2004).

| Release Name | Releasing Organization(s) | Reproductive Characteristics | Year of Release |
| --- | --- | --- | --- |
| Martin | USDA-NRCS | Vegetative | 2000 |
| St. Lucie | USDA-NRCS | Vegetative | 2000 |
| Texas Sue | Charles & Sue Lancaster, Lampasas, TX | | 2003 |
| Bumpers | USDA-NRCS | Apomictic tetraploid | 2005 |
| Dewald | Walter W. Robertson, Shippenville, PA | Apomictic tetraploid | 2005 |
| Verl | USDA-ARS, USDA-NRCS, and Oklahoma Agric. Exp. Sta. | Apomictic triploid | 2005 |
| Meadowcrest | USDA-NRCS, New York St. Col. Agric. & Life Sci., USDA-ARS | Apomictic tetraploid | 2006 |

cultivar "Verl," polyploid cultivars (Table 9.6) were derived from single plant selection from wild populations and all are tetraploids that reproduce through apomixis. Verl is a unique cultivar in that it is a fertile triploid that reproduces predominantly via apomixis. It was produced from a controlled pollination of a gynomonoecious sex form (GSF) diploid with a monoecious tetraploid. Verl was selected for its female fertility (seed set), male fertility (pollen stainability), and forage production attributes from 243 $F_1$ progeny resulting from the cross GSF-1 (PI 483447)/WW-1724 (Dewald and Dayton 1985a, 1985b). Verl has a high degree of apomictic reproduction in that 95% of its open-pollinated progeny were identical in appearance (Springer et al., 2006).

### 9.4.3 Future Goals

There is a great need to find and develop germplasm resistant for diseases and insects that affect eastern gamagrass. For example, Springer et al. (2004) estimated a forage dry matter yield reduction of 145 kg ha$^{-1}$ for the maize billbug and 475 kg ha$^{-1}$ for the southern cornstalk borer. Based on minimum and maximum insect numbers observed per plant Springer et al. (2004) estimated a minimum yield loss of 185 kg ha$^{-1}$ and a maximum 1320 kg ha$^{-1}$ for these two insects. In Maryland, Krizek et al. (2003) reported extensive plant damage due to the southern cornstalk borer that reduced plant vigor to the point of stand loss.

Unless the damage to eastern gamagrass plants was severe the cost of chemical control would be greater than the return from the forage and it is not known if chemical control would be effective. Controlling these insects may be more difficult because of the perennial habit of eastern gamagrass. An integrated approach to their control is likely the best. The first step is to develop eastern gamagrass cultivars with resistance to these pests. This would involve screening all available eastern gamagrass germplasms and possibly the acquisition of new germplasm. It could also involve using maize as a gene source, using other closely allied species, or using transgenics to develop *Bt* eastern gamagrass. Management strategies might include using trap crops, changing harvest dates to remove forage before insects bore into culms, or late fall burning or grazing to kill some larvae and remove thatch from plants which makes insects more vulnerable to freezing. Controlling plant diseases and insects will be important to the future of eastern gamagrass as a competitive biomass crop.

# References

Ahring RM, Frank H. Establishment of eastern gamagrass from seed and vegetative propagation. J Range Manage, 1968; 21: 27–30.

Aiken GE, Springer TL. Stand persistence and seedling recruitment for eastern gamagrass grazed continuously for different durations. Crop Sci, 1998; 38: 1592–1596.

Ainouche ML, Baumel A, Salmon A, Yannic G. Hybridization, polyploidy and speciation in Spartina (Poaceae). New Phytol, 2004; 161: 165–172.

Ainouche ML, Fortune PM, Salmon A, Parisod C, Grandbastien MA, Fukunaga K, Ricou M, Misset MT. Hybridization, polyploidy and invasion: lessons from Spartina (Poaceae). Biol Invasions, 2009; 11: 1159–1173.

Alderson J, Sharp WC. 1994. Grass varieties in the United States. USDA-SCS Agric Handbk No. 170. USDA. Washington, DC.

Anderson RC. Aspects of the germination ecology and biomass production of eastern gamagrass (*Tripsacum dactyloides* L.). Bot Gaz, 1985; 146: 353–364.

Anderson KL, Aldous AE. Improvement of Andropogon scoparius Michx. by breeding and selection. J Amer Soc Agron, 1938; 30: 862–869.

Anonymous. 2012. Southlow Michigan Germplasm big bluestem (http://www.plant-materials.nrcs.usda.gov/pubs/mipmcrb11253.pdf)

Barger GL. Total growing degree days. Weekly Weather Crop Bull, 1969; 56:10.

Barkworth ME, Anderton LK, Capels KM, Long S, Piep MB. (eds). 2007. Manual of grasses for North America north of Mexico. Utah St. Univ. Press.

Belt SV, Englert JM. (2008). Improved plant materials released by NRCS and cooperators through December 2007. USDA, NRCS National Plant Materials Center, Beltsville, MD.

Bennett MD. Variation in genomic form in plants and its ecological implications. New Phytol, 1987; 106: 177–200

Birkholz D. 1970. Variation in *Andropogon gerardii* Vitman. PhD thesis, Univ. Kansas, Lawrence (Diss. Abstr. 70-25305).

Blocker HD, Larsen KJ. A new leafhopper genus, *Cocrassana* (Homoptera: Cicadellidae), from Mexican *Tripsacinae* and a synopsis of related genera. J Kansas Entomol Soc, 1991; 64: 123–126.

Boe A, Bortnem R. Morphology and genetics of biomass in little bluestem. Crop Sci, 2009; 49: 411–418.

Boe A, Bortnem R, Kephart KD. Quantitative description of the phytomers of big bluestem. Crop Sci, 2000; 40: 737–741.

Boe A, Keeler KH, Norrmann GA, Hatch SL. The indigenous bluestems of the Western Hemisphere and gambagrass. In: Moser LE, Burson BL, Sollenberger LE (eds) *Warm-season ($C_4$) grasses*, 2004, pp. 873–908. Agron. Monogr. No. 45. ASA, Madison, WI.

Boe A, Lee DK. Genetic variation for biomass production in prairie cordgrass and switchgrass. Crop Sci, 2007; 47: 929–934.

Boe A, McDaniel B. *Conioscinella nuda* (Adams) and *Tetrastichus nebraskensis* (Girault) reared from big bluestem inflorescences in South Dakota. Prairie Nat, 1990; 22: 207–210.

Boe A, Owens V, Gonzalez-Hernandez J, Stein J, Lee DK, Koo BC. Morphology and biomass production of prairie cordgrass on marginal lands. GCB Bioenerg 2009; 1: 240–250.

Boe A, Peters DW, Ross JG. Accessory chromosomes in big bluestem (*Andropogon gerardii* Vit.) from eastern South Dakota collections. Proc S.D. Acad Sci, 1980; 59: 65–69.

Boe A, Robbins K, McDaniel B. Spikelet characteristics and midge predation of hermaphroditic genotypes of big bluestem. Crop Sci, 1989; 29: 1433–1435.

Boe A, Ross JG. Path coefficient analysis of seed yield in big bluestem. J Range Manage, 1983; 36: 652–653.

Boe A, Ross JG, Haas RJ, Tober DA. Registration of "Sunnyview" big bluestem. Crop Sci, 1999; 39:593.

Boe A, Ross JG, Wynia R. Pedicellate spikelet fertility in big bluestem from eastern South Dakota. J Range Manage, 1983; 36: 131–132.

Brejda JJ, Brown JR, Lorenz TE, Henry J, Lowry SR. Variation in eastern gamagrass forage yield with environments, harvests, and nitrogen rates. Agron J, 1997; 89: 702–706.

Brejda JJ, Brown JR, Lorenz TE, Henry J, Reid JL, Lowry SR. Eastern gamagrass responses to different harvest intervals and nitrogen rates in northern Missouri. J Prod Agric, 1996; 9: 130–135.

Brejda JJ, Kremer RJ, Brown JR. Indications of associative nitrogen fixation in eastern gamagrass. J Range Manage, 1994; 47: 192–195.

Brown WV, Emery WHP. Apomixis in the gramineae: *Panicoideae*. Am J Bot, 1958; 45: 253–263.

Bruckerhoff SB. 2004. OZ-70 Germplasm big bluestem (http://www.plant-materials.nrcs.usda.gov/pubs/mopmcpg5655.pdf)

Bruckerhoff SB, Cordsiemon RC. 2009. Refuge Germplasm big bluestem (http://www.plant-materials.nrcs.usda.gov/pubs/mopmcrb9068.pdf)

Burson BL, Voigt PW, Sherman RA, Dewald CL. Apomixis and sexuality in eastern gamagrass. Crop Sci, 1990; 30: 86–89.

Burton GW. Recurrent restricted phenotypic selection increases forage yields of Pensacola bahiagrass. Crop Sci, 1974; 14: 831–835.

Carter MC, Manglitz GR, Retwisch MD, Vogel KP. A seed midge pest of big bluestem. J Range Manage, 1988; 41: 253–254.

Casler MD, Pedersen JF, Eizenga GC, Stratton SD. Germplasm and cultivar development. In: Moser LE, Buxton DR, Casler MD (eds) *Cool-Season Forage Grasses*. 1996, pp. 413–470, Agron. Monogr. 34. ASA, Madison, WI.

Chakravarty T, Norcini JG, Aldrich JH, Kalmbacher RS. Plant regeneration of creeping bluestem [*Schizachyrium scoparium* (Michx.) Nash var. *stoloniferum* (Nash) J. Wipff] via somatic embryogenesis. In Vitro Cell Dev Biol-Plant 2001; 37: 550–554.

Chen CH, Boe A. Big bluestem (*Andropogon gerardii* Vitman), little bluestem [*Schizachyrium scoparium* (Michx.) Nash], and Indiangrass [*Sorghastrum nutans* (L.) Nash]. In: Bajaj YPS. *Biotechnology in Agric. and Forest.*, 1988,Vol. 6, pp. 444–457, Crops II. Springer-Verlag, Berlin, Germany.

Chen ZJ, Ni Z. Mechanisms of genomic rearrangements and gene expression changes in plant polyploids. BioEssays 2006; 28: 240–252.

Chen CH, Stenberg NE, Ross JG. Clonal propagation of big bluestem by tissue culture. Crop Sci, 1977; 17: 847–850.

Clark RB, Alberts EE, Zobel RW, Sinclair TR, Miller MS, Kemper WD, Foy CD. Eastern gamagrass (*Tripsacum dactyloides*) root penetration into and chemical properties of claypan soils. Plant Soil 1998; 200: 33–45.

Cordsiemon RC, Kaiser JK. 2010. Ozark Germplasm little bluestem (http://www.plant-materials.nrcs.usda.gov/pubs/mopmcrb9805.pdf)

Cornelius DR. Comparison of some soil-conserving grasses. J Amer Soc Agron, 1946; 38: 682–689.

Cornelius DR. The effect of source of little bluestem grass seed on growth, adaptation, and use in revegetation seedings. J Agric Res, 1947; 74: 133–143.

Cornelius DR. Seed production of native grasses under cultivation in eastern Kansas. Ecol Monogr, 1950; 20: 3–29.

Coyne PI, Bradford JA. Comparison of leaf gas exchange and water-use efficiency in two eastern gamagrass accessions. Crop Sci, 1985; 25: 65–75.

de Kroon H, Huber H, Stuefer JF, van Groenendael JM. A modular concept of phenotypic plasticity in plants. New Phytol, 2005; 166: 73–82.

Dewald CL, Dayton RS. A prolific sex form variant of eastern gamagrass. Phytologia 1985a; 57: 156.

Dewald CL, Dayton RS. Registration of gynomonoecious germplasm (GSF-I and GSF-II) of eastern gamagrass. Crop Sci, 1985b; 25: 715.

Dewald GW, Jalal SM. Meiotic behavior and fertility interrelationships in *Andropogon scoparius* and *A. gerardii*. Cytologia 1974; 39: 215–223.

Dewald CL, Kindiger B. Genetic transfer of gynomonoecy from diploid to triploid eastern gamagrass. Crop Sci, 1994; 34: 1259–1262.

Dewald CL, Kindiger B. Registration of FGT-1 eastern gamagrass germplasm. Crop Sci, 1996; 36: 219.

Dewald CL, Louthan VH. Sequential development of shoot system components in eastern gamagrass. J Range Manage, 1979; 32: 147–151.

Dewald CL, Sims PL. Seasonal vegetative establishment and shoot reserves of eastern gamagrass. J Range Manage, 1981; 34: 300–304.

Dewald CL, Taliaferro CM, Dunfield PC. Registration of four fertile triploid germplasm lines of eastern gamagrass. Crop Sci, 1992; 32:504.

de Wet JMJ, Harlan JR, Brink DE. Systematics of *Tripsacum dactyloides* (Gramineae). Am J Bot, 1982; 69: 269–276.

Durling JC, Leif JW, Burgdorf DW. Registration of Southlow Michigan little bluestem germplasm. J Plant Regist, 2007; 1:134.

Dwire, S. 2010. Evaluation of two native grass germplasm collections for their biomass potential. MS thesis, South Dakota State Univ., Brookings, SD.

Faix JJ, Kaiser CJ, Hinds FC. Quality, yield, and survival of Asiatic bluestems and an eastern gamagrass in southern Illinois. J Range Manage, 1980; 33: 388–390.

Fang X, Sugudhi PK, Venuto BC, Harrison SA. Mode of pollination, pollen germination, and seed set in smooth cordgrass (*Spartina alterniflora*, Poaceae). Int J Plant Sci, 2004; 165: 395–401.

Farquharson LI. Natural selection of tetraploids in a mixed colony of *Tripsacum dactyloides*. Proc Ind Acad Sci, 1954; 63: 80–82.

Farr DF, Bills GF, Chamuris GP, Rossman AY. *Fungi on plants and plant products in the United States*. 1989 APS Press, St. Paul, MN.

Fine GL, Barnett FL, Anderson KL, Lippert RD, Jacobson ET. Registration of "Pete" eastern gamagrass. Crop Sci, 1990; 30: 741–742.

Fine G, Thomassie G. 2000. "Vermilion" smooth cordgrass (http://www.plant-materials.nrcs.usda.gov/pubs/lapmcbr5830.pdf)

Finneseth CH. 2010. Evaluation and enhancement of seed lot quality in eastern gamagrass [*Tripsacum dactyloides* (L.) L.]. Ph.D. thesis, University of Kentucky, p 403.

Foy CD. Tolerance of eastern gamagrass to excess aluminum in acid soil and nutrient solution. J Plant Nutr, 1997; 20: 1119–1136.

Fu YB, Phan AT, Coulman B, Richards KW. Genetic diversity in natural populations and corresponding seed collections of little bluestem as revealed by AFLP markers. Crop Sci, 2004; 44: 2254–2260.

Gedye K, Gonzalez-Hernandez J, Ban Y, Ge X, Thimmapuram J, Sun F, Wright C, Ali S, Boe A, Owens V. Investigation of the transcriptome of prairie cordgrass, a new cellulosic biomass crop. Plant Genome 2010; 3: 69–80.

Gonzalez-Hernandez JL, Sarath G, Stein JM, Owens V, Gedye K, Boe A. A multiple species approach to biomass production from native herbaceous perennial feedstocks. In Vitro Cell Dev Biol-Plant 2009; 45: 267–281.

Gould FW. Chromosome counts and cytotaxonomic notes on grasses of the tribe Andropogoneae. Amer J Bot, 1956; 43: 395–404.

Gould FW, Shaw RB. 1983. *Grass systematics*. 2nd ed. Texas A & M Univ. Press, College Station.

Graves CR, Ellis FL, Bates GE. Growing eastern gamagrass. Tennessee Agric Sci, 1997; 181: 11–13.

Gross MF, Hardisky MA, Wolf PL, Klemas V. Relationship between aboveground and belowground biomass of *Spartina alterniflora* (smooth cordgrass). Estuaries, 1991; 14: 180–191.

Gustafson DJ, Gibson DJ, Nickrent DL. Random amplified polymorphic DNA variation among remnant big bluestem (*Andropogon gerardii* Vitman) from Arkansas Grand Prairie. Molec Ecol, 1999; 8: 1693–1701.

Gustafson DJ, Gibson DJ, Nickrent DL. Conservation genetics of two co-dominant grass species in an endangered grassland ecosystem. J Appl Ecol, 2004; 41: 389–397.

Gustafson DJ, Gibson DJ, Nickrent DL. Using local seeds in prairie restoration-data support the paradigm. Native Plants J, 2005; 6: 25–28.

Hall KE, George JR, Riedl RR. Herbage dry matter yields of switchgrass, big bluestem, and Indiangrass with N fertilization1. Agron J, 1982; 74: 47–51.

Hallauer AR, Miranda Filho JB. 1981. *Quantitative genetics in maize breeding*. Iowa St. Univ. Press, Ames, IA.

Handley MK, Kulakow PA, Henson J, Dewald CL. 1990. Impact of two foliar diseases on the growth and yield of eastern gamagrass. pp 31–39. In Eastern Gamagrass Conf. Proc., Poteau, OK. 23–25 Jan. 1989. The Kerr Center for Sustainable Agriculture, Poteau, OK.

Harlan JR, de Wet JMJ. On Ö. Winge and a prayer: the origins of polyploidy. Bot Rev, 1975; 41: 361–390.

Harper JL. 1985. Modules, branches, and capture of resources. In: Jackson JBC, Buss LW, Cook RE (eds) *Population Biology and Evolution of Clonal Organisms*. pp. 1–34, Yale Univ. Press. New Haven, CT.

Hendricks RC, Bushnell DM. 2008. Halophytes Energy Feedstocks: Back to Our Roots. Symposium on Transport Phenomena and Dynamics of Rotating Machinery, 17–22 February 2008, Honolulu, Hawaii.

Hitchcock AS. 1951. Manual of the grasses of the United States. 2nd ed. Revised by Agnes Chase. USDA Misc. Publ. 200. U.S. Govt. Print. Washington, DC.

Holechek JL, Pieper RD, Herbel CH. 2001. *Range Management*. 4th ed. Prentice Hall, Upper Saddle River, NJ.

Houck MJ. 2005a. Cottle County Germplasm sand bluestem (http://www.plant-materials.nrcs.usda.gov/pubs/txpmcpg6196.pdf)

Houck MJ. 2005b. OK Select Germplasm little bluestem (http://www.plant-materials.nrcs.usda.gov/pubs/txpmcpg6200.pdf)

Huff DR, Quinn JA, Higgins B, Palazzo AJ. Random amplified polymorphic DNA (RAPD) variation among native little bluestem [Schizachyrium scoparium (Michx.) Nash] populations from sites of high and low fertility in forest and grassland home. Molec Ecol, 1998; 7: 1591–1597.

Jacobson ET, Tober DA, Haas RJ, Darris DC. 1986. The performance of selected cultivars of warm-season grasses in the northern prairie and plains states. In: Clambey GK, Pemble RH (eds) Proc. 9th North Amer. Prairie Conf. 29 July–1 Aug. 1984. Moorhead, MN. Tri-college Univ. Ctr. Environ. Studies, NDSU, Fargo, ND.

Jacobson ET, Wark DB, Arnott RG, Haas RJ, Tober DA. Sculptured seeding: An ecological approach to revegetation. Restor Manage Notes, 1994; 12: 2–6.

Jefferson PG, McCaughey WP, May K, Woosaree J, Macfarlane L, Wright SMB. Performance of American native grass cultivars. Native Plants J, 2002; 3: 24–33.

Jefferson PG, McCaughey WP, May K, Woosaree J, McFarlan L. Potential utilization of native prairie grasses from western Canada as ethanol feedstock. Can J Pl Sci, 2004; 84: 1067–1075.

Jensen, N. K. 2006. Prairie cordgrass (*Spartina pectinata*) plant guide (http://www.plant-materials.nrcs.usda.gov/pubs/ndpmcpg5694.pdf)

Johnson SR, Knapp AK. Impact of *Ischnodemus falicus* (Hemiptera: Lygaeidae) on photosynthesis and production of *Spartina pectinata* wetlands. Environ Entomol, 1996; 25: 1122–1127.

Jung GL, Shaffer JA, Stout WL, Panciera MT. Warm-season grass diversity in yield, plant morphology, and nitrogen concentration and removal in northeastern USA. Agron J, 1990; 82: 21–26.

Kalmbacher RS, Dunavin LS, Martin FG. Fertilization and harvest season of eastern gamagrass at Ona and Jay, Florida. Soil Crop Sci Soc Fla Proc, 1990; 49: 166–173.

Keeler KH. Distribution of polyploidy variation in big bluestem (*Andropogon gerardii*, Poaceae) across the tallgrass prairie region. Genome, 1990; 33: 95–100.

Keeler KH, Davis GA. Comparison of common cytotypes of Andropogon gerardii (Andropogoneae, Poaceae). Amer J Bot, 1999; 86: 974–979.

Keeler KH, Williams CH, Vescio LS. Clone size of Andropogon gerardii Vitman (Big Bluestem) at Konza Prairie, Kansas. Am Midl Nat, 2002; 147: 295–304.

Kindiger, B. and Dewald CL. Genome accumulation in eastern gamagrass, *Tripsacum dactyloides* (L.) L. (Poaceae). Genetica, 1994; 92: 197–201.

Kindiger B, Dewald CL. The reproductive versatility of eastern gamagrass. Crop Sci, 1997; 37: 1351–1360.

King J. 2008. Hampton Germplasm big bluestem (http://www.plant-materials.nrcs.usda.gov/pubs/arpmcrb7986.pdf)

Klucas RV. Associative nitrogen fixation in plants. In: Dilworth M, Glenn A (eds) *Biology and Biochemistry of Nitrogen Fixation*. 1991, pp. 187–198, Elsevier Sci. Publ. Amsterdam.

Krizek DT, Alma Solis M, Touhey PA, Ritchie JC, Millner PD. Rediscovery of the southern cornstalk borer: a potentially serious pest of eastern gamagrass and strategies for mitigation. In: Burns JC. (ed) *Proc Eastern Native Grass Symp*. 3rd, Chapel Hill, NC. 1-3 Oct. 2002, 2003, Omni Press, Madison, WI.

Krupinsky JM, Tober DA. Leaf spot disease of little bluestem, big bluestem, and sand bluestem caused by *Phyllosticta andropononivora*. Plant Dis, 1990; 74: 442–445.

Kucera CL. Tall-grass prairie. In: Coupland RT (ed) *Ecosystems of the world 8A*, 1992, pp. 227–268, Natural grasslands. Elsevier, Amsterdam.

Law AG, Anderson KL. The effect of selection and inbreeding on the growth of big bluestem *Andropogon furcatus*, Muhl J Amer Soc Agron, 1940; 32: 931–944.

Leblanc O, Peel MD, Carman JG, Savidan Y. Megasporogenesis and megagametogenesis in several *Tripsacum* species. (Poaceae). Am. J. Bot, 1995; 82: 57–63.

Lee DK, Owens VN, Boe A, Koo BC. Biomass and seed yields of big bluestem, switchgrass, and intermediate wheatgrass in response to manure and harvest timing at two topographic positions. GCB Bioenerg 2009; 1: 171–179.

Leif J, Burgdorf D. 2005a. Prairie View Indiana Germplasm little bluestem (http://www.plant-materials.nrcs.usda.gov/pubs/mipmcrb6308.pdf)

Leif J, Burdgorf D. 2005b. Prairie View Indiana Germplasm big bluestem (http://www.plant-materials.nrcs.usda.gov/pubs/mipmcrb6309.pdf)

Leitch IJ, Bennett MD. Genome downsizing in polyploidy plants. J Linn Soc Bot, 2004; 82: 651–663.

Li Y, Gao J, Fei S. High frequency embryogenic callus induction and plant regeneration from mature caryopsis of big bluestem and little bluestem. Scient Horticult, 2009; 121: 348–352.

Maas DL, Springer TL, Arnold DC. Occurrence of the maize billbug, *Sphenophorus maidis*, in eastern gamagrass. Southwest Entomol, 2003; 28: 151–152.

Maas DL, Springer TL. Southern corn stalk borer, *Diatraea crambidoides* (Grote), feeding damage on eastern gamagrass in Oklahoma. Southwest Entomol, 2005; 30: 67–69.

Madakadze IC, Coulman BE, McElroy AR, Stewart KA, Smith DL. Evaluation of selected warm-season grasses for biomass production in areas with a short growing season. Bioresource Technol, 1998; 65: 1–12.

Mankin CJ. Diseases of grasses and cereals in South Dakota. S.D. Agric Exp Stn Tech Bull, 1969; 35.

Marchant CJ. Evolution in Spartina (Gramineae) III. Species chromosome numbers and their taxonomic significance. J Linn Soc, 1968; 60: 411–417.

McKone MJ, Lund CP, O'Brien JM. Reproductive biology of two dominant prairie grasses (*Andropogon gerardii* and *Sorghastrum nutans*, Poaceae): Male-biased sex allocation in wind-pollinated plants? Amer J Bot, 1998; 85: 776–783.

Missaoui AM, Fasoula VA, Bouton JH. The effect of low plant density on response to selection for biomass production in switchgrass. Euphytica, 2005; 142: 1–12.

Mobberley DG. Taxonomy and distribution of the genus *Spartina*. Iowa St Coll J Sci, 1956; 30: 471–574.

Montemayor MB, Price JS, Rochefort L, Boudreau S. Temporal variations and spatial patterns in saline and waterlogged peat fields. 1. Survival and growth of salt marsh graminoids. Environ Exper Bot, 2008; 62: 333–342.

Moser LE, Vogel KP. Switchgrass, big bluestem, and Indiangrass. In: Barnes RF et al. (ed) *Forages Vol.1: An Introduction to Grassland Agriculture*. 1995, pp. 409–420, Iowa St. Univ. Press, Ames, IA.

Moyer JL, Sweeney DW. 1995. Nitrogen rate and placement effects on eastern gamagrass under 1-cut and 2-cut harvest systems. pp. 41–43. Kansas Agric Exp. Stn. (Rep.) 733.

Mueller IM. An experimental study of rhizomes of certain prairie plants. Ecol Monogr, 1941; 11: 165–188. Madison, WI.

Mueller JP, Hall, TS, Spears JF, Penny BT. Winter establishment of eastern gamagrass in the southern Piedmont. Agron J, 2000; 92: 1184–1188.

Nault LR, DeLong DM, Triplehorn BW, Styer WE, Doebley JF. More on the association of *Dalbulus* (Homoptera: Cicadellidae) with Mexican *Tripsacum* (Poaceae), including the description of two new species of leafhoppers. Ann Entomol Soc Am, 1983; 76: 305–309.

Nelson RG, Langemeier MR, Ohlenbusch PD. Herbaceous energy crop production feasibility using conservation reserve program acreage. In: Burley SM, Arden ME, Campbell-Howe R, Wilkins-Crowder B. (eds), *Solar '94: Technical Papers*.1994, pp. 236–331. Am. Solar Energy Soc., Boulder, CO.

Newell LC. Effects of strain source and management practice on forage yields of two warm-season grasses. Crop Sci, 1968; 8: 205–210.

Newell LC, Peters LV. Performance of hybrids between divergent types of big bluestem and sand bluestem in relation to improvement. Crop Sci, 1961; 1: 370–373.

Norrmann GA, Quarin CL, Keeler KH. Evolutionary implications of meiotic chromosome behavior, reproductive biology, and hybridization in 6X and 9X cytotypes of *Andropogon gerardii* (Poaceae). Amer J Bot, 1997; 84: 201–207.

Owsley, M. 2009. "Flageo" marshhay cordgrass (http://www.plant-materials.nrcs.usda.gov/pubs/gapmcrb8429.pdf)

Phan AT, Smith SR. Seed yield variation in blue grama and little bluestem plant collections in southern Manitoba. Crop Sci, 2000; 40: 555–561.

Piper JK, Handley MK, Kulakow PA. Incidence of severity of viral disease symptoms on eastern gamagrass within monoculture and polyculture. Agric Ecosyst Environ, 1996; 59: 139–147.

Potter, L, Bingham MJ, Baker MG, Long SP. The potential of two perennial $C_4$ grasses and a perennial $C_4$ sedge as lingo-cellulosic fuel crops in N.W. Europe: crop establishment and yields in E. England Ann Bot, 1995; 76: 513–520.

Prasifka JR, Lee DK, Bradshaw JD, Parrish AS, Gray ME. Seed reduction in prairie cordgrass, *Spartina pectinata* Link., by the floret-feeding caterpillar *Aethes spartinana* (Barnes and McDunnough). Bioenerg Res, 2012; 5: 189–196.

Price D, Casler MD. 2010. Genetic diversity in natural populations of big bluestem from Wisconsin. ASA Abstr, 91–95. 2010 ASA Annual Meeting, Long Beach, CA.

Redmann RE. Production ecology of grassland plant communities in western North Dakota. Ecol Monogr, 1975; 45: 83–106.

Reeder JR. Chromosome numbers in western grasses. Amer J Bot, 1977; 64: 102–110.

Riley RD, Vogel KP. Chromosome numbers of released cultivars of switchgrass, Indiangrass, big bluestem, and sand bluestem. Crop Sci, 1982; 22: 1081–1083.

Salon PR, Earle ED. Chromosome doubling and mode of reproduction of induced tetraploids of eastern gamagrass. Plant Cell Rep, 1998; 17: 881–885.

Salon PR, Pardee WD. Registration of SG4X-1 germplasm of eastern gamagrass. Crop Sci, 1996; 36: 1425.

Schacht WH, Volesky JD, Bauer D, Smart AJ, Mousel EM. Plant community patterns of upland prairie in eastern Nebraska Sandhills. Prairie Natur, 2000; 32: 43–58.

Schieber, E. *Puccinia polysora* rust found on *Tripsacum laxum* in the jungle of Chiapas, Mexico. Plant Dis Rep, 1975; 59: 625–626.

Seifers DL, Handley MK, Bowden RL. Sugarcane mosaic virus strain maize dwarf mosaic virus B as a pathogen of eastern gamagrass. Plant Dis, 1993; 77: 335–339.

Seymour CP, Miller JW. The eradication of two potentially dangerous rusts, *Physopella pallescens* and *P. zeae*, from Florida. Proc Florida St. Hort Soc, 1974; 87: 124–125.

Shanmuganathan, N. Bacterial leaf blight of vetiver. FAO Plant Protect Bull, 1974; 22: 95–96.

Sherman RA, Voigt PW, Burson BL, Dewald CL. Apomixis in diploid x triploid *Tripsacum dactyloides* hybrids. Genome, 1991; 34: 528–532.

Silander JA, Antonovics J. The genetic basis of the ecological amplitude of Spartina patens I. Morphometric and physiological traits. Evolution, 1979; 33: 1114–1127.

Sims PL, Risser PG. Grasslands. In: Barbour MG, Billings WD (eds) *North American Terrestrial Vegetation*. 2nd ed. 2000, pp. 323–356, Cambridge Univ. Press, Cambridge.

Skaradek, W. 2006. "Avalon" saltmeadow cordgrass (http://www.plant-materials.nrcs.usda.gov/pubs/njpmcrb6484.pdf)

Skaradek, W. 2008a. Suther Germplasm big bluestem (http://www.plant-material.nrcs.usda.gov/pubs/njpmcrb8161.pdf)

Skaradek, W. 2008b. Suther Germplasm little bluestem (http://www.plant-materials.nrcs.usda.gov/pubs/njpmcrb8163.pdf)

Skinner RH, Zobel RW, van der Grinten M, Skaradek W. Evaluation of native warm-season grass cultivars for riparian zones. J Soil and Water Cons, 2009; 64: 413–422.

Snetselaar KM, Tiffany LH. A study of *Sphacelotheca occidentalis*, cause of kernel smut of big bluestem. J Iowa Acad Sci, 1991; 98: 145–152.

Songstad DD, Chen CH, Boe A. Plant regeneration in callus cultures derived from young inflorescences of little bluestem. Crop Sci, 1986; 26: 827–829.

Springer TL. Caryopsis size and germination of *Andropogon gerardii* pedicellate and sessile spikelets. Seed Sci Technol, 1991; 19: 461–468.

Springer TL, Dewald CL. Eastern gamagrass and other *Tripsacum* species. In: Moser LE, Sollenberger LE, Burson BL. (eds) *$C_4$ [Warm-season grasses]*. 2004, pp. 955–973, ASA, CSSA, and SSSA, Madison, WI.

Springer TL, Dewald CL, Aiken GE. Seed germination and dormancy in eastern gamagrass. Crop Sci, 2001; 41: 1906–1910.

Springer TL, Dewald CL, Sims PL, Gillen RL. How does plant population density affect the forage yield of eastern gamagrass? Crop Sci, 2003; 43: 2206–2211.

Springer TL, Dewald CL, Sims PL, Gillen RL, Louthan VH, Cooper WJ, Taliaferro CM, Maura C, Jr., Pfaff S, Wynia RL, Douglas JL, Henry J, Bruckerhoff SB, van der Grinten M, Salon PR, Houck MJ, Jr., Esquivel RG. Registration of "Verl" eastern gamagrass. Crop Sci, 2006; 46: 477–478.

Springer TL, Sims PL, Gillen RL. Estimate of forage yield loss in eastern gamagrass due to shoot boring insects. Proc Am Forage Grassl Counc Vol, 2004; 13: 333–336.

Springer TL, Taliaferro CM, McNew RW. Pollen size and pollen viability in big bluestem as related to spikelet type. Crop Sci, 1989; 29: 1559–1561.

Sumin K, Rayburn AL, Lee DK. Genome size and chromosome analyses in Pairie Cordgrass. Crop Sci, 2010; 50: 2277–2282.

Thornberry HH, Otterbacher AG, Thompson MR. Gamagrass (*Tripsacum dactyloides*): A new perennial host of maize dwarf mosaic virus. Plant Dis Rep, 1966; 50: 65–68.

Tian X, Knapp AD, Gibson LR, Struthers R, Moore KJ, Brummer EC, Bailey TB. Response of eastern gamagrass seed to gibberellic acid buffered below its pKa. Crop Sci, 2003; 43: 927–933.

Tian X, Knapp AD, Moore KJ, Brummer EC, Bailey TB. Cupule removal and caryopsis scarification improves germination of eastern gamagrass seed. Crop Sci, 2002; 42: 185–189.

Tober DA, Duckwitz W, Sieler S. 2007 Plant materials for Salt-Affected Sites in the Northern Great Plains. USDA, NRCS, (March 2007), Bismarck, ND, p. 8.

Tober DA, Jensen N, Duckwitz W, Knudson M. 2008. Big bluestem biomass trials in North Dakota, South Dakota, and Minnesota (http://www.plant-materials.nrcs.usda.gov/pubs/ndpmcpu7933.pdf).

U.S. Department of Agriculture. 2009. Summary Report: 2007 National Resources Inventory, National Resources Conservation Service, Washington, DC, and Center for Survey

Statistics and Methodology, Iowa State University, Ames, IA (http://www.nrcs.usda.gov/technical/NRI/2007/2007_NRI_Summary.pdf)

U.S. DOE. 2006. Breaking the biological barriers to cellulosic ethanol: A joint research agenda, DOE/SC-0095, U.S. Department of Energy Office of Science and Office of Energy Efficiency and Renewable Energy (www.doegenomestolife.org/biofuels/).

Vogel KP. Improving warm-season forage grasses using selection, breeding, and biotechnology. In: Moore KJ, Anderson BE (eds) *Native Warm-Season Grasses: Research Trends and Issues*. 2000, pp. 83–106, CSSA Spec. Pub. No. 30. ASA, Madison, WI.

Vogel KP, Mitchell RB, Klopfenstein TJ, Anderson BE. 2006a. Registration of "Goldmine" big bluestem. Crop Sci, 46: 2313–2314.

Vogel KP, Mitchell RB, Klopfenstein TJ, Anderson BE. 2006b. Registration of "Bonanza" big bluestem. Crop Sci, 46: 2314–2315.

Vogel KP, Pedersen JF. Breeding systems for cross-pollinated perennial grasses. Plant Breed Rev, 1993; 11: 251–274.

Warren RS, Baird LM, Thompson AK. Salt tolerance in cultured cells of *Spartina pectinata*. Plant Cell Rep, 1985; 4: 84–87.

Weaver JE. *North American Prairie*. 1954, Johnsen Publ. Co., Lincoln, NE.

Weaver JE. Extent of communities and abundance of the most common grasses in prairie. Bot Gaz, 1960; 122: 25–33.

Weaver JE, Albertson FW. Nature and degree of recovery of grassland from the great drought of 1933 to 1940. Ecol Monogr, 1944; 14: 393–479.

Weaver JE, Fitzpatrick TJ. Ecology and relative importance of the dominants of the tall-grass prairie. Bot Gaz, 1932; 93: 113–150.

Weimer PJ, Dien BS, Springer TL, Vogel KP. In Vitro gas production as a surrogate measure of the fermentability of cellulosic biomass to ethanol. Appl Microbiol Biotechnol, 2005; 67: 52–58.

Weimer PJ, Springer TL. Fermentability of eastern gamagrass, big bluestem and sand bluestem grown across a wide variety of environments. Bioresource Technol 2007; 98: 1615–1621.

Zeiders KE. Leaf spots of big bluestem, little bluestem, and Indiangrass caused by *Ascochyta brachypodii*. Plant Dis, 1982; 66: 502–505.

# Chapter 10
# Alfalfa as a Bioenergy Crop

Kishor Bhattarai, E. Charles Brummer, and Maria J. Monteros

*The Samuel Roberts Noble Foundation, 2510 Sam Noble Parkway, Ardmore, OK, USA*

## 10.1 Introduction

Increased global fuel and energy demands, depletion of fossil fuel reserves, political instability in some fossil-fuel-producing countries, and the desire for energy independence are driving the identification of alternative energy sources. The daily petroleum consumption in the United States for 2010 alone was 582 million gallons (United States Energy Information Administration, 2011). The use of fossil fuels releases sequestered carbon to the atmosphere and the increase in atmospheric carbon is one of the factors associated with global climate change. Efforts to reduce environmental impacts from global carbon emissions associated with global climate change (Brown, 2003; Farrell et al., 2006) are additional driving forces to identify alternative renewable energy resources.

In 2005, the U.S. Department of Agriculture (USDA) and the Department of Energy (DOE) issued a mandate to replace 30% of the transportation fuel consumption in the United States with biofuels by 2030, which translates to 90 billion gallons of ethanol. This "billion ton report" estimated that at least one billion tons of dry biomass will be required as feedstock to meet the energy mandate of the US government (USDA/DOE, 2005). This estimate included annual crop residues, forest residues, perennial crops, grains (mainly corn for ethanol and soybeans for biodiesel), animal and municipal waste, process residues and other materials. Aligned with the original biofuel goal, the US government passed the Energy Independence and Security Act (EISA) bill in 2007, which refined the biofuel target to 36 billion gallons by 2022. The EISA also put a cap on corn-grain-based ethanol production to 15 billion gallons per year to minimize large-scale land use changes. In 2010, 13.23 billion gallons of bioethanol were produced in the United States at 204 biorefineries (Renewable Fuels Association, 2011). Although most of these biorefineries used corn grain as feedstock, some also used sugar cane bagasse, cheese whey, barley, beer and potato waste, and wood residues as biofuel feedstock.

The search for alternative energy sources including bioenergy (both biofuel and biopower) is also driven by the global population growth with the desire for an increased standard of

---

*Bioenergy Feedstocks: Breeding and Genetics*, First Edition. Edited by Malay C. Saha, Hem S. Bhandari, and Joseph H. Bouton.
© 2013 John Wiley & Sons, Inc. Published 2013 by John Wiley & Sons, Inc.

living, as well as recent erratic weather patterns driven by global climate change (Vermerris, 2008). Bioenergy can contribute toward meeting energy demands jointly with other sources of alternative energy through optimization of feedstock production and processing, as well as advances in infrastructure to harvest, store, and transport biomass crops. Efforts to sustainably meet current and future energy demands using biofuels should be focused toward developing biomass crops that are more productive with lower agricultural inputs and have increased processing efficiency, specifically for lignocellulosic biomass crops including alfalfa (*Medicago sativa* L.).

## 10.2 Biomass for Biofuels

Bioenergy refers to energy produced from biological materials, specifically biosynthetic organisms (Vermerris, 2008). Photosynthesis is the process used by plants to convert solar energy into chemical energy in the form of sugar molecules from atmospheric carbon dioxide and water. When fuel is burned or utilized during the oxidation process, carbon dioxide and water molecules are recycled back into the atmosphere as fixed solar energy is released; therefore, biomass-based energy is a renewable source of energy. Fermentation and burning are oxidation reactions that release energy. Theoretically, 180 g of glucose can be converted into 92 g of ethanol through fermentation. A ton of dry herbaceous biomass could produce 100–110 gallons of ethanol (Dien, 2007) depending on the feedstock type. Bioenergy platforms include thermochemical or biochemical processes associated with pyrolysis or lignocellulosic conversion for the generation of electricity or liquid biofuels.

### 10.2.1 Lignocellulose-based Biofuels

Biomass crops can be used as a source of fermentable sugars to supply liquid transportation fuels including ethanol and other low-molecular-weight alcohols, depending on the microbial strain used for fermentation (Vermerris, 2008). Ethanol can be used as fuel in traditional engines when blended with gasoline. Ethanol blends with higher ethanol content require specialized engines such as those found in flexible fuel vehicles. Fermentable sugars needed for ethanol production can be obtained from multiple biomass crops and processing approaches including the hydrolysis of cellulose and hemicelluloses present in the plant cell wall. Ethanol obtained from cell walls is termed cellulosic ethanol (Vermerris, 2008).

The environmental and energy costs of producing lignocellulosic biomass vary between annual and perennial species (Hammerschlag, 2006). Annuals tend to maximize the aboveground biomass to enhance reproductive success in the current year, whereas perennials tend to devote a significant amount of carbon resources into their roots to maximize their persistence during the subsequent year. However, once established, perennials can utilize their extensive and deep root systems to survive in nutrient-depleted soils and water-limited environments. Perennials such as alfalfa tend to grow more in the second year after establishment (Ates and Tekeli, 2004; Rock et al., 2009). Additionally, perennials do not need annual reseeding as annuals do, thus spreading the cost and energy for seeding and establishment across multiple years. Perennials are also able to absorb nutrients efficiently, prevent soil erosion, increase soil tilth, and increase soil organic matter content compared to annual species.

Because the production of biofuels requires energy for plant growth, biomass harvest, transportation and conversion into biofuels, the net gain in energy and carbon balance depends on the biomass crop and the efficiency of these processes. The net energy gain obtained from corn-based ethanol is a contentious issue among scientists because of the various parameters used to estimate energy balance in their models. Although some studies have identified a negative energy balance in corn-based ethanol (Pimentel, 1991; Pimentel and Patzek, 2005), many other studies have shown a net positive energy balance (Shapouri et al., 2003; Farrell et al., 2006; Hammerschlag, 2006; Wang et al., 2007) in corn-based ethanol. Although the net energy balance of corn-based ethanol represents a short-term alternative (Sommerville, 2007), it does not address the concern of long-term environmental and energy sustainability. The fossil fuel ratio, determined by calculating the ratio of the energy delivered to a customer per unit of fossil energy consumed, shows higher efficiency (5.3) for lignocellulosic-based ethanol than that of corn-based ethanol (1.4) (Sheehan et al., 2003). The GREET model which stands for greenhouse gases, regulated emissions, and energy use in transportation, developed by Dr. Michael Wang at Argonne National Laboratory's Center for Transportation Research, shows that reduction of greenhouse gases for corn-based ethanol is 18–28% whereas cellulosic ethanol offers a reduction of greenhouse gases of 87%. Thus, cellulose-based ethanol represents a promising alternative to corn-based bioenergy, despite the fact that processing technologies are yet to be fully developed. Lignocellulosic biomass can be obtained from a wide range of sources (e.g., agricultural and forestry byproducts and municipal waste) and from species with significant biomass yield potential that are not part of the human food chain. Thus, they have the potential to address long-term environmental sustainability of food and fuel production.

### 10.2.2 Plant Cell Wall Components

The estimated biomass production by terrestrial plants is $170–200 \times 10^9$ Mg (Leith, 1975), out of which an estimated 70% is made up of plant cell walls (Duchesne and Larson, 1989). Plant cell wall constituents vary between species and tissues within a plant, across maturity within a tissue, and between cell wall layers within a cell. Within a species, leaf tissues generally have thinner cell walls to facilitate gas exchange and light absorption, whereas stem tissues generally have thicker cell walls at maturity because of their role in supporting plant weight and in conducting water, nutrients, and photosynthates to and from leaf and root tissues. Therefore, leaves have lower lignin concentration compared to stems. Stem tissues are a primary source of lignocelluloses in alfalfa. Markovic et al. (2007) reported that alfalfa leaves had 49 g kg$^{-1}$ of lignin content at the full flowering stage compared to 119.6 g kg$^{-1}$ of lignin in alfalfa stems. They also found a 17% increase in lignification of leaf tissues compared to 42% increase in lignification of stem tissues during the transition from bud stage to full flowering in alfalfa.

Cellulose, hemicellulose, pectin, and lignin are the major chemical components of alfalfa cell walls. Cellulose is the most dominant sugar molecule in the plant cell wall matrix, making it the most abundant biopolymer in the world. Cellulose is a major component in both primary and secondary cell walls in a plant cell and forms the scaffold of the cellular matrix. In alfalfa, cellulose contributes 27.5% of the total dry weight of the cell wall at the bud stage (Dien et al., 2006). Because cellulose is made of glucose units, the enzymes required to hydrolyze cellulose and to ferment subsequently released glucose molecules are simple and the technology is well developed. Hemicellulose is the second most abundant sugar molecule in nature and consists of pentose and hexose sugars such as xylose, mannose, fucose, arabinose, and galactose. Hemicellulose represents branched polysaccharides that cross-link cellulose microfibrils and lignin polymers and impart rigidity to the cell wall matrix. In alfalfa, hemicellulose contributes 10.5% of the total dry weight of the cell wall at the bud stage of growth (Dien et al., 2006). Hemicellulosic components vary among different plant species and

Table 10.1. Cell wall carbohydrate compositions in alfalfa harvested at bud and full flower stage of maturity.

|  | Bud | Full Flower (g kg$^{-1}$) |
| --- | --- | --- |
| Glucose | 275 | 306 |
| Xylose | 85 | 99 |
| Arabinose | 20 | 21 |
| Galactose | 17 | 17 |
| Mannose | 18 | 21 |
| Rhamnose | 6 | 5 |
| Fucose | 2 | 2 |
| Uronic acid | 82 | 76 |

Source: Dien et al., 2006.

within a species at different stages of maturity. As alfalfa stems mature, the concentration of xylose increases (Table 10.1) while the concentration of mannose and fucose decreases (Jung and Engels, 2002). The heterogeneous polymeric nature of hemicellulose has made it relatively difficult to use for biofuel production based on existing technologies. Hemicelluloses often degrade into inhibitory compounds during the biomass pretreatment processing and subsequent fermentation by commercially available yeast strains. Meanwhile, research to optimize the pretreatment process and use of different fermentation biota for converting pentose sugars into ethanol is ongoing (Walton and Van Heiningen, 2010).

Pectin is abundant in the primary cell walls and middle lamellas and consists of a group of polysaccharide molecules that are characterized by the presence of $\alpha$-1,4-linked galacturonic acid (Caffall and Mohnen, 2009). Pectin is described as structurally and functionally the most complex polysaccharide in plant cell walls (Mohnen, 2008), and has both pentose and hexose residues in the form of galactose, rhamnose, and arabinose as well as uronic acid. It is a branched chain of polysaccharides and joins cellulose microfibrils with other cell wall components. Dicot stems contain a significant amount of pectin in their cell walls (Jung and Engels, 2002), and more so than grasses (Cassida et al., 2007). Pectin contributes 12.5% of the total dry weight of cell walls in alfalfa at the bud growth stage (Dien et al., 2006). Pectin increases cell wall recalcitrance to enzymatic degradation for sugar release (Lionetti et al., 2010) and can increase the acid buffering capacity of alfalfa biomass (Dien et al., 2006). Therefore, higher amounts of acids are required during biomass pretreatment to release sugars. However, pectin can also be used directly as a source of sugar molecules for ethanol release (Peterson, 2003).

Lignin is the second most abundant biopolymer in nature that provides support for plant growth. The structural complexity of lignin makes it recalcitrant to digestion by bacterial enzymes (Eulgem, 2006). The hydrophobic property of lignin is important for water transportation in xylem vessels of alfalfa stems (Engels and Jung, 1998). Lignin is a complex phenylpropanoid heteropolymer that forms by polymerization of mainly three units: $p$-coumaryl, coniferyl, and sinapyl alcohols. When these units are arranged in lignin polymers, they are correspondingly designated as $p$-coumaryl (H), coniferyl (G), and syringyl (S) units. These units have the common basic structure of a phenyl ring with a propane chain, but they differ from each other at the $C_3$ and $C_5$ positions of the phenyl ring. The G unit has methoxylation at the $C_3$ position while the S unit has a methoxyl group at both the $C_3$ and $C_5$ positions, whereas the H unit has none. Concentrations of the G and S units vary between and within species during different stages of maturity. Lignin polymers that are rich in the G units have greater proportions of C–C bonds compared to lignin polymers that are rich in S units. Because the C–C bond requires higher energy to cleave, cell walls high in S lignin are less recalcitrant to enzyme degradation than cell walls high in G lignin (Huntley et al., 2003). In alfalfa, lignin contributes 15.8% of the total cell wall dry weight at the bud stage of growth (Dien et al., 2006).

Composition of cell walls vary among species as well as in various tissues and stages of maturity within species (Jung and Engels, 2002). Thus, efforts to modify alfalfa for biofuel production require an understanding of cell wall components and their variations in different alfalfa tissues. Thickening of cell walls in the epidermis, collenchyma, chlorenchyma, and cambial parenchyma in alfalfa is mainly due to the primary cell wall thickening and not due to lignification (Engels and Jung, 1998; Jung and Engels, 2002). The pith parenchyma in alfalfa exhibits centripetal lignification as it matures (Engels and Jung, 1998). The primary xylem parenchyma and phloem tissues in alfalfa do not undergo lignification until the complete elongation of the internodes. However, the primary xylem vessels begin to accumulate lignin from an early growth stage. The secondary xylem and secondary phloem tissues exhibit lignification as they mature. The primary phloem shows some lignification in the primary walls in the form of ring structures at later stages of maturity; however, the secondary walls do not show any lignification (Engels and Jung, 1998).

Maturity stage plays a role in alfalfa cell wall composition and the concentration of cellulose and hemicellulose increases slightly with maturity (Jung and Engels, 2002). In a regrowth study performed at the University of Minnesota, alfalfa stems contained 40% cellulose and roughly equal amounts of pectin, hemicelluloses, and lignin concentrations (Jung and Engels, 2002). As the alfalfa stem matured, the proportion of xylose in hemicelluloses increased and the proportions of mannose and fucose decreased. The composition of monolignols in alfalfa stems also changed with maturity as seen by an increase in the S/G ratio from 0.29 to 1.01 corresponding to an increase in S monolignol at later maturity stages. Thus, understanding cell wall composition in alfalfa can contribute to developing strategies that maximize access to fermentable sugars suitable for bioprocessing resulting in increased conversion efficiency and ethanol yields.

## 10.3 Why Alfalfa?

### 10.3.1 Background

Alfalfa is currently the fourth most widely grown crop in the United States. Over 8 million hectares of alfalfa were grown in 2010, during which alfalfa biomass production totaled 68 M tons, and ranged from 3.1 Mg ha$^{-1}$ in Vermont and New Hampshire, to 18.3 Mg ha$^{-1}$ in Arizona, with a US national average of 7.6 Mg ha$^{-1}$ (USDA/NASS, 2011). Alfalfa is a perennial forage legume that hosts nitrogen-fixing bacteria in its root nodules that enable atmospheric nitrogen fixation. As a result, alfalfa rarely requires nitrogen fertilizer for growth, thus reducing input costs for biomass production and results in high protein forage biomass.

Alfalfa is one among 87 species in the genus *Medicago*. Although cultivated alfalfa is an autotetraploid ($2n = 4x = 32$), the *M. sativa* complex includes various ploidy levels ranging from diploids to hexaploids with a basic genomic number $x = 8$ (except in five annual species where $x = 7$). Tetraploid alfalfa taxa display tetrasomic inheritance, meaning that each chromosome can pair and recombine genetically at meiosis with any of the other three homologous chromosomes. Nine subspecies in the *M. sativa* complex including *M. sativa* subspecies *sativa*,

*M. sativa* subspecies *falcata*, and *M. sativa* subspecies *glutinosa* are recognized (Gunn et al., 1978; Quiros and Bauchan, 1988). *M. sativa* subspecies *falcata* includes diploid and tetraploid forms and is characterized by yellow flowers with straight to sickle-shaped pods. *M. sativa* subspecies *sativa* is tetraploid and has flowers ranging in color from purple, violet, to lavender, and has coiled pods. *M. sativa* subspecies *caerulea* is similar to the *M. sativa* subspecies except it has a diploid genome. *M. sativa* ssp. *glutinosa* has both diploid and tetraploid forms with bright yellow to cream colored flowers that mature into yellow flowers. Pods are coiled and are covered with glandular hairs. Subspecies *falcata* is grown in northern climates, and thus tends to be more winter hardy than *sativa*.

Perennials in the *Medicago* genus likely evolved along the northern coast of the Mediterranean region, whereas annuals are thought to have evolved in relatively hotter and drier climates of the Mediterranean basin. The change in life form from perennials to annuals has been argued as the cause behind the evolution of seed dormancy and self-pollination systems in the annuals. Tetraploid alfalfa is distinguished from its diploid counterpart by the larger size of its flowers, pods, and seeds. Hybridization seems to be common among subspecies in the *M. sativa* complex. Lesins and Lesins (1964) proposed that diploid *M. glomerata* was the ancestor of present day alfalfa (including the subspecies *sativa*, *falcata,* and *glutinosa*). The hybridization of *M. sativa* ssp. *sativa* and *M. sativa* ssp. *falcata,* may have led to the cultivation of alfalfa in temperate zones (Lesins, 1976). Alfalfa is primarily outcrossing, and is pollinated mostly by various species of bees, including the alfalfa leafcutter bee (*Megachile rotundata*) (Pitts-Singer and Cane, 2011) and the alkali bee (*Nomia melanderi*) (Cane, 2008). The flower is complete, and self-pollination can occur in alfalfa, although some plants are self-incompatible. As a result, alfalfa often shows various degrees of inbreeding depression (Busbice and Wilsie, 1966; Li and Brummer, 2009).

### 10.3.2 Prospect as a Biofuel Feedstock

The Consortium for Alfalfa Improvement (CAI) was formed in 2003 and includes researchers from the Noble Foundation (Ardmore, OK), the U.S. Dairy Forage Research Center (USDFRC—Madison, WI), the Plant Science Research Unit (St. Paul, MN), Forage Genetics International Inc. (FGI—West Salem, WI), and Pioneer (Arlington, WI). The purpose of the consortium is to improve and "redesign alfalfa" by focusing on enhancing its yield and digestibility. Alfalfa has value as a dual-purpose forage for fuel and feed, based on utilization of the fibrous and less nutritious stems for biofuel (liquid fuels or electricity) and using the leaves as feed (Dale, 1983; Downing et al., 2005). However, alfalfa has more value as a high cost product for animal feed compared to the projected low value, low cost product of prospective bioenergy feedstocks. The potential rate of return differential may need to be addressed before alfalfa can play a significant role in the general biomass market.

Strategies to capture the dual role of alfalfa involve separation of leaves and stem portions for their use in different applications. Techniques used for alfalfa leaf and stem separation include: (1) dry fractionation, (2) wet fractionation, and (3) harvest fractionation (Shinners et al., 2007). Dry fractionation uses mechanical sieving and air separation of field-dried and chopped alfalfa samples but its feasibility depends on weather conditions. Wet fractionation extracts the juice from fresh-cut whole-plant alfalfa through maceration and is less weather dependent but has greater equipment and operational costs. The juice fraction can be later processed to obtain high-quality protein with less fiber and other co-products. After protein extraction from the alfalfa juice, the remnant juice can be used as fertilizer to increase the beneficial soil microflora and decrease pathogenic soil microflora (Doiphode and Mungikar, 2007). Harvest fractionation is less weather dependent and has lower operational costs (Shinners et al., 2007). Mechanical fractionation of alfalfa harvested in the field at 25% flowering into leaf and stem fractions resulted in 90% of the leaf tissue being separated from the stems (Shinners et al., 2007). The particle size of the harvested leaves was similar to the size of chopped whole alfalfa plants and thus no further size reduction was required for ensiling. The density of the stripped leaf was 11% greater than that of chopped whole plants. Stripped stems dried faster after harvest than nonstripped stems resulting in a reduction of weather-related sugar loss, thereby maximizing the forage and feedstock value of the stem fraction. The operational cost to separate alfalfa leaves and stems should be evaluated against the economic gain generated by reducing processing to make silage and condensed leaf material to ship as feed, and by limiting field drying that results in less weather-related loss of water-soluble molecules from the stem.

## 10.4 Breeding Strategies

Alfalfa is allogamous with inbreeding limited by a partial self-incompatibility system (Viands et al., 1988). Thus, populations are a highly heterogeneous mixture of genotypes, each of which can be heterozygous at many loci. Inbreeding severely depresses plant vigor and fertility in tetraploid alfalfa due to the loss of complementary gene interactions (Bingham et al., 1994; Osborn et al., 1997), and this inbreeding depression effectively prevents the development of inbred lines. Hence, most alfalfa cultivars have been developed by intercrossing a few to hundreds of highly heterozygous parental individuals to produce seed, which is increased by open pollination through two to four subsequent generations and released as a synthetic cultivar.

### 10.4.1 Germplasm Resources

The genetic diversity of alfalfa and species in the *M. sativa* complex is extensive, and it enables the selection of desirable traits for plant adaptation to a rapidly changing environment. Institutions with collections of *Medicago* species germplasm include: (1) the National Plant Germplasm System of the USDA whose Germplasm Resources Information Network (GRIN) includes extensive online documentation of characteristics of many of the accessions (http://www.ars-grin.gov/npgs, accessed August 20, 2011); (2) the N.I. Vavilov Institute of Plant Industry (VIR) in St. Petersburg, Russia (http://www.vir.nw.ru, accessed August 20, 2011); (3) the System-wide Information Network on Genetic Resources (SINGER), the germplasm information exchange network of the Consultative Group on International Agriculture Research (CGIAR) and its partners (http://www.singer.cgiar.org, accessed August 10, 2011); (4) the South Australia Research and Development Institute (SARDI) run by the Australian government; (5) the International Center for Agricultural Research in the Dry Areas (ICARDA), located in Aleppo, Syria, which maintains the Mediterranean and temperate forages; (6) a series of European gene banks (http://www.ecpgr.cgiar.org/germplasm_databases/international_multicrop_databases.html, accessed August 20, 2011); (7) the Canadian plant germplasm collection (http://pgrc3.agr.gc.ca/index_e.html, accessed September 11, 2011), which includes the collection of K. Lesins of the University of Alberta, Canada (Small, 2011). Core collections, a subsample of accessions representing the genetic or phenotypic diversity in the whole collection with minimal redundancy, are available for perennial *Medicago* (Basigalup et al.,

1995) and for annual species (Diwan et al., 1994). Some of the perennial species of *Medicago* hybridize with *M. sativa* and can serve as direct gene sources for cultivar improvement (McCoy and Bingham, 1988).

### 10.4.2 Cultivar Development

Alfalfa was first evaluated in the United States in the late 1700s, including at Monticello by Thomas Jefferson and at Mount Vernon by George Washington. By 1950, nine major germplasm introductions had been recognized, including "Falcata," "Varia," "Turkestan," "Flemish," "Ladak," "Chilean," "Peruvian," "Indian," and "African" (Barnes et al., 1977), and recurrent introductions from throughout the world continue today. Agricultural experiment stations and USDA-ARS initiated alfalfa breeding programs in the first half of the 1900s. Most early alfalfa cultivars were developed by the public sector, but by the 1970s, private companies captured the majority of the alfalfa cultivar market (Barnes et al., 1988).

### 10.4.3 Synthetic Cultivars and Heterosis

Conventional alfalfa breeding programs primarily use recurrent phenotypic selection to concentrate alleles for traits of interest into populations that will be marketed as synthetic cultivars. The breeding focus for much of the past 50 years has been on improving pest resistances, and consequently, today, most cultivars have very high levels of resistance to common diseases such as *Phytophthora*, *Aphanomyces*, Anthracnose, Bacterial blight, *Verticillium* wilt, and *Fusarium*, together with other pests (National Alfalfa & Forage Alliance, 2011). Breeding programs often develop improved populations from more than 100 parents to both capture desirable alleles for the suite of traits and to avoid inbreeding in advanced generations. Cultivars are marketed as synthetic varieties formed by two to three generations of open pollination following parental crossing. Broad-based populations reduce the amount of inbreeding depression that occurs due to sib mating during seed production compared to narrow-based populations (Busbice, 1969).

Complex, quantitatively inherited traits like yield could be more effectively selected using family-based methods. In alfalfa breeding, half-sib family recurrent selection, with half-sib families produced by intercrossing a group of plants either by hand or by using bees in a cage, is the most common method. Full-sib progenies and selfed family selection have been used to a very small extent, limited by the additional effort required to create the families. Interpopulation improvement, such as reciprocal recurrent selection, has not been used to any extent in commercial breeding programs. Improvements in genetic gain could potentially be realized through relatively minor changes in the specific breeding methods being used in many programs (Casler and Brummer, 2008).

The success of hybrid maize in the early 1900s led to interest in developing alfalfa hybrids (Tysdal and Kiesselbach, 1944). Heterosis for biomass yield has been documented in alfalfa on numerous occasions and reviewed in Brummer (1999), but the difficulty of producing inbred lines and the existence of progressive heterosis in tetraploids (Bingham et al., 1994; Li and Brummer, 2009) have precluded significant achievements in generating alfalfa hybrids. Male sterility and self-incompatibility systems exist in alfalfa and could be used to develop hybrids (Viands et al., 1988). A type of hybrid cultivar is currently marketed by Dairyland Seeds Inc. Semi-hybrid populations could partially exploit heterosis with relatively small changes needed from typical seed production practices (Brummer, 1999). However, lines with cytoplasmic male sterility may exhibit poor pollination due to selective avoidance by pollinators.

Heterosis in alfalfa can be exploited by improving populations with complementary alleles within and among loci (Bingham et al., 1994). Three existing germplasm pools in the United States have been proposed as heterotic groups to capture heterosis, including *M. sativa* ssp. *falcata*, dormant ssp. *sativa*, and nondormant ssp. *sativa* (Brummer, 2004). Heterosis between *M. sativa* ssp. *falcata* and ssp. *sativa* has been observed for biomass yield (Riday and Brummer, 2002). Factors affecting heterosis in elite *M. sativa* ssp. *falcata* hybrids include geographical origin and fall growth of the *M. sativa* ssp. *falcata* (Riday and Brummer, 2004). Semi- and nondormant derived *M. sativa* ssp. *sativa* populations did not exhibit heterosis for yield (Şakiroğlu and Brummer, 2007), but other nondormant populations may be better to generate high-yielding hybrids (Segovia-Lerma et al., 2004). Given the high levels of multiple pest resistances in modern germplasm, reinvestigation of some type of hybrid system and/or very narrow-based synthetic populations is warranted in order to capitalize on nonadditive and epistatic genetic variances known to be present in alfalfa for complex traits like yield (Dudley et al., 1960; Li and Brummer, 2009). The addition of transgenes makes the restriction of parental numbers important because it is easier to maintain populations in which high levels of individuals express the transgene.

### 10.4.4 Molecular Breeding

Molecular breeding encompasses various methods that use molecular markers to assist selection of phenotypic traits. A number of marker systems have been applied to alfalfa breeding, but the most widely used markers currently are simple sequence repeats (SSR). Single nucleotide polymorphism (SNP) markers are codominant markers and represent sequence variation at the nucleotide level. Compared to SSRs, SNPs are more amenable to high-throughput genotyping that allows samples to be genotyped faster and more economically than SSRs (Hurley et al., 2004). Alfalfa transcriptome sequencing has identified genome-wide SNP on all eight chromosomes (Han et al., 2011), and these markers have been used to fill in gaps in existing linkage maps. Due to the autotetraploid nature of alfalfa, SNP genotyping platforms require capabilities for allelic dosage detection. High-resolution melting (HRM) can be used for SNP genotyping and dosage identification in alfalfa (Han et al., 2012), as is the case for distinguishing each of the following allelic combinations: TTTT, TTTC, TTCC, TCCC, CCCC. Other platforms in the developmental pipeline include Illumina Infinium and GoldenGate arrays and Genotyping-by-Sequencing (GBS) (Elshire et al., 2011; Wei et al., 2011).

Markers can be used in a breeding program provided that sufficient levels of linkage disequilibrium (LD) exist between the marker and the desired trait allele. Initial mapping in alfalfa focused on diploids due to their simpler segregation patterns than autotetraploids. Since the development of diploid maps in alfalfa (Brummer et al., 1993; Echt et al., 1993; Kiss et al., 1993), the availability of a large number of SSR markers identified in the *Medicago truncatula* genome sequence, a model legume and close relative of alfalfa, has greatly enhanced mapping capabilities in tetraploid alfalfa (Eujayl et al., 2003; Julier et al., 2003; Sledge et al., 2005). Linkage maps in alfalfa at the tetraploid level were used to locate the genomic positions of quantitative trait loci (QTL) associated with aluminum tolerance (Khu et al., 2010b), winter hardiness, freezing injury, and fall growth (Brower et al., 2000), biomass yield (Robins et al., 2007c; Li et al., 2011), regrowth and plant height (Robins et al., 2007a), persistence (Robins et al., 2008), and selfed seed set (Brower et al., 2000; Robins et al., 2007b; Robins et al., 2008; Li et al., 2011). Mapping has generally been conducted in $F_1$ populations derived from noninbred parents to avoid the complications of segregation distortion in inbred populations resulting from uncovering deleterious recessive alleles during the inbreeding process (Osborn et al., 1997). Molecular markers can also be used for genome wide or genomic selection to incorporate desirable alleles at many loci that have small genetic effects based on their estimated breeding value (Heffner et al., 2009). Genomic selection may accelerate genetic gains by making predictions from marker-based models rather than requiring phenotypic evaluation, particularly useful for perennial species like alfalfa that require multi-year field evaluations.

Genomic resources developed in the model legume *M. truncatula*, including a large number of ESTs, genome sequence, gene chips, and mutant collections, can be applied to alfalfa breeding. However, characteristics of alfalfa that differ from *M. truncatula* (tetraploid vs. diploid, perennial vs. annual, and outcrossing vs. self-pollinating) make the model less useful for understanding the genetic mechanisms for traits like persistence, biomass yield, and winter hardiness. Currently, alfalfa transcriptome and genome sequencing are underway to develop a reference genome sequence that will be used to understand the gene content and genome structure in alfalfa. Organizations working toward this effort include The Samuel Roberts Noble Foundation, National Center for Genome Resources, and The J. Craig Venter Institute (USA), Keygene (The Netherlands), INRA (France), and the University of Toronto (Canada). The knowledge and tools developed from the sequencing project will be directly applied in breeding programs.

### *10.4.5 Trait Integration Through Biotechnology*

Genetic transformation of plant species has been used as a research tool and as a strategy for gene introgression (Parrott and Clemente, 2004). Transformation is useful to extend variation in a trait beyond what is available through existing natural genetic variation. Transformation of alfalfa using *Agrobacterium* began in the mid-1980s (Deak et al., 1986; Shahin et al., 1986). Transgenic approaches to modify traits include downregulating or upregulating individual genes or regulatory genes, including transcription factors (Chen and Dixon, 2007; Wang et al., 2010). Several economically important transgenes have been successfully incorporated in alfalfa to improve forage composition and abiotic stress responses (Jiang et al., 2009). Lignin concentration in alfalfa was reduced using transgenic techniques (Chen and Dixon, 2007; Wang et al., 2010), thus increasing the availability of sugar molecules for fermentation and processing. Reduced lignin alfalfa lines obtained by downregulating the caffeic acid 3-*O*-methyltransferase (COMT) and caffeoyl CoA 3-*O*-methyltransferase (CCOMT) genes from the lignin biosynthetic pathway are in the pipeline for commercial release by Forage Genetics International Inc. Suppression of genes in the lignin biosynthetic pathway also increased cell wall sugar concentration in switchgrass (*Panicum virgatum* L.) (Fu et al., 2011) and poplar (Hu et al., 1999). Bacterial *cp4-epsps* and *Bacillus thuringiensis* (Bt) genes were transferred to alfalfa to generate a cultivar tolerant to the herbicide glyphosate and a cultivar resistant to alfalfa weevil and/or clover root curculio, respectively (Mark McCaslin, personal communication, Forage Genetics International Inc.). McKersie et al. (2000) showed an increase in winter survival and forage yield in transgenic alfalfa with a superoxide dismutase (SOD) gene. Overexpression of malate dehydrogenase in transgenic alfalfa increased organic acid synthesis and conferred Al tolerance (Tesfaye et al., 2001).

Commercialization of transgenic alfalfa cultivars is hindered by expensive deregulation requirements, intellectual property protection, international trade considerations, and perceived risks associated with this technology. These risks may include the transfer of genes to nontarget species, the use of antibiotic genes as selectable markers, and the effect on local ecosystems and microbial communities. Prior to the deregulation and release of plant cultivars with transgene-derived traits, multiple field trials and extensive regulatory testing are performed to assess and mitigate any unintended effects or yield penalties associated with transgene insertion. The regulations in place for genetically engineered crops continue to evolve and regulatory approval is often required separately for each country in which the product will be marketed. The deregulatory process can be lengthy and expensive, as was the case for Roundup Ready alfalfa. Coexistence strategies are being implemented, including planting adequate buffer zones. In cases where a trait does not naturally exist in alfalfa, transgenic approaches represent a viable strategy. Given the costs associated with getting transgenic products to the market, only traits with significant value-added potential to seed companies have been targeted (Bouton, 2009). The success in market penetration and farmer adoption of Roundup Ready alfalfa in a very short time period highlights the value of biotechnology for trait improvement.

## 10.5 Breeding Targets

### 10.5.1 *Biomass Yield*

Increasing alfalfa yield is a key breeding target to increase the amount of forage or biofuel harvested per hectare. Alfalfa breeding programs in the United States began in the Great Plains during the 1930s and the alfalfa cultivars "Ranger" and "Buffalo" were released in 1942. Since then, genetic gain for quantitative traits such as biomass yield in alfalfa showed limited gains of 0.2–0.3% per year compared to yield increases in corn (2%) and white clover (1%) (Woodfield and Brummer, 2001). Observed yield increases were largely due to selection for disease and pest resistance rather than yield factors per se (Lamb et al., 2006). The fact that wide genetic variation is present within and among alfalfa germplasms for yield and that little direct effort has been placed on improving yield per se suggest that there is great potential to develop alfalfa germplasm with increased biomass production. By combining genetic modification with optimized management practices, alfalfa has shown a great potential for increasing biomass yield. Increases in alfalfa biomass production to more than 15.7 Mg ha$^{-1}$ in the upper Midwestern USA (producing over 750 gal ha$^{-1}$ of ethanol) were obtained by selecting for biomass-type alfalfa (Lamb et al., 2007). Population density and plant maturity during harvest affect alfalfa yields (Lamb et al., 2003). Specifically, alfalfa yields increase by increasing the plant density from 16 to 180 m$^{-2}$. However, planting more than 180 plants m$^{-2}$ actually decreases alfalfa yields when harvested at the green pod stage.

Strategies to increase biomass yields in alfalfa include marker-assisted selection and capturing heterosis, neither of which is fully exploited in breeding programs (Brummer, 2004). In a study evaluating nine high-yielding alfalfa accessions from the USDA-ARS National Plant Germplasm System for general combining ability (GCA) and heterosis for biomass yield, hybrids of parents with varying fall regrowth showed greater heterosis for biomass yield (Bhandari et al., 2007). Their findings indicate that selecting germplasm with high GCA and specific combining ability (SCA) could be a viable strategy for increasing alfalfa biomass yields. Gene expression profiling revealed that hybrids that showed heterosis for biomass yield had a higher number of genes showing nonadditive expression in the progeny compared to hybrids without biomass yield heterosis (Li et al., 2009). Field evaluations were used to identify genomic regions relevant to alfalfa biomass yield under irrigated and rainfed conditions (Han et al., 2007). Recently, Li et al. (2011) used association mapping with microsatellites to identify 15 alleles for biomass yield in alfalfa; one locus was previously associated with yield in an unrelated biparental population (Robins et al., 2007a, 2007b). Markers can be used to develop breeding strategies that maximize the frequency of positive alleles and of desirable allelic interactions in genes and regulators of gene networks that enhance forage/biomass production. The use of molecular markers can augment phenotypic selection, particularly on a genome-wide basis when the markers have been shown to accurately predict the phenotype.

### 10.5.2 Forage Quality and Composition

In addition to increased biomass yield, desirable traits in alfalfa for biofuel processing include long, thick stems, lodging tolerance, reduced lignin content, and higher concentrations of fermentable sugars. Finding all these desirable traits in a single germplasm or variety is very difficult; however, these traits can be incorporated into varieties through breeding programs. Several groups in the United States, including the Noble Foundation (Ardmore, OK), the USDFRC (Madison, WI), the Plant Science Research Unit (St. Paul, MN), and commercial breeding companies are working together to develop alfalfa varieties with altered cell wall chemistry for greater digestibility and processing efficiency.

Both selection and transgenesis can be used to alter lignin content in alfalfa. Lignin binds cellulose and thus, higher lignin content increases the recalcitrance of cell walls to hydrolytic degradation (Chen and Dixon, 2007). The genetic modification to reduce lignin biosynthesis in alfalfa improves forage quality and saccharification efficiency for bioethanol production. An evaluation of alfalfa populations divergently selected for lignin composition showed that reduced lignin populations had a greater leaf to stem ratio and lower yield compared to higher lignin populations (Kephart et al., 1989). Low lignin alfalfa populations had lower survival during a 2-year field trial at Ames, IA compared to high lignin alfalfa populations (Buxton and Casler, 1993). However, a slight negative correlation between lignin content and wood biomass has been shown in trees (Novaes et al., 2010).

Transgenes have been used to alter lignin content in alfalfa (Reddy et al., 2005; Chen and Dixon, 2007; Wang et al., 2010; Zhao et al., 2010). Alfalfa lines independently downregulated in the caffeic acid 3-*O*-methyltransferase (COMT) gene from the lignin biosynthesis pathway showed a larger reduction in Klason lignin at the early stage harvest (12%) than at the late stage harvest compared with the untransformed alfalfa control (Dien et al., 2011). Alfalfa lines suppressed for one of the two other genes from the monolignol biosynthesis pathway, cinnamyl CoA reductase (CCR) and cinnamyl alcohol dehydrogenase (CAD), had lower lignin content and significant increases in saccharification efficiency (Jackson et al., 2008). Many of the CCR downregulated lines had increased S/G ratio compared to CAD lines which had lower S/G ratios. Lower S/G ratios from poplar (*Populus* spp.) populations with varying lignin content and composition were shown to have higher theoretical ethanol yields (Davison et al., 2006). Poplar plants with greater efficiency at enzymatic degradation were obtained by downregulating the 4CL1 gene (Hu et al., 1999). The most efficient downregulation of lignin in alfalfa was observed when targeting the HCT (hydroxycinnamoyl CoA: shikimate hydroxycinnamoyl transferase) gene; however, it also results in reductions in plant height and biomass yield, changes in plant architecture, and delayed flowering (Shadle et al., 2007). Additional gene targets to directly engineer alfalfa plants with reduced lignin and normal growth were identified through screening a large collection of *M. truncatula* transposon insertion mutants for plants with altered lignification patterns. These efforts resulted in the identification of an NAC family transcription factor gene that encodes a master regulator of lignification in the fiber cells of the stem (Zhao et al., 2010). Knocking out expression of this gene reduced lignin levels specifically in the fiber cells, and increased cell wall sugar release efficiency, but did not negatively impact growth (Zhao et al., 2010). Comparative gene expression in alfalfa genotypes contrasting for cellulose and lignin concentrations were used to identify a number of genes with differential gene expression (Yang et al., 2010b), providing additional gene targets for evaluation.

The practical application resulting from the identification of lignin genes in alfalfa includes generation of progeny from crosses between lines independently downregulated in lignin genes and a common tester suitable for field testing. The HCT events had lower lignin levels and decreased plant height and biomass production throughout the growing season when grown in Oklahoma. Potential agronomic issues such as lodging of reduced lignin alfalfa can be addressed through backcrossing to elite lodging-tolerant alfalfa clones. These results highlight the importance of event sorting, field performance evaluations, and breeding of reduced lignin transgenic alfalfa to ensure the desired levels of biomass yield, disease resistance, and agronomic performance.

Biomass recalcitrance to enzymatic saccharification may be affected by other factors including lignin content. In poplar, although high lignin concentrations were responsible for low sugar release, the effect of lignin concentration on sugar release was minimal when the ratio of S to G lignin was greater than two (Studer et al., 2011). The levels of glucan release were dependent on both lignin concentration and S/G ratio whereas xylan release was only dependent on the S/G ratio. Pectin is another factor that affects recalcitrance for biochemical degradation (Jung and Engels, 2002). Lionetti et al. (2010) showed that the reduction of de-methyl-esterified homogalacturonan (HGA) increased the efficiency of enzymatic saccharification in *Arabidopsis*, wheat, and tobacco plants. Thus, optimizing alfalfa cell wall composition and bioprocessing technologies to incorporate pectin either as fermentable sugars or as valuable co-products represents another opportunity to use alfalfa as a biofuel feedstock.

### 10.5.3 Stress Tolerance

An increase in global population results in increased demands of water resources, changes in agricultural land use, and erratic weather patterns that will require modifications in farming practices and infrastructure to meet increased global demands for food and fuel. Along with agronomic and management-based approaches for enhancing crop productivity, improvements in a crop's ability to maintain yields with lower production inputs (water, nutrients, herbicides, and pesticides) will be critical (Tester and Langridge, 2010).

*Abiotic Stress*

Abiotic stress factors, including drought, are significant factors contributing to the reduction of agricultural productivity (Boyer, 1982; Rosegrant and Cline, 2003). All agricultural regions are likely to experience periods of limited water availability or drought throughout the growing season. Water levels of aquifers have declined due to lack of rainfall, and increased demands from agricultural and urban water use (McGuire, 2007). Efforts to improve and maintain crop yields under variable water supplies represent a critical goal for agriculture. Lindenmayer et al. (2011) proposed management practices to increase water use efficiency in alfalfa. By using less water for irrigation during the hot summer, seasonal water use efficiency (WUE) in alfalfa was increased. However, the use of less water for irrigation in the summer for alfalfa production reduced seasonal biomass yield overall and affected stand persistence. An alternative is to develop alfalfa varieties with enhanced WUE. Results from field evaluations for biomass yield and carbon isotope discrimination of thirty alfalfa half-sib families indicate the potential for improving WUE in alfalfa through breeding (Ray et al., 1999). Genomic regions relevant to biomass yield under drought conditions were identified in a population derived from *M. sativa* ssp. *sativa* var. Chilean (low WUE) and *M. sativa* ssp. *falcata*, var. Wisfal (high WUE). Further evaluations and screening of diverse alfalfa germplasm collected from arid regions is underway at the Noble Foundation to identify additional sources of drought tolerance. Overexpression of *WXP1*, a transcription factor gene, increases cuticular wax accumulation and enhances drought tolerance in transgenic alfalfa (Zhang et al., 2005).

Selection of root growth characteristics can also be used to increase drought tolerance in alfalfa. Roots from mature alfalfa plants grow as deep as 5–10 m from the soil surface depending on the soil type to reach water at the subsoil level. Root biomass of European alfalfa cultivars ranged from 8.2 to 9.3 Mg ha$^{-1}$ for the four best cultivars when soil down to 30 cm was sampled (Moghaddam et al., 2009). Another study estimated the amount of alfalfa root growth within 0–45 cm soil depth in Canada as 6.8 Mg ha$^{-1}$ for the first year and 14.4 Mg ha$^{-1}$ for the second year (Bolinder et al., 2002). They also reported greater root biomass for alfalfa between 30 and 45 cm from the soil surface compared to switchgrass during the second year harvest. The relationships between root and shoot biomass, drought tolerance, stand persistence, and nitrogen intake should be considered throughout the breeding process to develop alfalfa for biofuel. Plant roots represent a carbon sink that can affect the carbon neutrality of biofuel crops. A study to evaluate the carbon accumulation in the roots of reduced lignin alfalfa plants is currently underway at the Samuel Roberts Noble Foundation.

A significant portion of the world's arable lands is acidic (von Uexkull and Mutert, 1995). Aluminum (Al) toxicity in acid soils inhibits root growth, limiting nutrient and water uptake resulting in reduced crop productivity (Kochian, 1995). Lime applications to acidic soils only mitigate the issue at the soil surface and increase agricultural production input and costs. Genomic regions associated with Al tolerance in tetraploid alfalfa were identified (Khu et al., 2010a) and efforts are underway to identify mechanisms underlying aluminum tolerance in alfalfa, with the long-term goal of developing alfalfa cultivars that can be productive in acid and Al-toxic soils.

*Biotic Stress*

Pests and diseases in alfalfa can significantly reduce yields, stand life, and affect forage quality. The majority of alfalfa improvements have been targeted toward increasing resistance to insects, diseases, and nematodes. Most commercially available alfalfa cultivars have high resistance to major diseases (bacterial wilt, verticillium wilt, fusarium wilt, Phytophthora root rot, Aphanomyces root rot, and anthracnose), insects (aphids and potato leafhopper), and root knot nematodes (National Alfalfa & Forage Alliance, 2008). The increased tolerance to biotic stress factors resulted in higher yields and stand persistence in regions where these pests pose threats to alfalfa productivity. In addition to reducing biomass yields and persistence, biotic stress factors can also reduce alfalfa digestibility (Hutchins and Pedigo, 1990; Sulc et al., 2004). Molecular markers associated with the major biotic stresses in alfalfa are being developed through an initiative organized by the National Alfalfa & Forage Alliance.

### 10.5.4 Winter Hardiness

The simultaneous improvement of biomass yield and winter hardiness is hindered by the negative relationship that exists between these two traits (Castonguay et al., 2006). Alfalfa plants alter their growth physiology (plant height and biomass production) in response to temperature and photoperiod cues during autumn as a strategy to acclimate for the upcoming winter. Nondormant plants have a limited acclimation response and are able to continue growing during autumn compared to dormant plants. Dormancy is assessed indirectly based on plant height after the last harvest of the season (October), with shorter plants being more dormant (Brummer, 2004). Traditional breeding in alfalfa has been effective at improving winter hardiness of nondormant *M. sativa* ssp. *sativa* (Weishaar et al., 2005) and can be facilitated by the availability of molecular markers (Brower et al., 2000). Increasing yearly total biomass yield could be achieved by reducing autumn dormancy, aided by the identification of markers for dormancy control. Because effective evaluation and selection for winter hardiness require multiple field evaluations to ensure long-term persistence and sustained yield, genetic improvements accumulate slowly (Brummer, 2004). Transgenic approaches can be used to modify the dormancy response of nondormant *sativa* or *falcata* and enable utilization of germplasm with desired traits in plant improvement programs beyond its original area of adaptation.

## 10.6 Management and Production Inputs

Successful alfalfa production requires a series of necessary inputs to optimize product quality and the amount of profit. These inputs include decisions on soil amendments, tillage, planting practices, cultivar selection, cropping systems, post-planting disease, weed and insect management, and harvesting strategies, and are described in the Alfalfa Management Guide (Undersander et al., 2011). Alfalfa has been grown for several decades in a wide range of geographical locations in the United States assembling a solid foundation and infrastructure, machinery, technologies, and research programs aimed toward optimizing seeding, management, harvest, storage, transportation, and processing. Alfalfa is being managed for forage and hay, and is usually harvested at early maturity stages to obtain greater forage quality. In forage production systems, three to four harvests are recommended because as it matures, the proportion of fibers and stems increase in alfalfa herbage. Management systems aimed toward biomass production saw a 42% annual increase in biomass yield and nearly double potential ethanol production from alfalfa stems by harvesting at later flowering stages (Lamb et al., 2007). Delaying harvest benefits ethanol production from alfalfa due to increased biomass, altered lignin composition, and increases in both the S/G ratio and sugar accumulation (Jung and Engels, 2002). For biomass-type alfalfa in northern regions including Minnesota, two harvests per season were recommended with the potential for more harvests in locations with longer growing seasons such as the southern part of the United States.

Alfalfa is a perennial legume capable of symbiotic nitrogen fixation with rhizobia and thus does not require supplemental nitrogen (N) fertilization. Alfalfa can provide enough N for a subsequent crop in a crop rotation scheme and increased subsequent corn yields compared to continuous corn cropping or corn–soybean [*Glycine max* (L.)] rotations when no N fertilizer was applied (Sheaffer et al., 1989). The first and second year nitrogen credit to succeeding crop after alfalfa harvest was estimated at 0.17 Mg ha$^{-1}$ of N and 0.11 Mg ha$^{-1}$, respectively (2006 ISU Nitrogen Fertilizer Recommendation for corn in Iowa). The value of nitrogen credits increases as the price of N fertilizer increases. Increased alfalfa biomass may be obtained by optimizing the rhizobia strain used given the effect of the symbiotic host and subsequent biological N fixation on the rhizobia–host interaction (Yang et al., 2010a). A study utilizing the N-fixing bacteria *Rhizobium meliloti* genetically modified to contain an extra copy of the *nifA* regulatory nitrogen fixation gene and/or *dct* genes (genes involved in energy transport) found significant increases in alfalfa biomass in soils deficient in N and with low organic matter (Bosworth et al., 1994).

Advances in biotechnology coupled with plant breeding have resulted in the development of herbicide-tolerant (glyphosate-resistant) alfalfa cultivars. Glyphosate is a nonselective herbicide that kills most annual and perennial grass and broadleaf weeds and thus offers flexibility and options to manage weeds in alfalfa fields. These Roundup Ready® alfalfa cultivars can translate to reduced management decisions for producers associated with weed control in alfalfa. Additional opportunities to minimize inputs and maximize yields include the use of computers to track and manage field trials, and biometric methods for field-trial design and assessment of interactions between genotype, environment, and management strategies (Baenziger et al., 2006).

## 10.7 Processing for Biofuels

Alfalfa can be processed as a cellulosic feedstock for ethanol production, although cofiring with coal is also an option. Biochemical conversion involves saccharification and fermentation to generate liquid transportation fuels such as ethanol and possibly butanol, isobutanol, or short-chain alkanes. Thermochemical conversion is used to generate electricity, whereas gasification is used to generate syngas and subsequently electricity thorough combustion (Schubert, 2006), or alcohols and short-chain fatty acids generated by syngas-fermenting microorganisms (Henstra et al., 2007). Factors affecting conversion efficiency of alfalfa in various processing methods include cell wall constituents and cellulose and lignin concentration in plant stems (Boateng et al., 2008; Chen and Dixon, 2007). Lignin concentration inversely affects the saccharification and fermentation process (Boateng et al., 2008; Chen and Dixon, 2007; Rock et al., 2009). Alternatively, lignin content may be desirable for thermochemical conversion due to its high energy density (Hodgson et al., 2010) and thus is a good source of combustion fuel. Stems with higher lignin concentration produced higher syngas during pyrolysis compared to the stems with lower lignin concentrations (Boateng et al., 2008). Because generation of liquid fuels from syngas remains very inefficient (Sommerville, 2007), conversion technologies have focused on lignocellulosic biofuels.

Current technologies available for processing lignocellulosic biomass for ethanol production are more demanding than for corn starch processing due to the polysaccharides and lignin polymers in the plant cell walls. Lignin restricts access of enzymes to sugar molecules in cell walls and thus requires an additional pretreatment step to facilitate the release of fermentable sugars. Lignin content negatively affects the total glucose yield from dilute acid pretreatment and enzyme saccharification for several species including alfalfa and other biofuel feedstock (Dien et al., 2006). The cost of processing enzymes per gallon of ethanol produced has declined over the years as a result of increased enzyme production (Sommerville, 2007; Kalluri and Keller, 2010). Thus, the conversion component of the biofuel pipeline has significant opportunities for improvement to fully exploit the potential from lignocellulosic-based biofuels. These opportunities include engineering enzyme component mixtures, enhancing enzyme stability, and implementing synergistic processing alternatives (Ding et al., 2008). Additional efforts include developing enhanced microorganisms such as bacteria and fungi that are capable of efficiently fermenting lignocellulosic biomass into alternate fuels. Many questions remain as to the most optimal systems for harvesting, baling, storing, and transporting biofuel feedstocks. Net energy yields will vary by production inputs and can benefit from a consistent supply of locally produced biomass feedstock grown near processing facilities.

## 10.8 Additional Value from Alfalfa Production

### 10.8.1 Environmental Benefits

Alfalfa has been used for phytoremediation and to detoxify soil contaminants either by removing them from the soil via direct uptake or degrading them in soil. The alfalfa cultivar "Riley" was able to degrade petroleum hydrocarbons from the soil after 12 months of growth (Schwab et al., 2006). Symbiotic relationships between alfalfa and soil microbes can enhance microbe-mediated degradation of soil contaminants by providing a suitable environment for the growth of soil-detoxifying bacteria (Tang et al., 2010). Transgenic alfalfa was used to detoxify the herbicide atrazine from contaminated soil and water (Wang et al., 2005). This herbicide is widely used in corn fields, remains undegraded in the soil for months, and is reported to cause birth defects in babies. Alfalfa can also be used to manage N-contaminated soils (Russelle et al., 2007). Additionally, alfalfa was identified as the most biodiversity-friendly crop among 23 major crops evaluated (Montford and Small, 1999).

### 10.8.2 Alfalfa Co-products

High levels of proteins in alfalfa leaves contribute to its high value leaf meal used directly as a protein source or as a protein supplement to increase animal performance in cattle and aquatic animals. Tilapia fingerlings showed better growth when fed a fish meal diet in which 35% of the protein was replaced by alfalfa protein compared to a total fish meal diet (Olvera-Novoa et al., 1990). Alfalfa leaves were also used successfully to manufacture biodegradable plastics (Saruul et al., 2002). Nonpolysaccharide compounds are not amenable for ethanol production through fermentation and thus they can be extracted and used as valuable co-products. Commercially valuable co-products obtained from alfalfa include nutraceuticals such as lutein, phytoestrogen, cellulase, phytase, antibodies, and edible vaccines (Mueller et al., 2008). These high value co-products can increase profitability for growers and contribute to making alfalfa an economically viable crop for bioenergy.

## 10.9 Summary

Advancements in alfalfa breeding, ongoing research efforts to identify the genes and regulatory networks involved in lignin biosynthesis and mechanisms underlying abiotic and biotic stress tolerance in alfalfa, and technological innovations can be integrated to develop alfalfa cultivars that meet the demands of producers and consumers for feed and fuel. Efforts in alfalfa breeding are focused toward increasing biomass yields, enhancing its ability to adapt and thrive under environmental and biological stress conditions, and altering the forage composition to increase processing efficiency. Utilization of conventional and molecular breeding approaches integrated with an increasing number of genomic tools and genetic transformation approaches provide opportunities for increased efficiency in alfalfa improvement by capturing desirable agronomic and value-added traits. Expertise in alfalfa breeding and improvement programs, understanding the value of alfalfa for N fixation in crop rotations, generation of valuable co-products, and advances in bioprocessing technologies have positioned alfalfa as a valuable component of the biofuel equation. Biofuel industries will likely be local, and driven by the primary feedstock grown in a region. Alfalfa is adapted for growth in most of North America and can be used as a dual forage and biofuel feedstock. Although uncertainties remain about specific tonnage requirements and the economics of production and transportation, alfalfa has the potential to significantly contribute to an emerging biofuels industry. Profitability in alfalfa production systems for biofuels will be driven by implementation of strategies to reduce the cost per unit of biomass produced and/or increase the processing efficiency to increase ethanol yields. Investments in infrastructure associated with feedstock management to reduce inputs and transportation costs, processing plants, and well-defined regulatory environments will increase opportunities for the continued utilization of renewable feedstocks as energy sources in the biofuel industry.

## Acknowledgments

We thank Christy Motes, Shauna Smith, Will Chaney, Bonnie Farris, Dusty Pittman, and Brian Motes for supporting our program, and appreciate funding from the Oklahoma Bioenergy Center and the Oklahoma Center for the Advancement of Science & Technology (OCAST).

## References

Ates E, Tekeli AS. Assessing heritability and variance components of agronomic traits of four alfalfa (*Medicago sativa* L.) cultivars. Acta Agron Hungar, 2004; 52: 263–268.

Baenziger PS, Russell WK, Graef GL, Campbell BT. Improving lives: 50 years of crop breeding, genetics and cytology (C-1). Crop Sci, 2006; 46: 2230–2244.

Barnes DK, Bingham ET, Murphy RP, Hunt OJ, Beard DF, Skrdla WH, Teuber LR. *Alfalfa Germplasm in the United States: Genetic Vulnerability, Use, Improvement, and Maintenance*, 1977. Government Printing Office, Washington, DC.

Barnes DK, Goplen BP, Baylor JE. Highlights in the USA and Canada. In: Hanson AA, Barnes DK, Hills RR (eds) *Alfalfa and Alfalfa Improvement*, 1988, pp. 1–24. ASA-CSSA-CSSS, Madison, WI.

Basigalup DH, Barnes DK, Strucker RE. Development of a core collection for perennial *Medicago* plant introductions. Crop Sci, 1995; 35: 1163–1168.

Bhandari HS, Pierce CA, Murray LW, Ray IM. Combining abilities and heterosis for forage yield among high-yielding accessions of the alfalfa core collection. Crop Sci, 2007; 47: 665–673.

Bingham ET, Groose RW, Woodfield DR, Kidwell KK. Complementary gene interactions in alfalfa are greater in autotetraploids than diploids. Crop Sci, 1994; 34: 823–829.

Boateng AA, Weimer PJ, Jung HG, Lamb JFS. Response of thermochemical and biochemical conversion processes to lignin concentration in alfalfa stems. Energy Fuels, 2008; 22: 2810–2815.

Bolinder MA, Angers DA, Belanger G, Michaud R., Laverdiere MR. Root biomass and shoot to root ratios of perennial forage crops in eastern Canada. Can J Plant Sci, 2002; 82: 731–737.

Bosworth AH, Williams MK, Albrecht KA, Kwiatkowski R, Beynon J, Hankinson TR, Ronson CW, Cannon F, Wacek TJ, Triplett EW. Alfalfa yield response to inoculation with recombinant strains of *Rhizobium meliloti* with an extra copy of dctABD and/or modified nifA expression. Appl Environ Microbiol, 1994; 60: 3815–3832.

Bouton JH. Molecular breeding to improve forages for use in animal and biofuel production systems. In: Yamada T, German S (eds) *Molecular Breeding of Forage and Turf*, 2009, pp. 1–12. Springer, New York.

Boyer, JS. Plant productivity and environment. Science, 1982; 218: 443–448.

Brower DJ, Duke SH, Osborn TC. Mapping genetic factors associated with winter hardiness, fall growth, and freezing injury in autotetraploid alfalfa. Crop Sci, 2000; 40: 1387–1396.

Brown RC. *Biorenewable Resources: Engineering New Products from Agriculture*, 2003. Iowa State Press, Ames, IA.

Brummer EC. Capturing heterosis in forage crop cultivar development. Crop Sci, 1999; 39: 943–954.

Brummer EC. Applying genomics to alfalfa breeding programs. Crop Sci, 2004; 44: 1904–1907.

Brummer EC, Kochert G, Bouton JH. Development of an RFLP map in diploid alfalfa. Theor Appl Genet, 1993; 86: 329–332.

Busbice TH. Inbreeding in synthetic varieties. Crop Sci, 1969; 9: 601–604.

Busbice TH, Wilsie CP. Inbreeding depression and heterosis in autotetraploids with application to *Medicago sativa* L. Euphytica, 1966; 15: 52–67.

Buxton DR, Casler MD. Environmental and genetic factors affecting cell wall composition and digestibility. In: Jung HG, Buxton DR, Hartfield RD, Ralph J (eds) *Forage Cell Wall Structure and Digestibility*, 1993, pp. 685–714. American Society of Agronomy, Madison, WI.

Caffall KH, Mohnen D. The structure, function, and biosynthesis of plant cell wall pectic polysaccharides. Carbohydr Res, 2009; 344: 1879–1900.

Cane JH. A native ground-nesting bee (*Nomia melanderi*) sustainably managed to pollinate alfalfa across an intensively agricultural landscape. Apidologie, 2008; 39: 315–323.

Casler MD, Brummer EC. Theoretical expected genetic gains for among-and-within-family selection methods in perennial forage crops. Crop Sci, 2008; 48: 890–902.

Cassida KA, Turner KE, Hesterman OB. Comparison of detergent fiber analysis methods for forages high in pectin. Anim Feed Sci Technol, 2007; 135: 283–295.

Castonguay Y, Laberge S, Brummer EC, Volenec JJ. Alfalfa winter hardiness: A research retrospective and integrated perspective. Adv Agron, 2006; 90: 203–265.

Chen F, Dixon RA. Lignin modification improves fermentable sugar yields for biofuel production. Nat Biotechnol, 2007; 25: 759–761.

Dale, BE. Biomass refining: protein and ethanol from alfalfa. Ind Eng Chem Prod Res Dev, 1983; 22: 466–472.

Davison B, Drescher S, Tuskan G, Davis M, Nghiem N. Variation of S/G ratio and lignin content in a *Populus* family influences the release of xylose by dilute acid hydrolysis. Appl Biochem Biotechnol, 2006; 130: 427–435.

Deak M, Kiss GB, Koncz C, Dudits D. Transformation of *Medicago* by *Agrobacterium* mediated gene transfer. Plant Cell Rep, 1986; 5: 97–100.

Dien BS. Process review of lignocellulose biochemical conversion to fuel ethanol. In: *Frontiers of Engineering*, 2007, pp. 65–74. The National Academies Press, Washington, DC.

Dien BS, Jung HJG, Vogel KP, Casler MD, Lamb JFS, Iten L, Mitchell RB, Sarath G. Chemical composition and response to dilute-acid pretreatment and enzymatic saccharification of alfalfa, reed canarygrass, and switchgrass. Biomass Bioenerg, 2006; 30: 880–891.

Dien BS, Miller DJ, Hector RE, Dixon RA, Chen F, McCaslin M, Reisen P, Sarath G, Cotta MA. Enhancing alfalfa conversion efficiencies for sugar recovery and ethanol production by altering lignin composition. Bioresour Technol, 2011; 102: 6479–6486.

Ding SY, Xu Q, Crowley M, Zeng Y, Nimlos M, Lamed R, Bayer EA, Himmel ME. A biophysical perspective on the cellulosome: new opportunities for biomass conversion. Curr Opin Biotechnol, 2008; 19: 218–227.

Diwan N, Bauchan GR, McIntosh MS. A core collection for the United States annual *Medicago* germplasm collection. Crop Sci, 1994; 34: 279–285.

Doiphode DA, Mungikar AM. Rhizosphere and rhizoplane mycoflora under the influence of deproteinized leaf juice of lucerne. In: Gangawane L, Khilare VC (eds) *Sustainable Environmental Management: Dr. Jayashree Deshpande Festschrift Volume*, 2007, pp. 75–88. Daya Publishing House, New Delhi, India.

Downing M, Volk TA, Schmidt DA. Development of new generation cooperatives in agriculture for renewable energy research, development, and demonstration projects. Biomass Bioenerg, 2005; 28: 425–434.

Duchesne LC, Larson DW. Cellulose and the evolution of plant life. Bioscience, 1989; 39: 238–241.

Dudley JW, Busbice TH, Levings CS. Estimates of genetic variance in 'Cherokee' alfalfa (*Medicago sativa* L.). Crop Sci, 1960; 9: 228–231.

Echt CC, Kidwell KK, Knapp SJ, Osborn TC, McCoy TJ. Linkage mapping in diploid alfalfa (*Medicago sativa*). Genome, 1993; 37: 61–71.

Elshire RJ, Glaubitz JC, Sun Q, Poland JA, Kawamoto K, Buckler ES, Mitchell SE. A robust, simple genotyping-by-sequencing (GBS) approach for high diversity species. PLoS ONE, 2011; 6: e19379.

Engels FM, Jung HG. Alfalfa stem tissues: cell-wall development and lignification. Ann Bot, 1998; 82: 561–568.

Eujayl I, Sledge MK, Wang L, May GD, Chekhovskiy K, Zwonitzer JC, Mian MAR. *Medicago truncatula* EST-SSRs reveal cross-species genetic markers for *Medicago* spp. Theor Appl Genet, 2003; 108: 414–421.

Eulgem T. Dissecting the WRKY web of plant defense regulators. PLoS Pathogens, 2006; 2: 1028–1030.

Farrell AE, Plevin RJ, Turner BT, Jones AD, O'Hare M, Kammen DM. Ethanol can contribute to energy and environmental goals. Science, 2006; 311: 506–508.

Fu C, Mielenz JR, Xiao X, Ge Y, Hamilton CY, Rodriguez M, Chen F, Foston M, Ragauskas A, Bouton J, Dixon RA, Wang ZY. Genetic manipulation of lignin reduces recalcitrance and improves ethanol production from switchgrass. Proc Natl Acad Sci U S A, 2011; 108: 3803–3808.

Gunn CR, Skrdla WH, Spencer HC. *Classification of Medicago sativa L. using Legume Characters and Flowers Colors*, 1978. Government Printing Office, Washington, DC.

Hammerschlag R. Ethanol's energy return on investment: a survey of the literature 1990–present. Environ Sci Technol, 2006; 40: 1744–1750.

Han Y, Kang Y, Torres-Jerez I, Cheung F, Town C, Zhao P, Udvardi M, Monteros MJ. Genome-wide SNP discovery in tetraploid alfalfa using 454 sequencing and high resolution melting analysis. BMC Genomics, 2011; 12: 350.

Han Y, Khu DM, Monteros MJ. High-resolution melting analysis for SNP genotyping and mapping in tetraploid alfalfa (*Medicago sativa* L.). Mol Breed, 2012; 29: 489–501.

Han Y, Monteros MJ, Sledge MK, Ray IM, Bouton JH. Physiological characterization and QTL analysis of drought tolerance in alfalfa. *5th International Symposium on the Molecular Breeding of Forage and Turf*, July 1–6, 2007. Sapporo, Japan.

Heffner EL, Sorrells ME, Jannink JL. Genomic selection for crop improvement. Crop Sci, 2009; 49: 1–12.

Henstra AM, Sipma J, Rinzema A, Stams AJM. Microbiology of synthesis gas fermentation for biofuel production. Curr Opin Biotechnol, 2007; 18: 200–206.

Hodgson EM, Lister SJ, Bridgwater AV, Clifton-Brown J, Donnison IS. Genotypic and environmentally derived variation in the cell wall composition of Miscanthus in relation to its use as a biomass feedstock. Biomass Bioenerg, 2010; 34: 652–660.

Hu WJ, Harding SA, Lung J, Popko JL, Ralph J, Stokke DD, Tsai CJ, Chiang VL. Repression of lignin biosynthesis promotes cellulose accumulation and growth in transgenic trees. Nat Biotechnol, 1999; 17: 808–812.

Huntley SK, Ellis D, Gilbert M, Chapple C, Mansfield SD. Significant increases in pulping efficiency in C4H-F5H-transformed poplars: improved chemical savings and reduced environmental toxins. J Agric Food Chem, 2003; 51: 6178–6183.

Hurley JD, Engle LJ, Davis JT, Welsh AM, Landers JE. A simple, bead-based approach for multi-SNP molecular haplotyping. Nucleic Acids Res, 2004; 32: e186.

Hutchins SH, Pedigo LP. Phenological disruption and economic consequence of injury to alfalfa induced by potato leafhopper (Homoptera, Cicadellidae). J Econ Entomol, 1990; 83: 1587–1594.

Jackson L, Shadle G, Zhou R, Nakashima J, Chen F, Dixon R. Improving saccharification efficiency of alfalfa stems through modification of the terminal stages of monolignol biosynthesis. BioEnerg Res, 2008; 1: 180–192.

Jiang Q, Zhang JY, Guo X, Monteros MJ, Wang ZY. Physiological characterization of transgenic alfalfa (*Medicago sativa*) plants for improved drought tolerance. Int J Plant Sci, 2009; 170: 969–978.

Julier B, Flajoulot S, Barre P, Cardinet G, Santoni S, Huguet T, Huyghe C. Construction of two genetic linkage maps in cultivated tetraploid alfalfa (*Medicago sativa*) using microsatellite and AFLP markers. BMC Plant Biol, 2003; 3: 9.

Jung HG, Engels FM. Alfalfa stem tissues: cell wall deposition, composition, and degradability. Crop Sci, 2002; 42: 524–534.

Kalluri UC, Keller M. Bioenergy research: a new paradigm in multidisciplinary research. J Royal Soc Interface, 2010; 7: 1391–1401.

Kephart KD, Buxton DR, Hill RR. Morphology of alfalfa divergently selected for herbage lignin concentration. Crop Sci, 1989; 29: 778–782.

Khu DM, Reyno R, Brummer EC, Bouton JH, Monteros MJ. QTL mapping of aluminum tolerance in tetraploid alfalfa. In: Huyghe C (ed.) *Sustainable Use of Genetic Diversity in Forage and Turf Breeding*, 2010a, pp. 437–442. Springer, The Netherlands.

Khu DM, Reyno R, Brummer EC, Bouton JH, Han Y, Monteros MJ. QTL mapping of aluminum tolerance in tetraploid alfalfa. In: Huyghe C (ed.) *Sustainable Use of Genetic Diversity in Forage and Turf Breeding*, 2010b, pp. 437–442. Springer, The Netherlands.

Kiss GB, Csanadi G, Kalman K, Kalo P, Okresz L. Construction of a basic genetic map for alfalfa using RFLP, RAPD, isozyme and morphological markers. Mol Gen Genet, 1993; 238: 129–137.

Kochian LV. Cellular mechanisms of aluminum toxicity and resistance in plants. Annu Rev Plant Physiol Plant Mol Biol, 1995; 46: 237–260.

Lamb JFS, Jung HJG, Sheaffer CC, Samac DA. Alfalfa leaf protein and stem cell wall polysaccharide yields under hay and biomass management systems. Crop Sci, 2007; 47: 1407–1415.

Lamb JFS, Sheaffer CC, Rhodes LH, Sulc M, Undersander DJ, Brummer EC. Forage yield and quality of alfalfa cultivars released from the 1940's through the 1990's. Crop Sci, 2006; 46: 902–909.

Lamb JFS, Sheaffer CC, Samac DA. Population density and harvest maturity effects on leaf and stem yield in alfalfa. Agron J, 2003; 95: 635–641.

Leith, H. Modeling the primary productivity of the world, In: Leith H, Whittaker RH (eds) *Primary Productivity of the Biosphere*, 1975, pp. 237–263. Springer-Verlag, New York.

Lesins KA. Alfalfa, lucerne, *Medicago sativa* (Leguminosae-Papilionaceae). In: Simmonds NW (ed.) *Evolution of Crop Plants*, 1976, pp. 165–168. Longman Groups, London.

Lesins K, Lesins I. Diploid Medicago falcata L. Can J Genet Cytol, 1964; 6: 152–163.

Li X, Brummer EC. Inbreeding depression for fertility and biomass in advanced generations of inter- and intra-subspecies hybrids of tetraploid alfalfa. Crop Sci, 2009; 49: 13–19.

Li X, Wei Y, Moore KJ, Michaud R, Viands DR, Hansen JL, Acharya A, Brummer EC. Association mapping of biomass yield and stem composition in a tetraploid alfalfa breeding population. Plant Genome, 2011; 4: 24–35.

Li XH, Wei YL, Nettleton D, Brummer EC. Comparative gene expression profiles between heterotic and non-heterotic hybrids of tetraploid *Medicago sativa*. BMC Plant Biol, 2009; 9: 12.

Lindenmayer RB, Hansen NC, Brummer J, Pritchett JG. Deficit Irrigation of alfalfa for water-savings in the Great Plains and Intermountain West: a review and analysis of the literature. Agron J, 2011; 103: 45–50.

Lionetti V, Francocci F, Ferrari S, Volpi C, Bellincampi D, Galletti R, D'Ovidio R, De Lorenzo G, Cervone F. Engineering the cell wall by reducing de-methyl-esterified homogalacturonan improves saccharification of plant tissues for bioconversion. Proc Natl Acad Sci U S A, 2010; 107: 616–621.

Markovic J, Radovic J, Lugic Z, Sokolovic D. The effect of development stage on chemical composition of alfalfa leaf and steam. Biotechnol Anim Husb, 2007; 23: 383–388.

McCoy TJ, Bingham ET. Cytology and cytogenetics of alfalfa. In: Hanson AA et al. (eds) *Alfalfa and Alfalfa Improvement*, 1988, pp. 737–776. ASA-CSSA-SSSA, Madison, WI.

McGuire VL. Changes in water levels and storage in the high plains aquifer, predevelopment to 2005. US Geological Survey, 2007. Available at http://pubs.usgs.gov/fs/2007/3029/ (accessed March 24, 2011).

McKersie BD, Murnaghan J, Jones KS, Bowley SR. Iron-superoxide dismutase expression in transgenic alfalfa increases winter survival without a detectable increase in photosynthetic oxidative stress tolerance. Plant Physiol, 2000; 122: 1427–1438.

Moghaddam A, Pietsch G, Raza A, Vollmann J, Friedel JK. Root biomass of 18 alfalfa (*Medicago sativa* L.) cultivars in two different environments under organic management. In: *International Symposium on Root Research and Applications*, Boku-Vienna, Austria, 2009.

Mohnen D. Pectin structure and biosynthesis. Curr Opin Plant Biol, 2008; 11: 266–277.

Montford S, Small E. A comparison of the biodiversity friendliness of crops with special reference to hemp (*Cannabis sativa* L.). J Int Hemp Assoc, 1999. Available at http://www.druglibrary.net/olsen/HEMP/IHA/jiha6206.html (accessed March 24, 2011).

Mueller SC, Undersander DJ, Putnam DH. *Alfalfa for Industrial and Other Uses*, 2008. University of California, Division of Agricultural and Natural Resources, Oakland, CA.

National Alfalfa & Forage Alliance. Winter survival, fall dormancy and pest resistance ratings for alfalfa varieties, 2008.

National Alfalfa & Forage Alliance. Fall dormancy & pest resistance ratings for alfalfa varieties, 2011. Available at http://www.alfalfa.org/publications.html (accessed September 28, 2011).

Novaes E, Kirst M, Chiang V, Winter-Sederoff H, Sederoff R. Lignin and biomass: a negative correlation for wood formation and lignin content in trees. Plant Physiol, 2010; 154: 555–561.

Olvera-Novoa MA, Campos S, Sabido M, Palacios CAM. The use of alfalfa leaf protein concentrates as a protein source in diets for tilapia (*Oreochromis mossambicus*). Aquaculture, 1990; 90: 291–302.

Osborn TC, Brouwer D, McCoy TJ. Molecular marker analysis in alfalfa. In: McKersie BD, Brown DCW (eds) *Biotechnology and the Improvement of Forage Legumes*, 1997, pp. 91–109. CAB International, Guelph, Canada.

Parrott WA, Clemente TE. Transgenic soybean. In:Boerma HR, Specht JE (eds) *Soybeans: Improvement, Production and Uses*. Vol. 16, 3rd edn., 2004. ASA-CSSA-SSSA, Madison, WI.

Peterson JD. Pectin-rich biorefinery for production of ethanol and specialty chemicals. Ethanol Producer Magazine, 2003; 26–29.

Pimentel D. Ethanol fuels – energy security, economics, and the environment. J Agr Environ Ethic, 1991; 4: 1–13.

Pimentel D, Patzek TW. Ethanol production using corn, switchgrass, and wood; biodiesel production using soybean and sunflower. Nat Resour Res, 2005; 14: 65–76.

Pitts-Singer TL, Cane JH. The alfalfa leafcutting bee, *Megachile rotundata*: the world's most intensively managed solitary bee. Ann Rev Entomol, 2011; 56: 221–237.

Quiros CF, Bauchan GR. The genus *Medicago sativa* and the origin of the *Medicago sativa* complex. In: Hanson AA, Barnes DK, Hills RR (eds) *Alfalfa and Alfalfa Improvement*, Vol. 29, 1988, pp. 93–124. ASA-CSSA-SSSA, Madison, WI.

Ray IM, Townsend MS, Muncy CM, Henning JA. Heritabilities of water-use efficiency traits and correlations with agronomic traits in water-stressed alfalfa. Crop Sci, 1999; 39: 494–498.

Reddy MSS, Chen F, Shadle G, Jackson L, Aljoe H, Dixon RA. Targeted down-regulation of cytochrome P450 enzymes for forage quality improvement in alfalfa (*Medicago sativa* L.). Proc Natl Acad Sci U S A, 2005; 102: 16573–16578.

Renewable Fuels Association. R. ethanol facts, 2011. Available at http://www.ethanolrfa.org/pages/ethanol-facts. Accessed August 10, 2011.

Riday H, Brummer EC. Forage yield heterosis in alfalfa. Crop Sci, 2002; 42: 716–723.

Riday H, Brummer EC. Dissection of heterosis in alfalfa hybrids. In: Hopkins A, Wang ZY, Mian R, Sledge M, Barker RE (eds) *Molecular Breeding of Forage and Turf*, 2004, pp. 315–324. Kluwer, Dordrecht, The Netherlands.

Robins JG, Bauchan GR, Brummer EC. Genetic mapping forage yield, plant height, and regrowth at multiple harvests in tetraploid alfalfa (*Medicago sativa* L.). Crop Sci, 2007a; 47: 11–18.

Robins JG, Luth D, Campbell IA, Bauchan GR, He CL, Viands DR, Hansen JL, Brummer EC. Genetic mapping of biomass production in tetraploid alfalfa. Crop Sci, 2007b; 47 1–10.

Robins JG, Luth D, Campbell TA, Bauchan GR, He C, Viands DR, Hansen JL, Brummer EC. Mapping biomass production in tetraploid alfalfa (*Medicago sativa* L.). Crop Sci, 2007c; 47: 1–10.

Robins JG, Viands DR, Brummer EC. Genetic mapping of persistence in tetraploid alfalfa. Crop Sci, 2008; 48: 1780–1786.

Rock KP, Thelemann RT, Jung HJG, Tschirner UW, Sheaffer CC, Johnson GA. Variation due to growth environment in alfalfa yield, cellulosic ethanol traits, and paper pulp characteristics. Bioenerg Res, 2009; 2: 79–89.

Rosegrant MW, Cline SA. Global food security: challenges and policies. Science, 2003; 302: 1917–1919.

Russelle MP, Lamb JFS, Turyk NB, Shaw BH, Pearson B. Managing nitrogen contaminated soils: benefits of N-2-fixing alfalfa. Agron J, 2007; 99: 738–746.

Şakiroğlu M, Brummer EC. No heterosis between semidormant and nondormant derived alfalfa germplasm. Crop Sci, 2007; 47: 2364–2371.

Saruul P, Srienc F, Somers DA, Samac DA. Production of a biodegradable plastic polymer, poly-beta-hydroxybutyrate, in transgenic alfalfa. Crop Sci, 2002; 42: 919–927.

Schubert C. Can biofuels finally take center stage? Nat Biotechnol, 2006; 24: 777–784.

Schwab P, Banks MK, Kyle WA. Heritability of phytoremediation potential for the alfalfa cultivar Riley in petroleum contaminated soil. Water Air Soil Pollut, 2006; 177: 239–249.

Segovia-Lerma A, Murray LW, Townsend MS, Ray IM. Population-based diallel analyses among nine historically recognized alfalfa germplasms. Theor Appl Genet, 2004; 109: 1568–1575.

Shadle G, Chen F, Srinivasa Reddy MS, Jackson L, Nakashima J, Dixon RA. Down-regulation of hydroxycinnamoyl CoA: shikimate hydroxycinnamoyl transferase in transgenic alfalfa affects lignification, development and forage quality. Phytochemistry, 2007; 68: 1521–1529.

Shahin EA, Spielmann A, Sukhapinda K, Simpson RB, Yashar M. Transformation of cultivated alfalfa using disarmed *Agrobacterium tumefaciens*. Crop Sci, 1986; 26: 1235–1239.

Shapouri H, Duffield JA, Wang M. The energy balance of corn ethanol revisited. Trans ASAE, 2003; 46: 959–968.

Sheaffer CC, Barnes DK, Heichel GH. Alfalfa in crop rotation, 1989. Minnesota Agricultural Experiment Station, Station Bulletin 588.

Sheehan J, Aden A, Paustian K, Killian K, Brenner J, Walsh M, Nelson R. Energy and environmental aspects of using corn stover for fuel ethanol. Curr Biol, 2003; 7: 117–146.

Shinners KJ, Herzmann ME, Binversie BN, Digman MF. Harvest fractionation of Alfalfa. Trans ASABE, 2007; 50: 713–718.

Sledge M, Ray I, Jiang G. An expressed sequence tag SSR map of tetraploild alfalfa (*Medicago sativa* L.). Theor Appl Genet, 2005; 111: 980–992.

Small, E. *Alfalfa and Relatives: Evolution and Classification of Medicago*, 2011. NRC Research Press, Ottawa, Canada.

Sommerville C. Biofuels. Curr Biol, 2007; 17: R116.

Studer MH, Brethauer S, Demartini JD, McKenzie HL, Wyman CE. Co-hydrolysis of hydrothermal and dilute acid pretreated *Populus* slurries to support development of a high-throughput pretreatment system. Biotechnol Biofuels, 2011; 4: 19.

Sulc RM, Johnson KD, Sheaffer CC, Undersander DJ, van Santen E. Forage quality of potato leafhopper resistant and susceptible alfalfa cultivars. Agron J, 2004; 96: 337–343.

Tang J, Wang R, Niu X, Wang M, Zhou Q. Characterization on the rhizoremediation of petroleum contaminated soil as affected by different influencing factors. Biogeosci Discuss, 2010; 7: 4665–4688.

Tesfaye M, Temple SJ, Allan DL, Vance CP, Samac DA. Overexpression of malate dehydrogenase in transgenic alfalfa enhances organic acid synthesis and confers tolerance to aluminum. Plant Physiol, 2001; 127: 1836–1844.

Tester M, Langridge P. Breeding technologies to Increase crop production in a changing world. Science, 2010; 327: 818–822.

Tysdal HM, Kiesselbach TA. Hybrid alfalfa. J Amer Soc Agron, 1944; 36: 649–667.

Undersander D, Cosgrove D, Cullen E, Grau C, Rice ME, Renz M, Sheaffer CC, Shewmaker G, Sulc M. *Alfalfa Management Guide*, 2011. ASA-CSSA-SSSA, Madison, WI.

United States Energy Information Administration, E.I.A. Biodiesel overview, 2011. Available at http://www.eia.gov/totalenergy/data/monthly/pdf/sec10_8.pdf (accessed June 24, 2011).

USDA/DOE. Biomass as feedstock for a bioenergy and bioproducts industry: the technical feasibility of a billion-ton annual supply, 2005. Available at http://feedstockreview.ornl.gov/pdf/billion_ton_vision.pdf (accessed March 24, 2011).

USDA/NASS. Crop production 2010 summary, 2011. Available at http://www.usda01.library.cornell.edu/usda/nass/CropProdSu/2010s/2011/CropProdSu-01-12-2011_new_format.pdf (accessed May 16, 2011).

Vermerris W. *Genetic Improvement of Bioenergy Crops*, 2008. Springer Science+Business Media, LLC, New York.

Viands DR, Sun P, Barnes DK. Pollination control: mechanical and sterility. In: Hanson AA, Barnes DK, Hills RR(eds) *Alfalfa and Alfalfa Improvement*, pp. 931–960, 1988. ASA-CSSA-SSSA, Madison, WI.

von Uexkull HR, Mutert E. Global extent, development and economic impact of acid soils. Plant Soil, 1995; 171: 1–15.

Walton S, Van Heiningen A. Biological conversion of hemicellulose extracts from wood: production of fuel ethanol by Escherichia coli K011, 2010. http://forestbioproducts.umaine.edu/files/2010/10/EPSCOR-Poster_Walton.pdf (accessed August 10, 2011).

Wang HZ, Avci U, Nakashima J, Hahn MG, Chen F, Dixon RA. Mutation of WRKY transcription factors initiates pith secondary wall formation and increases stem biomass in dicotyledonous plants. Proc Natl Acad Sci U S A, 2010; 107: 22338–22343.

Wang M, Hong MW, Huo H. Life-cycle energy and greenhouse gas emission impacts of different corn ethanol plant types. Environ Res Lett, 2007; 2: 024001.

Wang L, Samac DA, Shapir N, Wackett LP, Vance CP, Olszewski NE, Sadowsky MJ. Biodegradation of atrazine in transgenic plants expressing a modified bacterial atrazine chlorohydrolase (atzA) gene. Plant Biotechnol J, 2005; 3: 475–486.

Wei Y, Li X, Brummer EC, Jiang Q, Kang J. Genotype-by-Sequencing (GBS) in alfalfa (*Medicago sativa*), 2011. ASA-CSSA-SSSA International Annual Meetings, San Antonio, Texas.

Weishaar MA, Brummer EC, Volenec JJ, Moore KJ, Cunningham S. Improving winter hardiness in nondormant alfalfa germplasm. Crop Sci, 2005; 45: 60–65.

Woodfield DR, Brummer EC. Integrating molecular techniques to maximize the genetic potential of forage legumes. In: Spangenberg G (ed.) *Molecular breeding of Forage Crops. Proceedings of the 2nd International Symposium on Molecular Breeding of Forage Crops, 19–24 November 2000, Lorne and Hamilton, Victoria, Australia*, 2001. Kluwer Academic Publishers, Dordrecht, The Netherlands.

Yang S, Tang F, Gao M, Krishnan HB, Zhu H. R gene-controlled host specificity in the legume-rhizobia symbiosis. Proc Natl Acad Sci U S A, 2010a; 107: 18735–18740.

Yang SS, Xu WW, Tesfaye M, Lamb JFS, Jung HJG, VandenBosch KA, Vance CP, Gronwald JW. Transcript profiling of two alfalfa genotypes with contrasting cell wall composition in stems using a cross-species platform: optimizing analysis by masking biased probes. BMC Genomics, 2010b; 11: 18.

Zhang J-Y, Broeckling CD, Blancaflor EB, Sledge MK, Sumner LW, Wang ZY. Overexpression of WXP1, a putative *Medicago truncatula* AP2 domain-containing transcription factor gene, increases cuticular wax accumulation and enhances drought tolerance in transgenic alfalfa (*Medicago sativa*). Plant J, 2005; 42: 689–707.

Zhao QA, Wang HZ, Yin YB, Xu Y, Chen F, Dixon RA. Syringyl lignin biosynthesis is directly regulated by a secondary cell wall master switch. Proc Natl Acad Sci U S A, 2010; 107: 14496–14501.

# Chapter 11
# Transgenics for Biomass

C. Frank Hardin and Zeng-Yu Wang

*Forage Improvement Division, The Samuel Roberts Noble Foundation, 2510 Sam Noble Parkway, Ardmore, OK 73401, USA*

## 11.1 Introduction

Global demand for energy is continuously increasing and currently these demands are primarily met through the use of fossil fuels. In 2007, 69% of these fossil fuels were used to meet transportation energy demands. Because fossil fuel is not renewable within a comprehensible time frame, continued dependence on fossil fuels is likely to push the fuel supply to the edge of extinction, dramatically declining over the course of this century, with the world's oil reserves estimated to be largely depleted by 2050 (Saxena et al., 2009). The burning of fossil fuel has led to climate change as a result of the accumulation of greenhouse gasses (e.g., $CO_2$), and pollutes the earth with acid rain ($SO_2$). Additionally, political tension between petroleum-rich countries and countries with large or growing energy demands has intensified due to unstable supply and the uneven geographic distribution of petroleum resources (Lu and Mosier, 2008). These problems, and others, have led to the exploration of renewable and sustainable energy sources.

### 11.1.1 Biomass for Biofuels

Biomass is a natural, renewable resource comprising of all organic material derived from plants including algae, trees, and crops. In green plants, photosynthesis converts solar energy into carbohydrates, the building blocks of plant materials (biomass). Energy in the form of chemical bonds between atoms in carbohydrates is released when these bonds are broken by digestion, combustion, or decomposition (McKendry, 2002). Biomass has long been exploited as a source of heat energy, used as forage crops providing sustenance to humans and animals that harvest the energy by digestion, as raw material (e.g., burning for energy) or, recently, as biofuel.

---

*Bioenergy Feedstocks: Breeding and Genetics*, First Edition. Edited by Malay C. Saha, Hem S. Bhandari, and Joseph H. Bouton.
© 2013 John Wiley & Sons, Inc. Published 2013 by John Wiley & Sons, Inc.

## 11.1.2 Biofuels

There are two primary transportation fuels that can be produced from biomass: biodiesel and bioethanol. Biodiesel is relatively easily produced from fat or oil via transesterification, yields 93% more energy than the amount of energy required for production, and produces less pollutants than either fossil fuel or bioethanol (Hill et al., 2006). However, because production of biodiesel is limited to crops such as soybean and palm, it directly competes with food supplies, and in 2006 only accounted for <0.2% of the amount of diesel used in the United States. Bioethanol is produced from the sugars that are extracted from the plant and fermented. In the United States, maize is the major source of bioethanol production; however, it only displaces a fraction of the demand for fossil fuel. In 2006, approximately 5 billion gallons of corn ethanol was produced, accounting for only 3.6% of the total amount of gasoline consumed that year (Li et al., 2008a). As with biodiesel, bioethanol is also a food-based fuel and has a damaging effect on food prices. Collectively, while it is unlikely that corn ethanol will ever meet the U S transportation fuel demand alone, the development and use of corn ethanol has jump-started the development of a new (second) generation of biofuel: cellulosic ethanol. Cellulosic ethanol, derived from lignocellulosic plant biomass is a promising, nonfood, nongrain alternative for renewable energy with the potential to meet much of the world's transportation fuel demand with less impact than fossil fuels (Lynd, 1996).

## 11.1.3 Lignocellulosic Biomass

Lignocellulosic plant biomass is a natural, renewable, highly abundant source of organic matter comprised primarily of cellulose, hemicellulose, and lignin. In the United States, over 180 million tons of lignocellulosic biomass are produced per year (Kim and Dale, 2004) and between 10 and 50 billion tons are produced worldwide (Greene, 2004). Lignocellulosic biomass can be divided into four main categories as the following: agricultural residues (i.e., sugarcane bagasse and corn stover), wood residues (wood chips, paper mill by-products), paper wastes, and dedicated energy crops such as poplar, *Miscanthus,* and switchgrass which recently has been chosen as a favorable feedstock for the production of cellulosic biofuels by the U.S. Department of Energy (McLaughlin and Kszos, 2005). In any of these forms, cellulose and hemicellulose may be converted into ethanol or other high-energy liquid fuels. However, challenges in producing ethanol from lignocellulosic biomass remain high due to the recalcitrant nature of plant cell walls. When producing ethanol from lignocellulosic biomass, cellulose and hemicellulose must first be depolymerized by hydrolysis to yield monomers (fermentable sugars) which can then be converted into liquid fuels. However, cellulose is embedded within a complex matrix that includes hemicellulose and lignin. Lignin is a structural component of the cell wall that acts as a mechanical barrier, preventing hydrolytic enzymes from breaking down the polysaccharide substrate, thereby limiting fermentable sugar release (Dien et al., 2006). It has been shown that sugar release from lignocellulosic biomass is inversely proportional to lignin content (Dien et al., 2006). Furthermore, lignin is capable of absorbing hydrolytic enzymes used to generate fermentable sugars, inhibiting fermentation processes (Keating et al., 2006). Therefore, only a very small proportion of monomeric glucose may be obtained by enzymatic hydrolysis of native, untreated lignocellulosic biomass (Mosier et al., 2005).

To improve fermentable sugar yields and processing efficiency, substrates are pretreated in an effort to hydrolyze hemicellulose, solubilize lignin, and increase the accessibility of cellulose-to-sugar conversion processes (EERE, 2012). Current strategies used to increase processing efficiency during biofuel production include pretreatment of the biomass with extreme heat and/or harsh chemicals to enhance hydrolysis of cellulose into fermentable sugars (Huang et al., 2011). Unfortunately, these processes produce by-products that inhibit hydrolytic enzymes and the fermentation process so the biomass must be detoxified and neutralized before it can be hydrolyzed by bacterial and fungal enzymes (expensively produced in bioreactors) and resulting sugars fermented into ethanol (Sticklen, 2008). The need for these processes drives the cost of producing the biofuel up to two- to threefold higher than the cost to produce starch-based ethanol (Abramson et al., 2010; Hisano et al., 2009; Sticklen, 2006; Sticklen, 2008). As a result, much effort is being made to overcome the high costs of producing the hydrolytic enzymes and pretreatment processes by further development of these pretreatment processes and identifying enzymes that can tolerate higher temperatures and broader pH ranges. To date, numerous cell-wall-deconstructing enzymes have been identified and characterized, and over the years, significant progress has been made toward genetically improving the enzymatic activity and microbes used in biofuel production. However, using current technologies and all available arable land to produce lignocellulosic biomass, it is estimated that this will only yield enough biofuel to displace 15% of transportation fuel (Gressel, 2008). Therefore, alternative approaches to reduce the cost of bioethanol production such as increasing biomass yield per unit area and modifying biomass composition are being explored. While plant biomass yield and quality can be improved through plant breeding, this process is likely to be time consuming. Transgenic approaches offer a viable alternative to conventional breeding, are capable of addressing issues that breeding cannot (i.e., expressing recombinant cellulases in the plant biomass itself, down-regulating enzymes involved in the lignin biosynthetic pathway), and are considered to be the most rapid and efficient solution to curbing the costs of ethanol production from lignocellulosic biomass (Gressel, 2008; Hisano et al., 2009; Sticklen, 2008). However, it is important to note that breeding will likely play an important role in the improvement of biomass for the production of cellulosic ethanol and should not be considered any less useful than transgenic approaches.

## 11.2 Transgenic Approaches

Genetic transformation of plants has become a useful tool for not only understanding basic biological processes but also for complementing conventional breeding approaches which has led to the development of transgenic cultivars in several cash crops (Vogel and Jung, 2001). Genetic transformation has also been used to improve forage and turf grasses (reviewed in Wang and Ge, 2006) and, more recently, to improve biofuel production from lignocellulosic biomass by introducing unique genetic variation that could not otherwise be achieved by breeding (Wang and Ge, 2006). There are two main methods currently used to produce transgenic plants: microprojectile bombardment or simply biolistics and *Agrobacterium*-mediated transformation.

### 11.2.1 Biolistics Transformation

Biolistics utilizes high-velocity metal particles to deliver exogenous DNA into the plant cells for stable transformation. Essentially, any plasmid DNA is coated onto approximately 1 $\mu$M particle of elemental heavy metal (usually gold or tungsten) and injected into plant cells. A major disadvantage of using this method is the integration of multiple copies of DNA which could lead to gene silencing, and may make sequencing of the transgene locus difficult. It also potentially causes problems with release of transgenics through regulatory agencies (Lee et al., 2008; Wang and Ge, 2006).

### 11.2.2 *Agrobacterium-mediated Transformation*

*Agrobacterium tumefaciens* is a soil bacterium that naturally infects plants and causes crown gall disease. Virulent *Agrobacterium* harbors a tumor-inducing (Ti) plasmid that contains several virulent (*vir*) genes. These genes encode Vir proteins that shuttle a segment of the Ti plasmid, referred to as transfer DNA or T-DNA (Chilton et al., 1977) that is defined by flanking border sequences called T regions (Yadav et al., 1982), from the *Agrobacterium* to the host plant cell, genetically transforming the host. Bacterial and host proteins then transport the T-DNA from the cytoplasm to the nucleus where it integrates into the plant genome. How and where T-DNA integrates into the plant genome is not well defined. *Agrobacterium*-mediated transformation is well reviewed in Gelvin (2003).

For the purposes of genetically engineering plants, the T-DNA is replaced with foreign genes intended for incorporation into the plant. Initially, introducing genes of interest (goi) into *Agrobacterium* involved homologous recombination between the T-DNA regions of the Ti plasmid with the goi of an exchange/co-integration vector (Fraley et al., 1985; Garfinkel et al., 1981; Zambryski et al., 1983). However, this difficult and complex genetic manipulation system was soon replaced by the simplified T-DNA binary vector system (de Framond et al., 1983; Hoekema et al., 1983). In this system, goi and *vir* genes are located on separate replicons; goi are maintained within the T-DNA region of a binary vector and *vir* genes are located on a separate vector (also known as the backbone vector sequence and *vir* helper, respectively). *Agrobacterium* strains containing a T-DNA and a *vir* helper are considered disarmed as long as they do not contain the oncogenes that may be transferred to plants. In general, when utilizing this system, goi are cloned into the T-DNA region of a binary vector, the construction is characterized and verified in *Escherichia coli* before it is amplified and isolated. The isolated vector is then transformed into an appropriate *Agrobacterium* strain carrying a *vir* helper region. *Agrobacterium* selected for the presence of the backbone vector is then used to transform plant explants. Numerous strains of disarmed *Agrobacterium* and backbone vectors have been constructed and many of which are summarized in Lee and Gelvin 2008. Today, backbone vectors commonly contain T-DNA left and right border sequences to define and delimit T-DNA, origins of replication to allow maintenance in both *Agrobacterium* and *E. coli*, bacterial selection genes, Gateway cloning (Hartley et al., 2000) for overexpression and RNA interference (RNAi), regulatory elements for expression of goi, plant selection cassettes, and reporter cassettes (Mann et al., 2010).

*Agrobacterium*-mediated transformation was originally used for dicotyledonous (dicot) plants such as tobacco, *Arabidopsis*, alfalfa, and poplar because *Agrobacterium* can naturally infect these plants. Starting from the mid-1990s, *Agrobacteria* have been successfully used to transform an array of monocotyledonous (monocot) plants including feedstock species such as maize (Ishida et al., 1996), rice (Hiei et al., 1994), sorghum (Zhao et al., 2000), wheat (Cheng et al., 1997), and switchgrass (Conger et al., 2002). An advantage of using *Agrobacterium*-mediated transformation over the biolistic approach is that *Agrobacterium*-mediated transformation allows stable integration of the transgene into the plant genome in a relatively lower copy number than biolistics generally leading to fewer rearrangements and more stable transgene expression over generations (Dai et al., 2001; Hu et al., 2003). From a regulatory perspective with respect to releasing transgenic plants into the field,

*Agrobacterium*-mediated transformation is the preferred method for introducing transgenes into biofuel crops (Conger et al., 2002).

## 11.3 Transgenic Approaches for Biomass Improvement

Producing cellulosic ethanol at this time is not economically viable, requiring expensive hydrolytic cellulases and pretreatment processes. Serious efforts to produce cheaper cellulosic ethanol are underway. Considering that many cellulosic feedstocks are essentially wild and have not benefited from decades of breeding as have food and feed crops, and that breeding for traits aimed at reducing the cost of producing cellulosic ethanol will take years, the development of transgenic plants for enhanced biofuel conversion is considered an effective solution for the production of ethanol from lignocellulosic biomass (Gressel, 2008).

The use of transgenic biomass for the improvement of biofuel production is very attractive. In theory, it is possible to increase biofuel yields without the need to increase land use or increase the amount of energy needed to produce the biofuel. Genetic engineering has already led to significant progress toward increasing the effectiveness of hydrolytic enzymes used for biofuel production. Several approaches aimed at improving biomass yield and modifying biomass composition of biofuel crops are underway. It is important to note that these efforts are still relatively new and as such, there is little literature available documenting the results of these studies. However, transgenic approaches directed toward improving biomass yield in food, feed, and model species is not new and much has been learned from those studies.

### *11.3.1 Improving Biomass Yield*

The bulk of plant biomass is stored in the form of polysaccharides which comprise approximately two-thirds of the plant secondary cell wall. This promises to be the most abundant source of renewable biomass for biofuel production (Carroll and Somerville, 2009; Pauly and Keegstra, 2010). On the basis of high biomass yields and low agronomic input, stover of conventional crops such as maize, rice, and sorghum along with several other plants, including *Miscanthus*, poplar, and switchgrass are considered sources of biomass for biofuel production. In the United States, switchgrass has been defined by the Department of Energy (DOE) as a major, dedicated bioenergy crop (McLaughlin and Kszos, 2005). The recent exploitation of biofuel as an alternative energy source to meet the world's increasing energy demands has led to an enormous effort to maximize fuel production from bioenergy crops. With this, there is an interest in understanding the molecular mechanisms that regulate biomass production and applying this toward improving plant biomass yield in biofuel crops. Transgenic approaches are expected to play an important role in increasing biomass yield. Little progress has been made in improving the biomass yield of dedicated bioenergy crops, but numerous strategies have been suggested to increase biomass yield in food, feed, and model plant species and are reviewed in Busov et al. (2008), Gonzalez et al. (2009), Rojas et al. (2010), Salas Fernandez et al. (2009). Some of the important strategies include the genetic modification of photosynthetic genes, transcription factors (TFs), cell-cycle machinery, hormone metabolism, and microRNAs. This knowledge may be applied directly toward improving biomass yield in biofuel feedstocks.

Photosynthesis supplies raw materials for vegetative growth and the efficiency of this process greatly influences a plant's biomass production (Demura and Ye, 2010). Most plants utilize the $C_3$ photosynthetic pathway. These plants contain a single type of chloroplast in which all reactions that convert light into chemical energy are used to fix $CO_2$ and synthesize reduced carbon compounds. In these plants, Rubisco (ribulose-1,5-bisphosphate carboxylase/oxygenase) catalyzes carbon fixation, converting $CO_2$ and ribulose-1,5-bisphosphate into two molecules of 3-phosphoglycerate, a three-carbon compound (hence, $C_3$). The $C_4$ pathway is a more complex adaptation to hot, arid climates. In $C_4$ plants, $CO_2$ fixation occurs in two steps in two specialized cell and chloroplast types; phosphoenolpyruvate carboxylase (PEPC) fixes $CO_2$ into four-carbon dicarboxylic acid oxaloacetate, which is then converted into malate or aspartate that are then decarboxylated. The $CO_2$ produced is then refixed by Rubisco. $C_4$ plants generally have higher photosynthetic rates at high temperatures and light intensity due to the increased efficiency of the photosynthetic carbon reduction (*PCR*) cycle (Hatch, 1987) and $C_4$ species such as switchgrass and *Miscanthus* are considered promising biofuel crops due to their high biomass yield.

Transgenic strategies involving modifying photosynthesis for improved biomass yield have been performed and mostly involve increasing the photosynthetic efficiency (Rojas et al., 2010). In rice, transgenic plants expressing maize PEPC or pyruvate, phosphate dikinase (PPDK) had enhanced photosynthetic capacity, improved biomass and grain yield than wild-type plants, primarily due to increased tiller number (Ku et al., 2007; Matsuoka et al., 2001). This is especially useful considering that the use of rice straw for ethanol production is very attractive given that it comprises a significant portion of the world's agronomic biomass (Sticklen, 2008). Similarly, in maize, the overexpression of sorghum PEPC improved $CO_2$ fixation and increased biomass yield under drought conditions (Jeanneau et al., 2002). In transgenic tobacco overexpressing two enzymes involved in the Calvin cycle; fructose-1,6-bisphosphatase and sedoheptulose-1,7-bisphosophatase improved dry-weight biomass yield (Lefebvre et al., 2005; Tamoi et al., 2006).

Another transgenic approach to improve biomass yield involves genetic modification of TFs. Besides their function at the cellular level, TFs also play important roles in plant architecture, growth stimulation, and regulation of photosynthetic capacity and cell cycle (Century et al., 2008). Some examples include members of the NAC (*N*AM, *A*TAF1/2, *C*UC2) family of TFs that stimulate growth and WRKY TFs (defined by a conserved *WRKY*GQK amino acid sequence at their N-terminal end) together with a novel zinc-finger-like motif (Rushton et al., 1995) that influence various physiological processes including pathogen defense, senescence, and trichome development (Eulgem et al., 2000). In *Arabidopsis*, plants overexpressing *NAC1* resulted in bigger plants with thicker stems and larger leaves, and more roots compared to control plants. The overexpression of the NAC-domain TF, ATAF2 also leads to increased biomass in *Arabidopsis*. Recently, it was shown that the disruption of WRKY TFs in *Arabidopsis* mutants leads to an approximately 50% increase in stem biomass. The loss of function of the WRKY TF resulted in the upregulation of genes that encode NAC- and CCCH-type TFs that activate secondary cell wall synthesis by stimulating the biosynthesis of xylan, cellulose, and lignin. Since the source of lignocellulosic biomass is the plant secondary cell wall, the discovery of negative regulators that affect secondary cell wall biosynthesis opens up the possibility of significantly increasing the biomass of bioenergy crops (Wang et al., 2010).

It has been demonstrated that modifying expression of cell-cycle-related genes improves biomass yield. During G1 phase of the cell cycle, cells increase mass in preparation for cell division. It is considered that the primary control point of the cell cycle and D-type cyclins (CycD) control responses to extracellular signals and the commitment to cell division during the G1 cycle (Cockcroft et al., 2000; Riou-Khamlichi et al., 1999; Soni et al., 1995). In tobacco, the overexpression of *Arabidopsis CYCD2* resulted in plants that were 35% taller than control plants and had increased growth and leaf initiation rates (Cockcroft et al., 2000). Another

*Arabidopsis* gene, *CDC27a,* that encodes a protein that plays a role in the E3 ligase Anaphase-promoting complex (APC) also increased tobacco biomass when overexpressed, resulting in plants that were 30% taller at flowering time (Rojas et al., 2009).

Plant hormones play an important role in growth and development by regulating processes such as meristem activity and cell elongation, both of which greatly contribute to biomass yield (Demura and Ye, 2010). Hormones such as gibberellins (GAs) have been shown to play a role in cell elongation. By altering GA metabolism and/or perception, it may be possible to genetically engineer biofuel crops with increased cell elongation and ultimately more biomass. Genes involved in GA biosynthesis have been characterized in rice and *Arabidopsis* (Hirano et al., 2008; Sakamoto et al., 2004) and overexpression of some of these genes has led to an increase in biomass production in tobacco and poplar. An elevation in GA production by overexpression of GA20-oxidase in poplar resulted in an increase in plant height and fiber length; thus dry stem weight is almost doubled that of wild-type plants (Eriksson et al., 2000). Furthermore, overexpression of the same gene in tobacco also led to a drastic increase in dry stem weight (Biemelt et al., 2004). Other hormones such as cytokinins and brassinosteroids have also been shown to be promising targets for genetic manipulation to increase plant biomass (Sakamoto et al., 2006).

Plant microRNAs (miRNA, *MIR* genes) are small 20–24 nucleotide RNA molecules that posttranscriptionally regulate gene expression. Primary miRNA transcripts (pri-miRNAs) are generally transcribed by RNA polymerase II and contain imperfect, self-complementary fold-back regions. DICER-LIKE 1 (DCL1) orchestrates processing (Bartel, 2004) in the nucleus resulting in an approximately 21 nucleotide, mature miRNA duplex that is then shuttled out of the nucleus where it associates with target mRNA sequences. The complementary base-pairing of miRNAs with their mRNA targets leads to the inhibition of gene expression by either targeting the mRNAs for degradation or inhibiting protein translation (Ambros, 2004). In plants, miRNAs play essential roles in regulating growth and development and are considered promising targets for gene manipulation toward increasing biomass. Plant miR156 is primarily expressed in juvenile plants and is involved in vegetative phase change. When the plants are young, miR156 expression is high and gradually decreases as the plants get older. MicroRNA156 targets genes belonging to the TF family SQUAMOSA PROMOTER BINDING PROTEIN-LIKE (SPL) (Rhoades et al., 2002). SPLs are involved in processes such as leaf development, flowering time, phase change, and shoot maturation. In *Arabidopsis*, the overexpression of miR156 reduced apical dominance, delayed flowering, and increased leaf number (10-fold) and biomass (Schwab et al., 2005). Overexpression of miR156 prolonged the vegetative phase and increased tiller numbers in maize and rice (Chuck et al., 2007; Xie et al., 2006). A miR156b precursor was overexpressed in switchgrass (Fu et al., 2012). The effects of miR156 overexpression on SPL genes were revealed by microarray and quantitative RT-PCR analyses. Morphological alterations, biomass yield, saccharification efficiency, and forage digestibility of the transgenic plants were characterized (Fu et al., 2012). Relatively low levels of miR156 overexpression were sufficient to increase biomass yield while producing plants with normal flowering time. Moderate levels of miR156 led to improved biomass but the plants were nonflowering. These two groups of plants produced 58–101% more biomass yield compared with the control. However, high miR156 levels resulted in severely stunted growth. The degree of morphological alterations of the transgenic switchgrass depends on miR156 level. Compared with floral transition, a lower miR156 level is required to disrupt apical dominance. The improvement in biomass yield was mainly because of the increase in tiller number. Targeted overexpression of miR156 also improved solubilized sugar yield and forage digestibility, and offered an effective approach for transgene containment (Fu et al., 2012).

Abiotic stresses are the primary causes of crop loss worldwide. While researchers have been working for years to improve tolerance to abiotic stresses in food and feed crops, conferring resistance to these stresses is especially important in biofuel species. Considering the food versus fuel debate, there is increasing concern about using productive land to produce biofuel feedstock as opposed to food or feed crops. Therefore, it is desirable to reserve productive land for food production and use marginal land for the development of biofuel crops. The use of transgenics to confer tolerance to abiotic stresses in biofuel feedstock would allow plants to grow better on marginal land, indirectly increasing the amount of land available for biofuel production. It would also address environmental stresses related to climate change and assure consistent and predictable biomass supplies for biofuel production (Wolt, 2009).

Plant adaptation to environmental stress is a complex system that begins with the sensing or perception of primary stresses such as drought, salinity, heat, and cold. These stresses are often interconnected and cause cellular damage and secondary stresses such as osmotic and oxidative stresses. Signal transduction cascades relay the perceived stress stimuli and invoke downstream transcription controls which activate stress-response mechanisms that act to maintain or re-establish homeostasis and preserve the integrity of membranes and proteins (Vinocur and Altman, 2005). These responses are genetically complex and generally do not involve a single gene but rather a multitude of genes and are therefore more difficult to control and engineer. Engineering strategies aimed at enhancing abiotic stress tolerance include the expression of genes involved in signaling and regulatory pathways, genes that encode proteins that confer stress tolerance, or enzymes that are involved in metabolite biosynthesis. Some of the recent advances include the overexpression of osmoprotectants, late embryogenesis abundant (LEA) proteins, and transporter genes (reviewed in Bhatnagar-Mathur et al. (2008), Jewell et al. (2009), Vinocur and Altman (2005). Many of these strategies may be applied to biofuel crops in an effort to increase biomass yield from marginal lands and ultimately biofuel production. However, given the complexity of stress-response mechanisms, genetic engineering alone may not be enough to improve abiotic stress tolerance and will be the most powerful when coupled with traditional breeding and classical physiology (Vinocur and Altman, 2005).

## 11.3.2 *Modifying Biomass Composition*

Cellulose is made up of bundles of microfibrils that consist of glucose monomers linked by glycosidic bonds and is the most abundant substance on earth. Hemicellulose is composed of monomeric, 5-carbon sugars such as xylose or arabinose and glucose. In the plant cell wall, hemicellulose cross-links cellulose microfibrils with lignin forming a complex cell wall network (Rubin, 2008). Cellulose and hemicellulose can be hydrolyzed to form fermentable sugars for biofuel production. However, as mentioned previously, the complex network of cellulose, hemicellulose, and lignin in the cell wall is largely recalcitrant to hydrolytic enzymes, primarily due to the association of lignin with cellulose and hemicellulose. This association not only necessitates pretreatment processes during ethanol production but also absorbs saccharification enzymes, driving the cost of cellulosic ethanol production up significantly (Sticklen, 2006; Sticklen, 2008).

While breeding approaches for reduced lignin content in plant biomass may solve this problem (Bouton, 2007), the process will likely take a while to achieve. In this case, transgenics offer a great alternative toward reducing the cost of cellulosic ethanol production (Gressel, 2008). Along with enhancing biomass yield to improve biofuel production, transgenic approaches that modify the biomass composition such as increasing the amount of cellulose by overexpressing genes involved in the biosynthesis of these compounds or producing recombinant cellulases *in planta* are viable alternatives. Another direct and effective approach is to modify lignin composition or reduce lignin content by down-regulating the enzymes involved in lignin biosynthesis (Hisano et al., 2009).

Lignin is a phenolic biopolymer synthesized in all plants and is considered critical for plant growth and development (Dixon et al., 2001; Rogers and Campbell, 2004). It is comprised of three monolignol precursors, p-coumaryl, coniferyl, and sinapyl alcohols which later undergo dehydrogenative polymerization to form p-hydroxyphenyl (H), guaiacyl (G), and syringyl (S) lignin, respectively (Weng et al., 2008). The current view of the general lignin biosynthetic pathway can be found in Li et al. (2008b). An important approach to reduce the cost of pretreatment during biofuel production from cellulosic biomass is to modify lignin composition or reduce lignin content by down-regulating enzymes involved in lignin biosynthesis (Ragauskas et al., 2006). Prior to the interest in modifying lignin in biofuel crops, the genetic modification of lignin biosynthesis was of interest for improving the pulping process in the paper industry and forage digestibility of forage crops (Li et al., 2008b), which include species such as maize and tall fescue, now considered important sources of biomass for cellulosic ethanol production (Hisano et al., 2009). The targets and outcomes of these experiments are summarized in Li et al. (2008b). In general, it was found that down-regulating genes early in the biosynthetic pathway such as PAL, C4H, HCT, and C3H were the most successful in reducing lignin content while F5H or COMT greatly reduced the lignin S/G ratio but had little effect on total lignin content. Despite the fact that many efforts have been made toward modifying lignin content, most were not aimed at specifically improving biofuel production. The first report specifically addressing the relationship between lignin content/composition and chemical/enzymatic saccharification was performed by Chen and Dixon (Chen and Dixon, 2007). These studies involved the down-regulation of lignin biosynthetic enzymes in alfalfa, a perennial forage crop that has also been proposed as a biofuel feedstock, and demonstrated that it is possible to overcome cell wall recalcitrance to bioconversion by genetically reducing lignin content. Furthermore, it may be possible to reduce or eliminate the need for pretreatment by using low-lignin biomass from transgenic plants, ultimately reducing the cost of biofuel production (Hisano et al., 2009).

Since these studies, it has recently been reported that the genetic manipulation of lignin reduces recalcitrance and improves ethanol production from the major energy crop, switchgrass (Fu et al., 2011). In these studies, the down-regulation COMT led to reduced lignin content and S/G ratio, improved forage quality, and up to a 38% increase in ethanol yield using conventional biomass fermentation processes. Furthermore, the transgenic lines required less severe pretreatment and less cellulase for equivalent ethanol fermentation compared to unmodified switchgrass. Together, these data suggest that transgenic switchgrass with reduced recalcitrance has the potential to significantly lower the costs of biofuel production. Furthermore, the requirement of less severe pretreatment allows for the opportunity to express recombinant hydrolytic enzymes *in planta*, potentially increasing processing efficiency and further reducing the cost of biofuel production (Chen and Dixon, 2007).

Currently, hydrolytic enzymes used in commercial biofuel production are expensively produced in microbial reactors. Due to the high costs of producing these enzymes in microbes, the DOE proposed a plan to reduce the cost of producing cellulases from \$5 gallon$^{-1}$ to \$0.10–0.15, so far this has not been achieved (Lee et al., 2008). A possible solution to curbing these costs is to produce cellulases within the feedstock biomass itself (Sticklen, 2006). Through the use of transgenics, hydrolytic enzymes may potentially be produced in all biofuel crops that will be used to produce cellulosic ethanol (Sticklen, 2008). It has been proposed that the ultimate transgenic feedstock contain enzyme levels high enough to ultimately perform autohydrolysis to release fermentable sugars or at least provide some of the enzymes necessary for hydrolysis (Sticklen, 2006). Toward this, hydrolytic enzymes have been expressed in numerous plants, including *Arabidopsis*, tobacco, alfalfa, rice, and maize (reviewed in Lee et al., 2008; Sticklen, 2008); however, expression in major biofuel feedstocks has yet to be achieved.

One final approach that can be used to improve biomass for biofuel production involves modifying fermentable sugar concentrations by overexpressing enzymes involved in cellulose and hemicellulose biosynthesis. While cellulose biosynthesis has been studied extensively (reviewed in Somerville, 2006), many of the steps of the pathway are not well understood and even less is known about hemicellulose biosynthesis. Future studies are necessary to improve our understanding the biosynthesis of these polysaccharides and how genetic modification can be used to increase polysaccharide concentrations in lignocellulosic biomass for improved biofuel production (Sticklen, 2008).

### 11.3.3 Regulatory Issues of Transgenic Bioenergy Crops

A major limitation for deployment of transgenics is the complicated GMO regulatory processes. Despite the wide adoption and the beneficial economic and environmental impacts of major transgenic crops (e.g., corn, soybean, cotton, canola), it has been extremely difficult to deregulate and commercialize other species. Many transgenic bioenergy crops (e.g., switchgrass, poplar) do not enter the food chain directly; their potential environmental or ecological impacts are the focus of risk assessment. Pollen-mediated transgene flow is a major concern for outcrossing species like switchgrass. The deregulation process has become so complicated and costly that only large companies with deep resources can afford to do it. The impacts of regulations on research and environmental studies of transgenic bioenergy crops are well summarized by Strauss et al. (2010). Several biological containment measures have been developed or proposed to control transgene flow. Such measures include male sterility, seed sterility, maternal inheritance, delayed flowering, etc. (Wang and Brummer, 2012). Nonflowering transgenic switchgrass plants have been produced by overexpression of miRNA156 (Fu et al., 2012). In addition, the intragenic or cisgenic approach may provide a cost-effective way for genetic engineering and commercialization of bioenergy species (Wang and Brummer, 2012).

## 11.4  Summary

Transgenics for improved biofuel production continues to evolve as a promising research tool that will undoubtedly play a significant role in developing improved feedstocks for cellulosic energy production. Overcoming recalcitrance is perhaps the most significant achievement that is necessary to reduce our dependence on transportation fossil fuels by at least 30% prior to 2030. It is likely that a combination of strategies including transgenic approaches and breeding efforts targeting the development of feedstocks with increased biomass yield and more efficient processing technologies can contribute toward the goal of developing a more sustainable and economically viable alternative for the production of liquid biofuels.

## Acknowledgments

This work was supported by the BioEnergy Science Center and The Samuel Roberts Noble Foundation. The BioEnergy Science Center is a U.S. Department of Energy Bioenergy Research

Center supported by the Office of Biological and Environmental Research in the DOE Office of Science.

## References

Abramson M, Shoseyov O, Shani Z. Plant cell wall reconstruction toward improved lignocellulosic production and processability. Plant Sci, 2010; 178: 61–72. doi:10.1016/j.plantsci.2009.11.003.

Ambros V. The functions of animal microRNAs. Nature, 2004; 431: 350–355. doi:10.1038/nature02871.

Bartel DP. MicroRNAs: genomics, biogenesis, mechanism, and function. Cell, 2004; 116: 281–297.

Bhatnagar-Mathur P, Vadez V, Sharma K. Transgenic approaches for abiotic stress tolerance in plants: retrospect and prospects. Plant Cell Rep, 2008; 27: 411–424. doi:10.1007/s00299-007-0474-9.

Biemelt S, Tschiersch H, Sonnewald U. Impact of altered gibberellin metabolism on biomass accumulation, lignin biosynthesis, and photosynthesis in transgenic tobacco plants. Plant Physiol, 2004; 135: 254–265. doi:10.1104/pp.103.036988.

Bouton JH. Molecular breeding of switchgrass for use as a biofuel crop. Curr Opin Genet Dev, 2007; 17: 553–558.

Busov VB, Brunner AM, Strauss SH. Genes for control of plant stature and form. New Phytologist, 2008; 177: 589–607. doi:10.1111/j.1469-8137.2007.02324.x.

Carroll A, Somerville C. Cellulosic biofuels. Ann Rev Plant Biol, 2009; 60: 165–182. doi:10.1146/annurev.arplant.043008.092125.

Century K, Reuber TL, Ratcliffe OJ. Regulating the regulators: the future prospects for transcription-factor-based agricultural biotechnology products. Plant Physiol, 2008; 147: 20–29. doi:10.1104/pp.108.117887.

Chen F, Dixon RA. Lignin modification improves fermentable sugar yields for biofuel production. Nat Biotechnol, 2007; 25: 759–761.

Cheng M, Fry JE, Pang S, Zhou H, Hironaka CM, Duncan DR, Conner TW, Wan Y. Genetic Transformation of Wheat Mediated by Agrobacterium tumefaciens. Plant Physiol, 1997; 115: 971–980.

Chilton MD, Drummond MH, Merio DJ, Sciaky D, Montoya AL, Gordon MP, Nester EW. Stable incorporation of plasmid DNA into higher plant cells: the molecular basis of crown gall tumorigenesis. Cell, 1977; 11: 263–271.

Chuck G, Cigan AM, Saeteurn K, Hake S. The heterochronic maize mutant Corngrass1 results from overexpression of a tandem microRNA. Nat Genet, 2007; 39: 544–549. doi:10.1038/ng2001.

Cockcroft CE, den Boer BGW, Healy JMS, Murray JAH. Cyclin D control of growth rate in plants. Nature, 2000; 405: 575–579.

Conger BV, Somleva MN, Tomaszewski Z. Agrobacterium-mediated genetic transformation of switchgrass. Crop Sci, 2002; 42: 2080–2087.

Dai S, Zheng P, Marmey P, Zhang S, Tian W, Chen S, Beachy RN, Fauquet C. Comparative analysis of transgenic rice plants obtained by Agrobacterium-mediated transformation and particle bombardment. Mol Breed, 2001; 7: 25–33. doi:10.1023/a:1009687511633.

de Framond AJ, Barton KA, Chilton M-D, Mini-Ti. A new vector strategy for plant genetic engineering. Nat Biotech, 1983; 1: 262–269.

Demura T, Ye ZH. Regulation of plant biomass production. Curr Opin Plant Biol, 2010; 13: 299–304. doi:10.1016/j.pbi.2010.03.002.

Dien BS, Jung HJG, Vogel KP, Casler MD, Lamb JFS, Iten L, Mitchell RB, Sarath G. Chemical composition and response to dilute-acid pretreatment and enzymatic saccharification of alfalfa, reed canarygrass, and switchgrass. Biomass Bioenerg, 2006; 30: 880–891. doi:10.1016/j.biombioe.2006.02.004.

Dixon RA, Chen F, Guo D, Parvathi K. The biosynthesis of monolignols: a "metabolic grid", or independent pathways to guaiacyl and syringyl units? Phytochemistry, 2001; 57: 1069–1084.

EERE, USDE. *Biomass Program*, 2012. www1.eere.energy.gov/biomass/about.html

Eriksson ME, Israelsson M, Olsson O, Moritz T. Increased gibberellin biosynthesis in transgenic trees promotes growth, biomass production and xylem fiber length. Nature Biotechnol, 2000; 18: 784–788. doi:10.1038/77355.

Eulgem T, Rushton PJ, Robatzek S, Somssich IE. The WRKY superfamily of plant transcription factors. Trends Plant Sci, 2000; 5: 199–206. doi:10.1016/s1360-1385(00)01600-9.

Fraley RT, Rogers SG, Horsch RB, Eichholtz DA, Flick JS, Fink CL, Hoffmann NL, Sanders PR. The SEV System: A new disarmed Ti plasmid vector system for plant transformation. Nat Biotech, 1985; 3: 629–635.

Fu C, Mielenz JR, Xiao X, Ge Y, Hamilton CY, Rodriguez M, Jr, Chen F, Foston M, Ragauskas A, Bouton J, Dixon RA, Wang ZY. Genetic manipulation of lignin reduces recalcitrance and improves ethanol production from switchgrass. Proc Natl Acad Sci U S A, 2011; 108: 3803–3808. doi:10.1073/pnas.1100310108.

Fu C, Sunkar R, Zhou C, Shen H, Zhang JY, Matts J, Wolf J, Mann D, Stewart CN, Tang Y, Wang ZY. Overexpression of miR156 in switchgrass (*Panicum virgatum* L.) results in various morphological alterations and leads to improved biomass production. Plant Biotechnol J, 2012; 10: 443–452.

Garfinkel DJ, Simpson RB, Ream LW, White FF, Gordon MP, Nester EW. Genetic analysis of crown gall: fine structure map of the T-DNA by site-directed mutagenesis. Cell, 1981; 27: 143–153.

Gelvin SB. Agrobacterium-mediated plant transformation: the biology behind the "gene-jockeying" tool. Microbiol Mol Biol Rev, 2003; 67: 16–37.

Gonzalez N, Beemster GT, Inze D. David and Goliath: what can the tiny weed Arabidopsis teach us to improve biomass production in crops? Curr Opin Plant Biol, 2009; 12: 157–164. doi:10.1016/j.pbi.2008.11.003.

Greene N. *Growing Energy. How Biofuels Can Help End America's Oil Dependence*, 2004. Natural Resources Defence Council, New York.

Gressel J. Transgenics are imperative for biofuel crops. Plant Sci, 2008; 174: 246–263. doi:10.1016/j.plantsci.2007.11.009.

Hartley JL, Temple GF, Brasch MA. DNA cloning using in vitro site-specific recombination. Genome Res, 2000; 10: 1788–1795.

Hatch MD. C4 Photosynthesis: A unique blend of modified biochemistry, anatomy and ultrastructure. Biochim Biophys Acta, 1987; 895: 81–106.

Hiei Y, Ohta S, Komari T, Kumashiro T. Efficient transformation of rice (Oryza sativa L.) mediated by Agrobacterium and sequence analysis of the boundaries of the T-DNA. Plant J, 1994; 6: 271–282.

Hill J, Nelson E, Tilman D, Polasky S, Tiffany D. Environmental, economic, and energetic costs and benefits of biodiesel and ethanol biofuels. Proc Natl Acad Sci U S A, 2006; 103: 11206–11210. doi:10.1073/pnas.0604600103.

Hirano K, Ueguchi-Tanaka M, Matsuoka M. GID1-mediated gibberellin signaling in plants. Trends Plant Sci, 2008; 13: 192–199. doi:10.1016/j.tplants.2008.02.005.

Hisano H, Nandakumar R, Wang ZY. Genetic modification of lignin biosynthesis for improved biofuel production. In Vitro Cell Dev Biol-Plant, 2009; 45: 306–313. doi:10.1007/s11627-009-9219-5.

Hoekema A, Hirsch PR, Hooykaas PJJ, Schilperoort RA. A binary plant vector strategy based on separation of vir- and T-region of the Agrobacterium tumefaciens Ti-plasmid. Nature, 1983; 303: 179–180.

Hu T, Metz S, Chay C, Zhou HP, Biest N, Chen G, Cheng M, Feng X, Radionenko M, Lu F, Fry J. Agrobacterium-mediated large-scale transformation of wheat (Triticum aestivum L.) using glyphosate selection. Plant Cell Rep, 2003; 21: 1010–1019. doi:10.1007/s00299-003-0617-6.

Huang R, Su R, Qi W, He Z. Bioconversion of lignocellulose into bioethanol: process intensification and mechanism research. Bioenerg Res, 2011; 4: 225–245.

Ishida Y, Saito H, Ohta S, Hiei Y, Komari T, Kumashiro T. High efficiency transformation of maize (Zea mays L.) mediated by Agrobacterium tumefaciens. Nature Biotechnol, 1996; 14: 745–750. doi:10.1038/nbt0696-745.

Jeanneau M, Vidal J, Gousset-Dupont A, Lebouteiller B, Hodges M, Gerentes D, Perez P. Manipulating PEPC levels in plants. J Exp Bot, 2002; 53: 1837–1845.

Jewell MC, Campbell BC, Godwin ID. Transgenic plants for abiotic stress resistance. In: Kole C (ed.) *Transgenic Crop Plants. Volume 2, Utilization and Biosafety*, 2009, pp. 67–132. Springer, Berlin, London.

Keating JD, Panganiban C, Mansfield SD. Tolerance and adaptation of ethanologenic yeasts to lignocellulosic inhibitory compounds. Biotechnol Bioeng, 2006; 93: 1196–1206. doi:10.1002/bit.20838.

Kim S, Dale BE. Global potential bioethanol production from wasted crops and crop residues. Biomass Bioenerg, 2004; 26: 361–375. doi:10.1016/j.biombioe.2003.08.002.

Ku MSB, Cho D, Li X, Jiao D-M, Pinto M, Miyao M, Matsuoka M. *Introduction of Genes Encoding C4 Photosynthesis Enzymes into Rice Plants: Physiological Consequences, Novartis Foundation Symposium 236 – Rice Biotechnology: Improving Yield, Stress Tolerance and Grain Quality*, 2007, pp. 100–116. John Wiley & Sons Ltd.

Lee D, Chen A, Nair R. Genetically engineered crops for biofuel production: regulatory perspectives. Biotechnol Genet Eng Rev, 2008; 25: 331–361.

Lee LY, Gelvin SB. T-DNA binary vectors and systems. Plant Physiol, 2008; 146: 325–332. doi:10.1104/pp.107.113001.

Lefebvre S, Lawson T, Fryer M, Zakhleniuk OV, Lloyd JC, Raines CA. Increased sedoheptulose-1,7-bisphosphatase activity in transgenic tobacco plants stimulates photosynthesis and growth from an early stage in development. Plant Physiol, 2005; 138: 451–460. doi:10.1104/pp.104.055046.

Li X, Weng JK, Chapple C. Improvement of biomass through lignin modification. Plant J, 2008a; 54: 569–581. doi:10.1111/j.1365-313X.2008.03457.x.

Li X, Weng JK, Chapple C. Improvement of biomass through lignin modification. Plant J, 2008b; 54: 569–581. doi:10.1111/j.1365-313X.2008.03457.x.

Lu Y, Mosier NS. Current technologies for fuel ethanol production from lignocellulosic plant biomass. In: Vermerris W (ed.) *Genetic Improvement of Bioenergy Crops*, 2008, pp. 161–182. Springer, New York.

Lynd LR. Overview and evaluation of fuel ethanol from cellulosic biomass: technology, economics, the environment, and policy. Ann Rev Energ Environ, 1996; 21: 403–465. doi:10.1146/annurev.energy.21.1.403.

Mann DGJ, LaFayette PR, Abercrombie LL, Parrott WA, Stewart CN. *pANIC: A Versatile Set of Gateway-Compatible Vectors for Gene Overexpression and RNAi-Mediated down-Regulation in Monocots, Plant Transformation Technologies*, 2010, pp. 161–168. Wiley-Blackwell, Oxford.

Matsuoka M, Furbank RT, Fukayama H, Miyao M. Molecular engineering of c4 photosynthesis. Ann Rev Plant Physiol Plant Mol Biol, 2001; 52: 297–314. doi:10.1146/annurev.arplant.52.1.297.

McKendry P. Energy production from biomass (part 1): overview of biomass. Bioresource Technol, 2002; 83: 37–46. doi:10.1016/s0960-8524(01)00118-3.

McLaughlin SB, Kszos LA. Development of switchgrass (Panicum virgatum) as a bioenergy feedstock in the United States. Biomass Bioenerg, 2005; 28: 515–535. doi:10.1016/j.biombioe.2004.05.006.

Mosier N, Wyman C, Dale B, Elander R, Lee YY, Holtzapple M, Ladisch M. Features of promising technologies for pretreatment of lignocellulosic biomass. Bioresource Technol, 2005; 96: 673–686. doi:10.1016/j.biortech.2004.06.025.

Pauly M, Keegstra K. Plant cell wall polymers as precursors for biofuels. Curr Opin Plant Biol, 2010; 13: 305–312. doi:10.1016/j.pbi.2009.12.009.

Ragauskas AJ, Williams CK, Davison BH, Britovsek G, Cairney J, Eckert CA, Frederick WJ, Jr, Hallett JP, Leak DJ, Liotta CL, Mielenz JR, Murphy R, Templer R, Tschaplinski T. The path forward for biofuels and biomaterials. Science, 2006; 311: 484–489. doi:10.1126/science.1114736.

Rhoades MW, Reinhart BJ, Lim LP, Burge CB, Bartel B, Bartel DP. Prediction of plant microRNA targets. Cell, 2002; 110: 513–520.

Riou-Khamlichi C, Huntley R, Jacqmard A, Murray JAH. Cytokinin activation of arabidopsis cell division through a D-type cyclin. Science, 1999; 283: 1541–1544. doi:10.1126/science.283.5407.1541.

Rogers LA, Campbell MM. The genetic control of lignin deposition during plant growth and development. New Phytologist, 2004; 164: 17–30. doi:10.1111/j.1469-8137.2004.01143.x.

Rojas CA, Eloy NB, Lima Mde F, Rodrigues RL, Franco LO, Himanen K, Beemster GT, Hemerly AS, Ferreira PC. Overexpression of the Arabidopsis anaphase promoting complex subunit CDC27a increases growth rate and organ size. Plant Mol Biol, 2009; 71: 307–318. doi:10.1007/s11103-009-9525-7.

Rojas CA, Hemerly AS, Ferreira PCG. Genetically modified crops for biomass increase. Genes and strategies. GM Crops, 2010; 1: 137–142.

Rubin EM. Genomics of cellulosic biofuels. Nature, 2008; 454: 841–845. doi:10.1038/nature07190.

Rushton PJ, Macdonald H, Huttly AK, Lazarus CM, Hooley R. Members of a new family of DNA-binding proteins bind to a conserved cis-element in the promoters of alpha-Amy2 genes. Plant Mol Biol, 1995; 29: 691–702.

Sakamoto T, Miura K, Itoh H, Tatsumi T, Ueguchi-Tanaka M, Ishiyama K, Kobayashi M, Agrawal GK, Takeda S, Abe K, Miyao A, Hirochika H, Kitano H, Ashikari M, Matsuoka M. An overview of gibberellin metabolism enzyme genes and their related mutants in rice. Plant Physiol, 2004; 134: 1642–1653. doi:10.1104/pp.103.033696.

Sakamoto T, Morinaka Y, Ohnishi T, Sunohara H, Fujioka S, Ueguchi-Tanaka M, Mizutani M, Sakata K, Takatsuto S, Yoshida S, Tanaka H, Kitano H, Matsuoka M. Erect leaves caused by brassinosteroid deficiency increase biomass production and grain yield in rice. Nat Biotech, 2006; 24: 105–109. doi:http://www.nature.com/nbt/journal/v24/n1/suppinfo/nbt1173_S1.html.

Salas Fernandez MG, Becraft PW, Yin Y, Lubberstedt T. From dwarves to giants? Plant height manipulation for biomass yield. Trends Plant Sci, 2009; 14: 454–461. doi:10.1016/j.tplants.2009.06.005.

Saxena RC, Adhikari DK, Goyal HB. Biomass-based energy fuel through biochemical routes: a review. Renew Sust Energy Rev, 2009; 13: 167–178. doi:10.1016/j.rser.2007.07.011.

Schwab R, Palatnik JF, Riester M, Schommer C, Schmid M, Weigel D. Specific effects of microRNAs on the plant transcriptome. Dev Cell, 2005; 8: 517–527. doi:10.1016/j.devcel.2005.01.018.

Somerville C. Cellulose synthesis in higher plants. Annual Rev Cell Dev Biol, 2006; 22: 53–78. doi:10.1146/annurev.cellbio.22.022206.160206.

Soni R, Carmichael JP, Shah ZH, Murray JA. A family of cyclin D homologs from plants differentially controlled by growth regulators and containing the conserved retinoblastoma protein interaction motif. Plant Cell, 1995; 7: 85–103. doi:10.1105/tpc.7.1.85.

Sticklen M. Plant genetic engineering to improve biomass characteristics for biofuels. Curr Opin Biotechnol, 2006; 17: 315–319.

Sticklen MB. Plant genetic engineering for biofuel production: towards affordable cellulosic ethanol. Nat Rev Genet, 2008; 9: 433–443. doi:10.1038/nrg2336.

Strauss SH, Kershen DL, Bouton JH, Redick TP, Tan H, Sedjo RA. Far-reaching deleterious impacts of regulations on research and environmental studies of recombinant DNA-modified perennial biofuel crops in the United States. Bioscience, 2010; 60: 729–741.

Tamoi M, Nagaoka M, Miyagawa Y, Shigeoka S. Contribution of fructose-1,6-bisphosphatase and sedoheptulose-1,7-bisphosphatase to the photosynthetic rate and carbon flow in the Calvin cycle in transgenic plants. Plant Cell Physiol, 2006; 47: 380–390. doi:10.1093/pcp/pcj004.

Vinocur B, Altman A. Recent advances in engineering plant tolerance to abiotic stress: achievements and limitations. Curr Opin Biotechnol, 2005; 16: 123–132. doi:10.1016/j.copbio.2005.02.001.

Vogel KP, Jung HJG. Genetic modification of herbaceous plants for feed and fuel. Crit Rev Plant Sci, 2001; 20: 15–49. doi:10.1080/20013591099173.

Wang H, Avci U, Nakashima J, Hahn MG, Chen F, Dixon RA. Mutation of WRKY transcription factors initiates pith secondary wall formation and increases stem biomass in dicotyledonous plants. Proc Natl Acad Sci U S A, 2010; 107: 22338–22343. doi:10.1073/pnas.1016436107.

Wang ZY, Brummer EC. Is genetic engineering ever going to take off in forage, turf and bioenergy crop breeding? Ann Bot, 2012. doi:10.1093/aob/mcs027.

Wang ZY, Ge Y. Recent advances in genetic transformation of forage and turf grasses. In Vitro Cell Dev Biol-Plant, 2006; 42: 1–18. doi:10.1079/ivp2005726.

Weng JK, Li X, Bonawitz ND, Chapple C. Emerging strategies of lignin engineering and degradation for cellulosic biofuel production. Curr Opin Biotechnol, 2008; 19: 166–172. doi:10.1016/j.copbio.2008.02.014.

Wolt JD. Advancing environmental risk assessment for transgenic biofeedstock crops. Biotechnol Biofuels, 2009; 2: 27. doi:10.1186/1754-6834-2-27.

Xie K, Wu C, Xiong L. Genomic organization, differential expression, and interaction of SQUAMOSA promoter-binding-like transcription factors and microRNA156 in rice. Plant Physiol, 2006; 142: 280–293. doi:10.1104/pp.106.084475.

Yadav NS, Vanderleyden J, Bennett DR, Barnes WM, Chilton MD. Short direct repeats flank the T-DNA on a nopaline Ti plasmid. Proc Natl Acad Sci U S A, 1982; 79: 6322–6326.

Zambryski P, Joos H, Genetello C, Leemans J, Montagu MV, Schell J. Ti plasmid vector for the introduction of DNA into plant cells without alteration of their normal regeneration capacity. EMBO J, 1983; 2: 2143–2150.

Zhao ZY, Cai T, Tagliani L, Miller M, Wang N, Pang H, Rudert M, Schroeder S, Hondred D, Seltzer J, Pierce D. Agrobacterium-mediated sorghum transformation. Plant Mol Biol, 2000; 44: 789–798.

# Chapter 12
# Endophytes in Low-input Agriculture and Plant Biomass Production

Sita R. Ghimire[1,2] and Kelly D. Craven[2]

[1]*Center for Agricultural and Environmental Biotechnology, RTI International, Research Triangle Park, NC 27709, USA*
[2]*Plant Biology Division, The Samuel Roberts Noble Foundation, 2510 Sam Noble Parkway, Ardmore, OK 73401, USA*

## 12.1 Introduction

This chapter provides information on a group of plant-associated microbes termed endophytes and describes fungal endophytes of cool season grasses ($C_3$), warm season grasses ($C_4$), and woody plants, and the fitness benefits these endosymbionts can contribute to their host plants. Two important members of the fungal order Sebacinales, *Piriformospora indica* and *Sebacina vermifera,* have shown tremendous potential for enhancement of biomass production and other plant characteristics in a wide range of plant species. The possibility of combining several microbial endophytes into groups, or consortia, with synergistic attributes that may facilitate improved crop production even in a low-input agriculture system and the importance of fungal endophytes for nonagricultural applications will also be discussed.

## 12.2 What are Endophytes?

The term endophyte was originally coined by the pioneering plant pathologist, Anton de Bary to refer to any organisms occurring within plant tissues (De Bary, 1866). Since then, endophytes have been defined in several ways with the addition of new information and recognition of the seemingly ubiquitous nature of these microbes (Hyde and Soytong, 2008; Ghimire and Hyde, 2004). Perhaps the most commonly used definition is "all those organisms inhabiting plant organs that at some time in their life, can colonize internal plant tissues without causing apparent harm to the host" (Petrini, 1991). The strength of this definition lies in its inclusivity of all microbes that are present *asymptomatically* within plant shoot and root tissues. However, this can also be seen as a weakness due to the inclusion of latent plant pathogens or saprotrophs. At the heart of the issue is whether the length of time, or the proportion of the microbes' lifestyle spent in this endophytic state should be taken into account, as well as whether the organism must reside exclusively within plant tissues. For example, should a saprotrophic fungus (feeds

---

*Bioenergy Feedstocks: Breeding and Genetics*, First Edition. Edited by Malay C. Saha, Hem S. Bhandari, and Joseph H. Bouton.
© 2013 John Wiley & Sons, Inc. Published 2013 by John Wiley & Sons, Inc.

on dead plant material) that has had the good fortune of landing a spore on a still-living plant tissue, where it can lie in a quiescent state until the plant material naturally senesces, be considered endophytic? What if the spore germinates to form initial invasive hyphae but is constrained to a small, localized infection, albeit an asymptomatic one? Issues such as these are discussed in greater detail below.

A large body of research on plants and their fungal endophytes suggests that most, if not all, plants in natural ecosystems are symbiotic with mycorrhizal fungi and/or fungal endophytes (Petrini, 1986). Mycorrhizae present an additional challenge to the definition of endophyte, as they seemingly spend a great majority of their life cycle associated with asymptomatic plant tissues, but here the discrepancy is more of a spatial question. Much of the fungal mycelium (the fungal "body") is indeed internalized in the host root tissues, but some, perhaps most, extends into the soil matrix. At least part of the fungus seems to fulfill Petrini's definition. A further distinction can be made as to whether an endophytic relationship is beneficial to both interacting organisms or not. Endophytism does not reflect costs or benefits to the interacting partners nor does it reflect the level of integration, be it completely internalized or partially so, localized or systemic.

An endophyte can be systemically infecting or localized; external as long as some symbiont material is internal; and even pathogenic, as long as some part of the relationship is asymptomatic. Some may argue that most microbial interactions with plants can then be considered at least partially endophytic. The key point is that from a mechanistic and utilitarian standpoint, the focus should remain on why the interactions manifest as asymptomatic in the first place. Perhaps latent pathogens or even initial asymptomatic infections by pathogenic fungi utilize similar mechanisms to manipulate host defense responses or, alternatively, that those microbes maintained in a symbiotic state for longer periods of time are coerced to do so by active plant responses. The current evidence suggests that even the most "intimate" of fungal symbionts can be turned into pathogens by the disruption of any one of a number of genes (Eaton et al., 2010; Tanaka et al., 2006). Further, transfer of some of even the most beneficial endophytes into new hosts results in either the death of the plant or the loss of the symbiont (Latch and Christensen, 1985). It is becoming increasingly apparent that many of these symbioses are finely tuned, requiring an intimate interplay between both partners. It is observed that a great number, perhaps even the majority of microbes, are endophytic in some plant species, but may be pathogenic in others. This phenomenon may even extend across kingdom boundaries, where microbes that are endophytic in plants are potential pathogens in humans (Barac et al., 2004; Guo et al., 2002). The remainder of this review will focus on fungal endophytes, where the majority of studies have been focused and the "currencies" mediating the exchange have been best defined. Emphasis has been made in the pleiomorphic and dynamic nature of these interactions and how they may fundamentally change over time and space.

Fungal endophytes are broadly categorized into two major groups as clavicipitaceous endophytes (C-endophytes) and nonclavicipitaceous endophytes (NC-endophytes) based on both phylogeny and life history characteristics. The C-endophytes are associated with a great number of cool season grasses and several warm season grasses (Bischoff and White, 2005), whereas NC-endophytes are associated with diverse plant species that range from nonvascular plants to angiosperms (Arnold and Lutzoni, 2007). In a recent review, fungal endophytes were grouped into four functional classes based on host range, types, and extent of host tissue(s) colonization, *in planta* endophyte diversity, mode of transmission and the fitness benefits conferred to their host plants (Rodriguez et al., 2009). In this classification, C-endophytes are regarded as Class 1 endophyte, whereas NC-endophytes are further divided into three functional groups as Class 2, Class 3, and Class 4 endophytes.

## 12.3 Endophytes of Cool Season Grasses

The clavicipitaceous fungi of genus *Epichloë* (anamorphs: *Neotyphodium*; together termed the epichloae) form endophytic, and often highly mutualistic relationships with cool season grasses (subfamily: Pooideae). These symbionts are systemic within the aerial tissues of the plant, and span a continuum including asexual mutualists (that are vertically transmitted through seeds) to exclusively sexual pathogens (that are horizontally transmitted), and a mixed strategy, where the endophyte exhibits both mutualistic and pathogenic properties (Moon et al., 2004; Schardl and Clay, 1997).

The mutualistic benefits provided by these endophytes are primarily protective in nature, enhancing host fitness against a number of biotic and abiotic stresses. Protection from biotic stresses is mediated by the production of one or more of at least four distinct classes of alkaloids by these endophytes (Clay and Schardl, 2002). Two of these classes, the loline alkaloids and peramine, are insecticidal and feeding deterrents, respectively (Clay, 1990; Rowan and Gaynor, 1986; Patterson et al., 1991), while the ergot alkaloids and indole-diterpenes are toxic to grazing mammalian herbivores (Gentile et al., 2005; Li et al., 2004; Bacon et al., 1977). Besides their alkaloid production ability, *Epichloë* endophytes can alter phytohormone production in plant tissues (De Battista et al., 1990), which can affect a variety of morphological and physiological responses including greater growth of endophyte-infected plants than their uninfected conspecific (Clay and Schardl, 2002). Further, endophyte-infected plants often have reduced fungal disease incidence (Gwinn and Gavin, 1992; Clarke et al., 2006) and nematode infection (West et al., 1988; Kimmons et al., 1990). Endophyte-infected plants are also typically more competitive than their uninfected conspecifics, and exhibit enhanced tolerance to drought (Arachevaleta et al., 1989) and salt stresses (Sabzalian and Mirlohi, 2010). The multitude of beneficial attributes ascribed to these fungal endophytes likely explains their prevalence within the cool season grasses, as well as the high infection levels often seen in grass populations, often approaching 100% (Siegel et al., 1984). In fact, it is highly likely that this grass subfamily coevolved with the epichloae fungi (Schardl et al., 2008).

Besides the epichloae endophytes, cool season grasses harbor a variety of other fungal endophytes that are typically nonsystemic and not transmitted vertically through seeds. Indeed, the presence of numerous NC-endophytes seems to be a general characteristic of many cool season grasses (Neubert et al., 2006; Sanchez Marquez et al., 2007, 2008).

## 12.4 Endophytes of Warm Season Grasses

In contrast to the cool season grass endophytes, understanding of warm season grass endophytes is very limited. Although the epichloae are not found here, warm season grasses are infected with both C-endophytes and NC-endophytes. The C-endophytes infecting warm season grasses belong to the genera *Balansia*, *Balansiopsis*, and *Myriogenospora* (Cheplick and Clay, 1988; Clay, 1984; Kelley and Clay, 1987). Of these, species of *Balansia* appear to have the broadest host range, infecting over a hundred warm season grass and sedge species (Clay, 1989). Similar to the epichloae, *Balansia epichloë* and *Balansia henningsiana* are known to produce alkaloids and protect plants from insects and mammalian herbivores (Bacon et al., 1981).

Perhaps the best-studied system involving warm season grasses and C-endophytes involves an interaction between *B. henningsiana* and its host *Panicum agrostoides*, a close relative of an important bioenergy crop switchgrass (*Panicum virgatum*). Like the epichloae, this symbiotic fungus grows in the intercellular spaces of all aboveground plant tissues with hyphae particularly concentrated in the base of the tillers (Diehl, 1950). Endophyte-associated damage to the host is negligible (fruiting structures of the fungus emerge without significant host tissue damage) and is greatly outweighed by the beneficial qualities imparted by the fungus. Intriguingly, infected plants have been shown to yield two to three times more vegetative growth than uninfected conspecifics in natural populations, and produce up to two times as many tillers. Further, infected plants in the field appear to be more resistance to infection by the fungal pathogen *Alternaria triticina* (Clay et al., 1989). Follow-up greenhouse studies showed that infected plants produced up to 25% more shoot biomass and 50% more vegetative tillers. Several other *Panicum* species, including switchgrass, have been reported as natural hosts of *B. henningsiana* (Diehl, 1950). However, we have thus far been unable to recover this fungal endophyte from natural populations of switchgrass (see later) and have been unable to artificially inoculate strains from *P. agrostoides* or *Panicum anceps* into switchgrass (S. R. Ghimire and K. D. Craven, unpublished data), suggesting some level of host specificity.

In a 2-year survey of native stands of switchgrass growing in the prairies of northern Oklahoma, a remarkable diversity in the fungal endophyte communities isolated from sterilized shoot and root tissues was documented (Ghimire et al., 2011). It was hypothesized that centers of host diversity should also represent centers of diversity for their microbial endophytes and suggested that these endophytes belonged to 51 and 58 operational taxonomic units (OTUs), respectively, representing at least 18 taxonomic orders (Ghimire et al., 2011). The most prevalent strains recovered belonged to genera such as *Acremonium*, *Alternaria*, *Codinaeopsis*, *Gibberella*, *Fusarium*, *Phoma*, *Periconia*, and *Sporisorium*. Interestingly, none of these fungal isolates from switchgrass belonged to the family Clavicipitaceae. However, several fungi belonging to the same order (Hypocreales) were common, including isolates of species which have been shown to promote plant growth and development. For example, several *Acremonium* and *Fusarium* spp. were isolated from these grasses, and have been used previously as biocontrol agents in both food crops and forage $C_4$ grasses (Kelemu et al., 2001; Dongyi and Kelemu, 2004; Horinouchi et al., 2007; Kaur et al., 2010). *Acremonium strictum* is an effective mycoparasite, preventing the colonization and subsequent disease of the host plant caused by at least five different plant pathogenic fungi (Choi et al., 2009). A natural endophyte of maize, *Acremonium zeae*, is used to effectively eliminate mycotoxin accumulation in the kernels caused by *Fusarium verticillioides* and *Aspergillus flavus* (Wicklow et al., 2005). Further, *A. zeae* produces a full complement of hemicellulolytic enzymes capable of hydrolyzing arabinoxylans from several industrially important feedstocks, demonstrating its potential for application in the bioconversion of lignocellulosic biomass into fermentable sugars (Bischoff et al., 2009). Indeed, initial experiments suggest that at least one of the *Acremonium* species we isolated enhances growth when artificially inoculated back into switchgrass (S. R. Ghimire, unpublished results).

Biomass gains, disease resistance, and stress tolerance are some common benefits to plant hosts from their NC-endophytes. Biomass gains may sometimes result from induction of plant hormones by the endophyte or production of analogs by the endophyte itself (Tudzynski and Sharon, 2002). The plant hormone class known as gibberellins is so named because it was first found to be produced by an endophytic fungus called *Gibberella fujikuroi*, causing "bakanae or foolish seedling" disease in rice (Sun and Snyder, 1981). Disease protection conferred to plants infected with NC-endophytes may be a consequence of niche exclusion of plant pathogens by endophytes for available resources (Rodriguez et al., 2009). The stress tolerance of host plants infected with these endophytes seems to be associated with reactive oxygen species (ROS) production and increased water use efficiency (Rodriguez et al., 2008; Baltruschat et al., 2008).

## 12.5 Endophytes of Woody Angiosperms

Fungal endophytes are present, primarily as localized infections, in the aerial tissues of all trees and shrubs studied to date (Petrini, 1991; Arnold et al., 2003). However, our current understanding of these endophytes and the role they may play in the ecology of woody angiosperms is limited to a small number of plant species. Immature leaves are typically endophyte-free while unfurling from the bud and infection density tends to increase with leaf age. These endophytes appear to be nonsystemic and are typically transmitted horizontally by spores (Helander et al., 1993; Wilson and Carroll, 1994). Several studies in trees, shrubs, and ferns have shown that individual species and even individual plants typically harbor scores of fungal species (Saikkonen et al., 1998; Carroll and Carroll, 1978; Espinosa-Garcia and Langenheim, 1990; Faeth and Hammon, 1997; McCutcheon et al., 1993; Lodge et al., 1996). The species composition of tree endophytes typically comprises a few dominant species and numerous sporadically detected species (Saikkonen, 2007). These dominant fungal species are usually presumed to be specific to the host tree species or closely related species (Gennaro et al., 2003). In general, the nature of the relationship between woody plants and their foliar endophytes is not mutualistic as the fungus is localized, not tightly coupled with the host reproductive system, or even highly dependent on the host (Stone, 1987; Barklund, 1987; Danti et al., 2002). In fact, many of the endophytes of woody plants appear to be closely related to pathogenic fungi (Saikkonen et al., 1998) suggesting that while endophytic, they may in many cases be quiescent pathogens or saprotrophs. Although their role in protecting hosts from herbivores and abiotic stresses is not common (Carroll, 1995), some of these horizontally transmitted endophytes have been reported to enhance host defense (Arnold et al., 2003).

## 12.6 Other Fungal Endophytes

Two members of the newly defined basidiomycete order Sebacinales, *P. indica*, and its close relative *S. vermifera* have stimulated considerable attention over the past several years. These fungi form an endophytic association with the roots of several mono- and dicotyledonous plants in a manner similar to those of mycorrhizal fungi (Weiss et al., 2011; Waller et al., 2008). Indeed, these and related species are known to form several types of mycorrhizal relationships including orchidaceous, ericoid, and jungermannioid (Weiss et al., 2004). Both species are axenically cultivable, possess plant-growth-promoting characteristics, and contribute several other benefits to host plants (Deshmukh et al., 2006; Varma et al., 1999; Waller et al., 2008; Barazani et al., 2005). *S. vermifera* promotes growth and fitness of *Nicotiana attenuata* by impairing ethylene biosynthesis (Barazani et al., 2007). Similarly, *P. indica* is involved in auxin metabolism and the production of high quantities of cytokinin in some plants (Sirrenberg et al., 2007; Vadassery et al., 2008).

Root colonization of switchgrass by *S. vermifera* enhances biomass production up to 113%, and also significantly improves seed germination (Ghimire et al., 2009). More recent work investigating biomass production under water-limiting conditions confirmed enhancement of switchgrass biomass (with the best combinations yielding as high as 300% increase in biomass) (Figure 12.1) and surprisingly, cocultivated plants under drought stress produced significantly higher biomass than their mock-inoculated conspecifics under normal watering conditions (Ghimire and Craven, 2011). Further, both studies reveal that plants colonized by *S. vermifera* exhibit a significant increase in both shoot and root biomass, indicating that aboveground biomass gains are not simply a consequence of reallocated carbohydrate. Indeed, cocultivated

**Figure 12.1.** Effect of *Sebacina vermifera* on early growth of switchgrass (after 6 weeks of cocultivation). One-month-old rooted switchgrass clones of NF/GA-993 were cocultivated with two different strains (MAFF 305828 and MAFF 305830) of *S. vermifera*. (*For color details, see color plate section.*)

plants consistently produced higher root biomass than mock-inoculated plants suggesting a greater potential to sequester carbon and hold soils, both highly desired properties in a crop grown under a low-input regime.

## 12.7 Endophytes in Biomass Crop Production

Unprecedented global population growth led to the green revolution of the 20th century that in turn resulted in research, development, and technology transfer initiatives that increased agriculture production around the globe (Hazell, 2009). The use of high-yielding crop varieties, development of hybrid cultivars, expansion of irrigation infrastructures, use of fertilizers and pesticides, and modernization of crop management practices were the foundation for this revolution. In other words, the green revolution was the outcome of an intensive use of agricultural resources to feed an ever-growing world population that would have led to mass starvation in many regions had it not succeeded. However, the intensive use of agricultural resources, particularly water and fertilizer, now and into the foreseeable future is exceedingly expensive and/or environmentally damaging, and cannot remain indefinitely because of overwhelming demand for water for nonagricultural residential uses and the skyrocketing energy cost of fertilizer production.

There has been a steady rise in agricultural production since the green revolution, but the scope of high-input agriculture with current technologies and available crop plants will not be sufficient to feed the rapidly growing world population in the context of a dwindling supply of agricultural inputs (Den Herder et al., 2010). This is particularly true in developing countries, where population growth will be the greatest but access to agronomic inputs will be the most limited. Extensive utilization of endophytic microbes, particularly those that are readily cultivable on minimal resources to produce large amounts of plant inoculum, could play a significant role in feeding an ever-burgeoning world population.

Future plant breeding efforts need to be focused on belowground characteristics of plants such as root architecture including rooting depth, spread of roots, root biomass, root to shoot biomass ratio, water and nutrient uptake, and nitrogen fixation. This is critical in the context of significant climate change, where plants will be grown on exceedingly more marginal lands impacted by drought, salinity, and mineral and nutrient depletion, all while likely facing an increased incidence of pests and diseases. Moreover, there is increasing pressure on arable land due to land degradation, erosion, and the increasing demand for biofuel feedstock production. Merging high-yielding agriculture with ecologically sensitive land use will require a multifaceted approach that combines conventional plant breeding, agronomy, molecular biology, genetic engineering, and other disciplines to develop plants with traits that can deal with diminishing agricultural resource availability (Scherr and McNeely, 2008).

Shifting society's dependence away from petroleum toward renewable biomass resources is generally viewed as an important contributor to the development of a sustainable industrial society and the effective management of greenhouse gas emissions (Ragauskas et al., 2006). To facilitate this shift, the U.S. Department of Energy has set a goal of replacing 30% of current U.S. petroleum consumption with bioethanol by 2030. Similarly, the biofuel directive of the European Union has targeted the replacement of 5.75% of all petrol and diesel transportation fuel with biomass-derived fuels by December 2010. Achieving these goals requires billions of tons of biomass annually (Perlack et al., 2005). Theoretically, biomaterials-based biofuel production seems achievable as the current sustainable global biomass energy potential has been estimated at $\sim 10^{20}$ J/year (Parikka, 2004). However, competition between biomass crops and food crops for finite resources (including land use) and the adverse environmental effects of agrochemical overuse and misuse are major challenges for feedstock-based biofuels to emerge as a viable energy alternative in the years ahead. Achieving high amounts of biomass on marginal lands, under adverse environmental conditions and with minimum agronomic inputs should therefore define the "ideal" biofuel crop. To bring a renewable biofuel-based economy to fruition, a concerted, multidisciplinary effort is necessary to maximize not only the biomass yield of bioenergy crops, but also the sustainability of land and other resource use. Only then we will be able to produce enough feedstock for biorefineries for generations to come.

The agricultural resources that limit crop production mostly include water and nutrients, especially nitrogen (N) and phosphorous (P). Perhaps fortuitously (or not), the mycorrhizal fungi that are known to form associations with more than 80% of plant species, often enhance N and P uptake (Abbott and Robson, 1984; George et al., 1995), water uptake, and drought tolerance (Safir and Boyer, 1971; Sylvia and Williams, 1992), and can even provide protection or tolerance to rhizosphere pathogens in a broad variety of plants (Linderman and Hendrix, 1982). Despite these obvious benefits, the use of mycorrhizal fungi (particularly arbuscular–vesicular, or AM mycorrhizae) in agriculture is limited, as mass production of these fungi in the absence of host plant is not possible (Diop et al., 1994; Singh et al., 2000). The use of species such as *Sebacina* and *Piriformaspora* is particularly attractive as these fungi are amenable to axenic culture, form symbiotic associations with a wide range of plant species (Singh et al., 2000; Waller et al., 2005; Clay et al., 1989), and contributing pleiotropic benefits to their host plants (see earlier sections). Furthermore, it is observed that the general strategy of isolating endophytic fungi from wild relatives of the target crop of interest holds great promise for enhancement of several growth characteristics, whether the intended purpose is for biomass or food production. The use of endosymbionts as a complementary approach to breeding and transgenic efforts can maximize the utility of existing and future crop varieties toward economical and sustainable biomass production.

## 12.8 The Use of Fungal Endophytes in Bioenergy Crop Production Systems

Given the potential of fungal endophytes to enhance plant health and productivity, how would such microorganisms be used and deployed in a bioenergy crop production system? This largely depends upon the type of fungal organism utilized. For example, most of the cool season grass endophytes mentioned previously are capable of nondestructively infecting the embryonic tissues of the developing seed ovule and are incorporated into the resulting seed at a very high frequency (Schardl, 2010). This is a perfect delivery vehicle, as it packages both host and endophyte together in a single, fairly stable unit that can be readily bulked up and distributed with relative ease. Unfortunately, the large majority of useful fungal endophytes, at least those currently recognized, are not seed-transmissible. One reason for this is that many of the useful fungi do not colonize the aerial parts of the plant, much less the seed. Arbuscular mycorrhizae (AM) and the aforementioned *P. indica* and *S. vermifera*, being restricted to root tissues cannot be transmitted internally in seed tissues. Dissemination of the former fungi (AM) is further complicated by the fact that they are obligate symbionts, meaning that they require a living host plant to grow and develop. Methods employed to overcome these hurdles involve producing fungal inoculum in pot cultures in the presence of suitable plants (Feldman and Idczak, 1994) or, more recently, in *Agrobacterium*-transformed root cultures (Declerck et al., 1996). In this manner, AM spores are produced that can then be mixed with the soil substrate. The latter fungi such as *S. vermifera* can be cultured *in vitro* and grown on various substrates in the absence of a host plant (Ghimire and Craven, 2011). This greatly simplifies the process and allows the potential for large-scale inoculum production. As with AM fungi, the inoculum could then be incorporated into the soil matrix as a biofertilizer and protectant. As both of these root-infecting endophytic mycorrhizae have seemingly broad host ranges and are documented to impart a variety of fitness benefits to their host plants, incorporation of one or both should prove very beneficial in production systems involving bioenergy crops, or any crop for that matter.

## 12.9 Endophyte Consortia

It has been shown that plant diversity fosters soil microbial diversity (Zak et al., 2003; Bardgett and Shine, 1999; Millard and Singh, 2010). Host diversity may have similar effects on fungal endophyte assemblages, such that endophyte diversity is expected to be lower in monoculture stands of genetically uniform crop varieties than in mixture of genetically diverse crop varieties. It is becoming increasingly apparent that diverse soil communities are critical for sustainable biomass production (Solis-Dominguez et al., 2011) as well as protection from plant pathogens (Weller et al., 2002; Alabouvette, 1999; Hornby, 1983). Monocultures have historically shown increased susceptibility to disease-causing microbes, as a pathogen that can infect one plant genotype can infect the entire population (Zhu et al., 2000; Browning and Frey, 1969; Wolfe, 1985). This notion is also supported by observational evidence, wherein there is considerably less disease prevalence in genetically diverse plant populations than in monoculture stands. These findings suggest the possibility that at least part of the resistance to perturbation (such as a diseased state) exhibited by diverse plant assemblages is due to their associated microbial communities. We propose that this same phenomenon can be applied to endophytes, and consortia of such microbes could be devised that would promote various aspects of plant health and productivity. For example, some member(s) of the consortium could contribute to water and nutrient uptake while others may improve disease resistance, drought tolerance, seed germination, and so on. The synergetic interactions among microbial consortium members may enhance survival and fitness of symbionts (Porras-Alfaro et al., 2008), and the impact of the consortium may often be observed in the ecosystem (Clay and Holah, 1999).

Besides fungal endophytes, plants typically harbor several endophytic bacteria. These endophytic bacteria have also shown tremendous potential for enhancing the growth and stress tolerances of numerous plants, including domesticated cereals like corn (Lalande et al., 1989; McInroy and Kloepper, 1995). Like the plant-growth-promoting rhizobacteria (PGPR), which typically reside in the immediate vicinity of the plant roots, endophytic bacteria can enhance nutrient availability to their host plant. Some fix atmospheric nitrogen while others can solubilize minerals such as phosphorus and iron, thereby increasing their availability. Further, like endophytic fungi and PGPR, endophytic bacteria can enhance plant fitness by suppressing pathogenic microbes. Like their fungal counterparts, the synthesis of a wide range of phytohormones has been demonstrated for various endophytic bacteria. These endophytically derived compounds can affect plant growth and development, such as stimulation of root development, or induce important phenotypes, such as drought tolerance and systemic resistance against pathogens in their host plant (Cho et al., 2008; Han et al., 2006; Ryu et al., 2003; Ryu et al., 2004).

We promote the notion that future agronomic strategies incorporate studies of natural systems that bringing these eukaryotic and prokaryotic organisms together may have positive and synergistic effects on the overall fitness of potentially any plant species, including those grown for biomass production. As we promote the use of marginal lands for the production of biomass crops, we have chosen switchgrass as a model to explore the creation of endophytic consortia to facilitate low-input agriculture. This strategy reduces the competition with food crops for arable land, and if designed intelligently, can also serve to begin rebuilding damaged and depleted soils to increase arable lands in the future. Understanding endophytic microbial diversity in a given host (or its relatives), selection of microbial candidates for combining into consortia and subsequent evaluation of these consortia on fitness attributes of target plant species under challenging environments are critical steps toward utilizing these microbes in agriculture. Although this approach is nontransgenic (thus perhaps enhancing its attractiveness to certain target groups), it is perfectly complementary to a transgenic approach and may even alleviate any deficiencies introduced as the unintended consequences of genetic modification.

## 12.10  Source of Novel Compounds

Endophytes are novel natural products that have pharmaceutical, agricultural, and industrial importance (Strobel and Daisy, 2003; Gunatilaka, 2006). As endophytes are presumed to be in a more or less constant metabolic and environmental interaction with their hosts (Schulz et al., 2002), they may be more capable of producing a wide range of bioactive metabolites with varied properties, including metabolites with antimicrobial, anticancer, antioxidant, and insecticidal properties (Gunatilaka, 2006; Huang et al., 2007). For example, the promising cancer drug taxol was originally isolated from the Pacific Yew tree (Stierle et al., 1993). Since that time, it has been recognized that much of the taxol production is accomplished by endophytic symbionts of the Yew tree. Many other fungi have subsequently been shown to produce taxol (Bashyal et al., 1999; Li et al., 1996; Strobel et al., 1996), several of which can be grown in batch culture, thus alleviating the need to extract the taxol from the bark of the tree (Stierle et al., 1993).

In general, fungi are expert secondary metabolite producers, and these metabolites quite often have bioactive properties. Endophytes in particular are viewed as an outstanding source of bioactive natural products because many of them occupying literally millions of unique biological niches (higher plants) growing in so many unusual environments (Strobel and Daisy, 2003). Thus, mining fungal endophytes for potentially useful compounds holds tremendous promise for agriculture and medicine. As mentioned earlier, some of the C-endophytes in cool season grasses produce ergot alkaloids that cause toxicosis in grazing livestock. However, these same compounds (which are potent vasoconstrictors) have shown great promise in treating human conditions ranging from migraines and blood loss during child birth to Parkinson's disease. The number of review articles on natural products from endophytic microorganisms that have appeared over the past decade demonstrates to this interest (Gunatilaka, 2006; Strobel and Daisy, 2003; Gusman and Vanhaelen, 2000; Tan and Zou, 2001; Schulz et al., 2002; Strobel, 2003).

## 12.11 Endophyte in Genetic Engineering of Host Plants

Endophytic microbes offer the potential to introduce desirable traits into the plant of interest by genetic manipulation of the endophyte itself (Pamphile et al., 2004). Often, these microbes are easier to manipulate genetically, so this can serve as complementary approach to traditional plant genetic manipulation. This concept was successfully used in phytoremediation, where endophytic bacteria were engineered to degrade organic contaminants, after which their inoculation into Poplar trees allowed the plants to grow and thrive in toxic soils that would otherwise kill them in the absence of the symbiotic microbes (Barac et al., 2004; Taghavi et al., 2005; Weyens et al., 2009). This concept could potentially be expanded to capitalize on the systemic or localized nature of the endophyte, as well as the differential nature in expression profiles exhibited between plant and microbial symbiont. In this regard, the endophyte, whether it is fungal or bacterial in nature, can be viewed as an additional means of integration into the host plant.

## 12.12 Conclusions

"Endophyte" refers to all organisms inhabiting plant organs that at some time in their life can colonize internal plant tissues without causing apparent harm to the host. This is a definition that is still evolving today. Early research was focused on C-endophytes because of their role in agriculture and livestock production as well as the very limited knowledge on the ecological significance of NC-endophytes. Both the C- and NC-endophytes not only impact the fitness of their hosts but also affect plant ecology and community structure, thus suggest a synergism between plants and their associated microbes. As the role of endosymbionts in host defense, stress tolerance, nutrient acquisition, and water use has become better defined, a new research arena has opened with regard to their utilization in low-input agriculture. Efforts that facilitate low-input agriculture are a necessity to meet the demand for food and biomass crops in the future. We envision the use of endosymbionts as a complementary approach to maximize the utility of existing and future crop varieties toward economical and sustainable biomass production.

## Acknowledgments

This research was supported by the BioEnergy Science Center and the Samuel Roberts Nobel Foundation. The BioEnergy Science Center is a U.S. Department of Energy Bioenergy Research Center supported by the Office of Biological and Environmental Research in the DOE Office of Science.

## References

Abbott LK, Robson AL. The effect of VA mycorrhizae on plant growth. In: Powell CL, Bagyaraj DJ (eds) *VA Mycorrhiza*, 1984, pp. 113–130. CRC Press, Boca Raton, FL.

Alabouvette C. Fusarium wilt suppressive soils: an example of disease-suppressive soils. Aust Plant Pathol, 1999; 28: 57–64.

Arachevaleta M, Bacon CW, Hoveland CS, Radcliffe DE. Effect of the tall fescue endophyte on plant response to environmental stress. Agron J, 1989; 81: 83–90.

Arnold AE, Lutzoni F. Diversity and host range of foliar fungal endophytes: are tropical leaves biodiversity hotspots? Ecology, 2007; 88: 541–549.

Arnold AE, Mejia LC, Kyllo D, Rojas EI, Maynard Z, Robbins N, Herre EA. Fungal endophytes limit pathogen damage in a tropical tree. Proc Natl Acad Sci U S A, 2003; 100: 15649–15654.

Bacon CW, Porter JK, Robbins JD. Ergot alkaloid biosynthesis by isolates of *Balansia epichloë* and *Balansia henningsiana*. Can J Bot, 1981; 59: 2534–2538.

Bacon CW, Porter JK, Robbins JD, Luttrell ES. *Epichloe typhina* from toxic tall fescue grasses. Appl Environ Microbiol, 1977; 34: 576–581.

Baltruschat H, Fodor J, Harrach BD, Niemczyk E, Barna B, Gullner G, Janeczko A, Kogel KH, Schafer P, Schwarczinger I, Zuccaro A, Skoczowski A. Salt tolerance of barley induced by the root endophyte *Piriformospora indica* is associated with a strong increase in antioxidants. New Phytol, 2008; 180: 501–510.

Barac T, Taghavi S, Borremans B, Provoost A, Oeyen L, Colpaert JV, Vangronsveld J, van der Lelie D. Engineered endophytic bacteria improve phytoremediation of water-soluble, volatile, organic pollutants. Nat Biotechnol, 2004; 22: 583–588.

Barazani O, Benderoth M, Groten K, Kuhlemeier C, Baldwin IT. *Piriformospora indica* and *Sebacina vermifera* increase growth performance at the expense of herbivore resistance in *Nicotiana attenuata*. Oecologia, 2005; 146: 234–243.

Barazani O, Von Dahl CC, Baldwin IT. *Sebacina vermifera* promotes the growth and fitness of *Nicotiana attenuata* by inhibiting ethylene signaling. Plant Physiol, 2007; 144: 1223–1232.

Bardgett RD, Shine A. Linkages between plant litter diversity, soil microbial biomass and ecosystem function in temperate grasslands. Soil Biol Biochem, 1999; 31: 317–321.

Barklund P. Occurrence and pathogenicity of *Lophodermium piceae* appearing as an endophyte in needles of *Picea abies*. Trans Brit Mycol Soc, 1987; 89: 307–313.

Bashyal B, Li JY, Strobel G, Hess WM, Sidhu R. *Seimatoantlerium nepalense*, an endophytic taxol producing coelomycete from Himalayan yew (*Taxus wallachiana*). Mycotaxon, 1999; 72: 33–42.

Bischoff JF, White JF. Evolutionary development of the Clavicipitaceae. In: Dighton JF., White Jr, Oudemans P (eds) *The Fungal Community: Its Organization and Role in the Ecosystem*, 2005, pp. 505–518. Taylor & Francis, Boca Raton, FL.

Bischoff KM, Wicklow DT, Jordan DB, de Rezende ST, Liu SQ, Hughes SR, Rich JO. Extracellular hemicellulolytic enzymes from the maize endophyte *Acremonium zeae*. Curr Microbiol, 2009; 58: 499–503.

Browning JA, Frey KJ. Multiline cultivars as a means of disease control. Ann Rev Phytopathol, 1969; 7: 355–382.

Carroll G. Forest endophytes – pattern and process. Can J Bot, 1995; 73: S1316–S1324.

Carroll GC, Carroll FE. Studies on the incidence of coniferous needle endophytes in the pacific northwest. Can J Bot, 1978; 56: 3034–3043.

Cheplick GP, Clay K. Acquired chemical defenses in grasses—the role of fungal endophytes. Oikos, 1988; 52: 309–318.

Cho SM, Kang BR, Han SH, Anderson AJ, Park JY, Lee YH, Cho BH, Yang KY, Ryu CM, Kirn YC. 2R,3R-butanediol, a bacterial volatile produced by Pseudomonas chlororaphis O6, is involved in induction of systemic tolerance to drought in Arabidopsis thaliana. Mol Plant Microbe Int, 2008; 21: 1067–1075.

Choi GJ, Kim JC, Jang KS, Nam MH, Lee SW, Kim HT. Biocontrol activity of *Acremonium strictum* BCP against *Botrytis* diseases. Plant Pathol J, 2009; 25: 165–171.

Clarke BB, White JF, Hurley RH, Torres MS, Sun S, Huff DR. Endophyte-mediated suppression of dollar spot disease in fine fescues. Plant Dis, 2006; 90: 994–998.

Clay K. The effect of the fungus *Atkinsonella hypoxylon* (Clavicipitaceae) on the reproductive system and demography of the grass *Danthonia spicata*. New Phytol, 1984; 98: 165–175.

Clay K. Clavicipitaceous endophytes of grasses – their potential as biocontrol agents. Mycol Res, 1989; 92: 1–12.

Clay K. Fungal endophytes of grasses. Ann Rev Ecol Syst, 1990; 21: 275–297.

Clay K, Cheplick GP, Marks S. Impact of the fungus *Balansia henningsiana* on *Panicum agrostoides*: frequency of infection, plant growth and reproduction, and resistance to pests. Oecologia, 1989; 80: 374–380.

Clay K, Holah J. Fungal endophyte symbiosis and plant diversity in successional fields. Science, 1999; 285: 1742–1744.

Clay K, Schardl C. Evolutionary origins and ecological consequences of endophyte symbiosis with grasses. Am Nat, 2002; 160: S99–S127.

Danti R, Sieber TN, Sanguineti G. Endophytic mycobiota in bark of European beech (*Fagus sylvatica*) in the Apennines. Mycol Res, 2002; 106: 1343–1348.

De Bary A. *Morphologie und Physiologie der Pilze, Flechten, und Myxomyceten*, 1866; W. Engelmann, Leipzig.

De Battista JP, Bacon CW, Severson R, Plattner RD, Bouton JH. Indole acetic acid production by the fungal endophyte of tall fescue. Agron J, 1990; 82: 878–880.

Declerck S, Strullu DG, Plenchette C. In vitro mass-production of the arbuscular mycorrhizal fungus, *Glomus versiforme*, associated with Ri T-DNA transformed carrot roots. Mycol Res, 1996; 100: 1237–1242.

Den Herder G, Van Isterdael G, Beeckman T, De Smet I. The roots of a new green revolution. Trends Plant Sci, 2010; 15: 600–607.

Deshmukh S, Hueckelhoven R, Schaefer P, Imani J, Sharma M, Weiss M, Waller F, Kogel KH. The root endophytic fungus Piriformospora indica requires host cell death for proliferation during mutualistic symbiosis with barley. Proc Natl Acad Sci U S A, 2006; 103: 18450–18457.

Diehl WW. Balansia and Balansiae in America, 1950. Agriculture Monograph No. 4, United States Department of Agriculture.

Diop TA, Plenchette C, Strullu DG. Dual axenic culture of sheared root inocula of vesicular arbuscular mycorrhizal fungi associated with tomato roots. Mycorrhiza, 1994; 5: 17–22.

Dongyi H, Kelemu S. *Acremonium implicatum*, a seed-transmitted endophytic fungus in *Brachiaria* grasses. Plant Disease, 2004; 88: 1252–1254.

Eaton CJ, Cox MP, Ambrose B, Becker M, Hesse U, Schardl CL, Scott B. Disruption of signaling in a fungal-grass symbiosis leads to pathogenesis. Plant Physiol, 2010; 153: 1780–1794.

Espinosa-Garcia FJ, Langenheim JH. The endophytic fungal community in leaves of a costal redwood population, diversity and spatial patterns. New Phytol, 1990; 116: 89–97.

Faeth SH, Hammon KE. Fungal endophytes in oak trees: Long-term patterns of abundance and associations with leafminers. Ecology, 1997; 78: 810–819.

Feldman F, Idczak E. Inoculum production of VA mycorrhizal fungi. In: Norris JR, Read DJ, Varma AK (eds) *Techniques for Mycorrhizal Research*, 1994, pp. 799–817. Academic Press, San Diego.

Gennaro M, Gonthier P, Nicolotti G. Fungal endophytic communities in healthy and declining *Quercus robur* L. and *Q. cerris* L. trees in northern Italy. J Phytopathol, 2003; 151: 529–534.

Gentile A, Rossi MS, Cabral D, Craven KD, Schardl CL. Origin, divergence, and phylogeny of *Epichloë* endophytes of native Argentine grasses. Mol Phylogenet Evol, 2005; 35: 196–208.

George E, Marschner H, Jakobsen I. Role of arbuscular mycorrhizal fungi in uptake of phosphorus and nitrogen from soil. Crit Rev Biotechnol, 1995; 15: 257–270.

Ghimire SR, Craven KD. The ectomycorrhizal fungus *Sebacina vermifera*, enhances biomass production of switchgrass (*Panicum virgatum* L.) under drought conditions. Appl Environ Microbiol, 2011; 77(19): 7063–7067.

Ghimire SR, Charlton ND, Bell JD, Krishnamurthy YL, Craven KD. Biodiversity of fungal endophyte communities inhabiting switchgrass (*Panicum virgatum* L.) growing in the native tallgrass prairie of northern Oklahoma. Fungal Divers, 2011; 47: 19–27.

Ghimire SR, Charlton ND, Craven KD. The mycorrhizal fungus, *Sebacina vermifera*, enhances seed germination and biomass production in switchgrass (*Panicum virgatum* L). Bioenerg Res, 2009; 2: 51–58.

Ghimire SR, Hyde KD. Fungal Endophytes. In: Varma A, Abbott L, Werner D, Hampp R (eds) *Plant Surface Microbiology*, 2004, pp. 281–292. Springer, Berlin, Heidelberg.

Gunatilaka AAL. Natural products from plant-associated microorganisms: Distribution, structural diversity, bioactivity, and implications of their occurrence. J Nat Prod, 2006; 69: 509–526.

Guo X, van Iersel MW, Chen JR, Brackett RE, Beuchat LR. Evidence of association of salmonellae with tomato plant grown hydroponically in inoculated nutrient solution. Appl Environ Microbiol, 2002; 68(7): 3639–3643.

Gusman J, Vanhaelen M. Endophytic fungi: an underexploited source of biologically active secondary metabolites. Recent Res Dev Phytochem, 2000; 4: 187–206.

Gwinn KD, Gavin AM. Relationship between endophyte infestation level of tall fescue seed lots and *Rhizoctonia zea* seedling disease. Plant Dis, 1992; 76: 911–914.

Han SH, Lee SJ, Moon JH, Park KH, Yang KY, Cho BH, Kim KY, Kim YW, Lee MC, Anderson AJ, Kim YC. GacS-dependent production of 2R, 3R-butanediol by Pseudomonas chlororaphis O6 is a major determinant for eliciting systemic resistance against Erwinia carotovora but not against Pseudomonas syringae pv. tabaci in tobacco. Mol Plant Microbe In, 2006; 19: 924–930.

Hazell PBR. *The Asian Green Revolution*, 2009, p. 34. International Food Policy Research Institute, Washington, DC.

Helander ML, Neuvonen S, Sieber T, Petrini O. Simulated acid rain affects birch leaf endophyte populations. Microb Ecol, 1993; 26: 227–234.

Horinouchi H, Muslim A, Suzuki T, Hyakumachi M. *Fusarium equiseti* GF191 as an effective biocontrol agent against Fusarium crown and root rot of tomato in rock wool systems. Crop Prot, 2007; 26: 1514–1523.

Hornby D. Suppressive soils. Ann Rev Phytopathol, 1983; 21: 65–85.

Huang WY, Cai YZ, Xing J, Corke H, Sun M. A potential antioxidant resource: Endophytic fungi from medicinal plants. Econ Bot, 2007; 61: 14–30.

Hyde KD, Soytong K. The fungal endophyte dilemma. Fungal Divers, 2008; 33: 163–173.

Kaur R, Kaur J, Singh RS. Nonpathogenic *Fusarium* as a biological control agent. Plant Pathol J, 2010; 9: 88–100.

Kelemu S, White JF, Munoz F, Takayama Y. An endophyte of the tropical forage grass Brachiaria brizantha: isolating, identifying, and characterizing the fungus, and determining its antimycotic properties. Can J Microbiol, 2001; 47: 55–62.

Kelley SE, Clay K. Interspecific competitive interactions and the maintenance of genotypic variation within two perennial grasses. Evolution, 1987; 41: 92–103.

Kimmons CA, Gwinn KD, Bernard EC. Nematode reproduction on endophyte infected and endophyte free tall fescue. Plant Dis, 1990; 74: 757–761.

Lalande R, Bissonnette N, Coutlee D, Antoun H. Identification of rhizobacteria from maize and determination of their plant growth promoting potential. Plant Soil, 1989; 115: 7–11.

Latch GCM, Christensen MJ. Artificial infection of grasses with endophytes. Ann Appl Biol, 1985; 107: 17–24.

Li CJ, Nan ZB, Paul VH, Dapprich PD, Liu Y. A new *Neotyphodium* species symbiotic with drunken horse grass (*Achnatherum inebrians*) in China. Mycotaxon, 2004; 90: 141–147.

Li JY, Strobel G, Sidhu R, Hess WM, Ford EJ. Endophytic taxol-producing fungi from bald cypress, *Taxodium distichum*. Microbiology, 1996; 142: 2223–2226.

Linderman RG, Hendrix JW. Evaluation of plant response to colonization by vascular-arbuscular mycorrhizal fungi: A. Host variables. In: Schenck NC (ed.) *Methods and Principles of Mycorrhizal Research*, 1982, pp. 69–76. American Phytopathological Society, St. Paul, MN.

Lodge DJ, Fisher PJ, Sutton BC. Endophytic fungi of *Manilkara bidentata* leaves in Puerto Rico. Mycologia, 1996; 88: 733–738.

McCutcheon TL, Carroll GC, Schwab S. Genotypic diversity in populations of a fungal endophyte from Douglas fir. Mycologia, 1993; 85: 180–186.

McInroy JA, Kloepper JW. Population dynamics of endophytic bacteria in field grown sweet corn and cotton. Can J Microbiol, 1995; 41: 895–901.

Millard P, Singh BK. Does grassland vegetation drive soil microbial diversity? Nutr Cycl Agroecosys, 2010; 88: 147–158.

Moon CD, Craven KD, Leuchtmann A, Clement SL, Schardl CL. Prevalence of interspecific hybrids amongst asexual fungal endophytes of grasses. Mol Ecol, 2004; 13: 1455–1467.

Neubert K, Mendgen K, Brinkmann H, Wirsel SGR. Only a few fungal species dominate highly diverse mycofloras associated with the common reed. Appl Environ Microbiol, 2006; 72: 1118–1128.

Pamphile JA, da Rocha CL, Azevedo JL. Co-transformation of a tropical maize endophytic isolate of *Fusarium verticillioides* (synonym *F. moniliforme*) with gusA and nia genes. Genet Mol Biol, 2004; 27: 253–258.

Parikka M. Global biomass fuel resources. Biomass Bioenerg, 2004; 27: 613–620.

Patterson CG, Potter DA, Fannin FF. Feeding deterrency of alkaloids from endophyte-infected grasses to Japanese beetle grubs. Entomol Exp Appl, 1991; 61: 285–289.

Perlack RD, Wright LL, Turhollow AF, Graham RL, Stokes BJ, Erbach DC. Biomass as Feedstock for Bioenergy and Bioproducts Industry: the Technical Feasibility of a Billion-ton Annual Supply, 2005, U.S. Department of Energy and U.S. Department of Agriculture.

Petrini O. Taxonomy of endophytic fungi of aerial plant tissues. In: Fokkema NJ, van del Huevel J (eds) *Microbiology of the Phyllosphere*, 1986, pp. 175–187. Cambridge University Press, Cambridge.

Petrini O. Fungal endophytes in tree leaves. In: Andrews JH, Hirano SS (eds) *Microbial Ecology of Leaves*, 1991, pp. 179–197. Springer, New York.

Porras-Alfaro A, Herrera J, Sinsabaugh RL, Odenbach KJ, Lowrey T, Natvig DO. Novel root fungal consortium associated with a dominant desert grass. Appl Environ Microbiol, 2008; 74: 2805–2813.

Ragauskas AJ, Williams CK, Davison BH, Britovsek G, Cairney J, Eckert CA, Frederick WJ, Hallett JP, Leak DJ, Liotta CL, Mielenz JR, Murphy R, Templer R, Tschaplinski T. The path forward for biofuels and biomaterials. Science, 2006; 311: 484–489.

Rodriguez RJ, Henson J, Van Volkenburgh E, Hoy M, Wright L, Beckwith F, Kim YO, Redman RS. Stress tolerance in plants via habitat-adapted symbiosis. Isme J, 2008; 2: 404–416.

Rodriguez RJ, White JF, Arnold AE, Redman RS. Fungal endophytes: diversity and functional roles. New Phytol, 2009; 182: 314–330.

Rowan DD, Gaynor DL. Isolation of feeding deterrents against Argentine stem weevil from ryegrass infected with the endophyte *Acremonium loliae*. J Chem Ecol, 1986; 12: 647–658.

Ryu CM, Farag MA, Hu CH, Reddy MS, Wei HX, Pare PW, Kloepper JW. Bacterial volatiles promote growth in Arabidopsis. Proc Natl Acad Sci U S A, 2003; 100: 4927–4932.

Ryu CM, Murphy JF, Mysore KS, Kloepper JW. Plant growth-promoting rhizobacteria systemically protect Arabidopsis thaliana against Cucumber mosaic virus by a salicylic acid and NPR1-independent and jasmonic acid-dependent signaling pathway. Plant J, 2004; 39: 381–392.

Sabzalian MR, Mirlohi A. *Neotyphodium* endophytes trigger salt resistance in tall and meadow fescues. J Plant Nutr Soil Sci, 2010; 173: 952–957.

Safir GR, Boyer JS. Mycorrhizal enhancement of water transport in soybean. Science, 1971; 172: 581–583.

Saikkonen K. Forest structure and fungal endophytes. Fungal Biol Rev, 2007; 21: 67–74.

Saikkonen K, Faeth SH, Helander M, Sullivan TJ. Fungal endophytes: A continuum of interactions with host plants. Annu Rev Ecol Sys, 1998; 29: 319–343.

Sanchez Marquez S, Bills GF, Zabalgogeazcoa I. The endophytic mycobiota of the grass Dactylis glomerata. Fungal Divers, 2007; 27: 171–195.

Sanchez Marquez S, Bills GF, Zabalgogeazcoa I. Diversity and structure of the fungal endophytic assemblages from two sympatric coastal grasses. Fungal Divers, 2008; 33: 87–100.

Schardl CL. The Epichloae, symbionts of the grass family Poöideae. Ann Missouri Bot Gard, 2010; 97: 646–665.

Schardl CL, Clay K. Evolution of mutualistic endophytes from plant pathogen. In: Carroll GC, Tudzynski P (eds) *The Mycota V: Plant Relationships Part B*, 1997, pp. 221–238. Springer, Berlin.

Schardl CL, Craven KD, Speakman S, Stromberg A, Lindstrom A, Yoshida R. A novel test for host-symbiont codivergence indicates ancient origin of fungal endophytes in grasses. Systematic Biol, 2008; 57: 483–498.

Scherr SJ, McNeely JA. Biodiversity conservation and agricultural sustainability: towards a new paradigm of 'ecoagriculture' landscapes. Philos Trans R Soc Lond B Biol Sci, 2008; 363: 477–494.

Schulz B, Boyle C, Draeger S, Rommert AK, Krohn K. Endophytic fungi: a source of novel biologically active secondary metabolites. Mycol Res, 2002; 106: 996–1004.

Siegel MR, Johnson MC, Varney DR, Buckner RC. Incidence and dissemination of the tall fescue fungal endopyte. Phytopathology, 1984; 74: 856–856.

Singh A, Sharma J, Rexer KH, Varma A. Plant productivity determinants beyond minerals, water and light: Piriformospora indica—a revolutionary plant growth promoting fungus. Curr Sci, 2000; 79: 1548–1554.

Sirrenberg A, Goebel C, Grond S, Czempinski N, Ratzinger A, Karlovsky P, Santos P, Feussner I, Pawlowski K. *Piriformospora indica* affects plant growth by auxin production. Physiol Plantarum, 2007; 131: 581–589.

Solis-Dominguez FA, Valentin-Vargas A, Chorover J, Maier RM. Effect of arbuscular mycorrhizal fungi on plant biomass and the rhizosphere microbial community structure of mesquite grown in acidic lead/zinc mine tailings. Sci Total Environ, 2011; 409: 1009–1016.

Stierle A, Strobel G, Stierle D. Taxol and taxane production by taxomyces andreanae, an endophytic fungus of Pacific Yew. Science, 1993; 260: 214–216.

Stone JK. Initiation and development of latent infection by Rhabdocline parkeri on Douglas fir. Can J Bot, 1987; 65: 2614–2621.

Strobel G, Daisy B. Bioprospecting for microbial endophytes and their natural products. Microbiol Mol Biol Rev, 2003; 67: 491–502.

Strobel G, Yang XS, Sears J, Kramer R, Sidhu RS, Hess WM. Taxol from *Pestalotiopsis microspora*, an endophytic fungus of *Taxus wallachiana*. Microbiology, 1996; 142: 435–440.

Strobel GA. Endophytes as sources of bioactive products. Microbes Infect, 2003; 5: 535–544.

Sun SK, Snyder WC. The bakanae disease of the rice plant. In: Nelson PE, Toussoun TA, Cook PJ (eds), *Fusarium: Diseases, Biology and Taxonomy*, 1981, pp. 104–113. The Pennsylvania State University Press, University Park.

Sylvia DM, Williams SE. Vesicular-arbuscular mycorrhizae and environmental stress. In: Bethlenfalvay GJ, Linderman RG (eds) *Mycorrhizae in Sustainable Agriculture*, 1992, pp. 101–124. American Society of Agronomy, Madison, Wisconsin.

Taghavi S, Barac T, Greenberg B, Borremans B, Vangronsveld J, van der Lelie D. Horizontal gene transfer to endogenous endophytic bacteria from poplar improves phytoremediation of toluene. Appl Environ Microbiol, 2005; 71: 8500–8505.

Tan RX, Zou WX. Endophytes: a rich source of functional metabolites. Natural Product Reports, 2001; 18: 448–459.

Tanaka A, Christensen MJ, Takemoto D, Park P, Scott B. Reactive oxygen species play a role in regulating a fungus-perennial ryegrass mutualistic interaction. Plant Cell, 2006; 18: 1052–1066.

Tudzynski B, Sharon A. Biosynthesis, biological role and application of fugal-phytohormones. In: Osiewacz HD (ed.) *The Mycota X. Industrial applications*, 2002, pp. 183–212. Springer, Berlin.

Vadassery J, Ritter C, Venus Y, Camehl I, Varma A, Shahollari B, Novak O, Strnad M, Ludwig-Muller J, Oelmuller R. The role of auxins and cytokinins in the mutualistic interaction between Arabidopsis and *Piriformospora indica*. Mol Plant Microbe Interact, 2008; 21: 1371–1383.

Varma A, Verma S, Sudha, Sahay N, Butehorn B, Franken P. *Piriformospora indica*, a cultivable plant-growth-promoting root endophyte. Appl Environ Microbiol, 1999; 65: 2741–2744.

Waller F, Achatz B, Baltruschat H, Fodor J, Becker K, Fischer M, Heier T, Huckelhoven R, Neumann C, von Wettstein D, Franken P, Kogel KH. The endophytic fungus *Piriformospora indica* reprograms barley to salt-stress tolerance, disease resistance, and higher yield. Proc Natl Acad Sci U S A, 2005; 102: 13386–13391.

Waller F, Mukherjee K, Deshmukh SD, Achatz B, Sharma M, Schaefer P, Kogel KH. Systemic and local modulation of plant responses by *Piriformospora indica* and related Sebacinales species. J Plant Physiol, 2008; 165: 60–70.

Weiss M, Selosse MA, Rexer KH, Urban A, Oberwinkler F. Sebacinales: a hitherto overlooked cosm of heterobasidiomycetes with a broad mycorrhizal potential. Mycol Res, 2004; 108: 1003–1010.

Weiss M, Sykorova Z, Garnica S, Riess K, Martos F, Krause C, Oberwinkler F, Bauer R, Redecker D. Sebacinales everywhere: previously overlooked ubiquitous fungal endophytes. PLoS One, 2011; 15: e16793. doi:10.1371/journal.pone.0016793.

Weller DM, Raaijmakers JM, Gardener BBM, Thomashow LS. Microbial populations responsible for specific soil suppressiveness to plant pathogens. Ann Rev Phytopathol, 2002; 40: 309–348.

West CP, Izekor E, Oosterhuis DM, Robbins RT. The effect of Acremonium coenophialum on the growth and nematode infestation of tall fescue. Plant Soil, 1988; 112: 3–6.

Weyens N, Van Der Lelie D, Artois T, Smeets K, Taghavi S, Newman L, Carleer R, Vangronsveld J. Bioaugmentation with engineered endophytic bacteria improves contaminant fate in phytoremediation. Environ Sci Technol, 2009; 43: 9413–9418.

Wicklow DT, Roth S, Deyrup ST, Gloer JB. A protective endophyte of maize: *Acremonium zeae* antibiotics inhibitory to *Aspergillus flavus* and *Fusarium verticillioides*. Mycol Res, 2005; 109: 610–618.

Wilson D, Carroll GC. Infection studies of *Discula quercina*, an endophyte of *Quercus garryana*. Mycologia, 1994; 86: 635–647.

Wolfe MS. The current status and prospects of multiline cultivars and variety mixtures for disease resistance. Ann Rev Phytopathol, 1985; 23: 251–273.

Zak DR, Holmes WE, White DC, Peacock AD, Tilman D. Plant diversity, soil microbial communities, and ecosystem function: Are there any links? Ecology, 2003; 84: 2042–2050.

Zhu YY, Chen HR, Fan JH, Wang YY, Li Y, Chen JB, Fan JX, Yang SS, Hu LP, Leung H, Mew TW, Teng PS, Wang ZH, Mundt CC. Genetic diversity and disease control in rice. Nature, 2000; 406: 718–722.

# Index

Aberystwyth spaced plant diversity trial, 71, 72f, 73f, 76f, 78
Abiotic stress tolerance, 240
   of alfalfa, 219–20
   of bluestems, 187
   of eastern gamagrass, 193–94
   of maize stover, 157
   in prairie cordgrass, 179
   sorghum breeding for, 99
*Acremonium zeae*, 252
Adati, S, 47, 51, 55–56
*Aegilops tauschii*, 44
AFEX. *See* Ammonia fiber expansion
AFLP. *See* Amplified fragment length polymorphism
Africa sorghum evolution, 85
*Agrobacterium*, 23
*Agrobacterium*-mediated transformation, 236–37
   of *Miscanthus*, 61
"Alamo" switchgrass, 7, 20, 34, 40
Albert-Thenet, JR, 139
Alexander, AG, 134
Alfalfa, 207–24
   background, 211–12
   biofuel feedstock prospect, 212–13
   biofuels processing, 222
   co-products, 223
   environmental benefits, 223
   introduction, 207–8
Alfalfa biomass
   lignocellulose-based biofuels, 208–9
   plant cell wall components, 209–11
Alfalfa breeding strategies
   cultivar development, 214
   germplasm resources, 213–14

   molecular breeding, 215–16
   synthetic cultivars and heterosis, 214–15
   trait integration, 216–17
Alfalfa breeding targets
   biomass yield, 217–18
   forage quality and composition, 218–19
   management and production inputs, 221–22
   stress tolerance, 219–20
   winter hardiness, 220–21
*Alternaria triticina*, 252
Aluminum toxicity, 102
American Recovery and Reinvestment Act, 34
Ammonia fiber expansion (AFEX), 153
Amplified fragment length polymorphism (AFLP), 18–19, 56
An, GH, 51
Anas, A, 102
Anderson, I, 91
*Andropogoneae*, 124
*Andropogon gerardii*. *See* Big bluestem
*Andropogon hallii*. *See* Sand bluestem
*Arabidopsis*, 39
   mutations, 238
Artificial hybridization of *Miscanthus*, 57
*Aspergillus flavus*, 252
Association mapping in switchgrass genetics, 22
Atienza, SG, 60
Azhar, FM, 102

BAC-by-BAC approach, 34
Badaloo, MGH, 139
*Balansia epichlo*ë, 251–52
*Balansia henningsiana*, 251–52
BBSRC. *See* Biotechnology and Biological Sciences Research Council
Best linear unbiased predictors (BLUPs), 136

*Bioenergy Feedstocks: Breeding and Genetics*, First Edition. Edited by Malay C. Saha, Hem S. Bhandari, and Joseph H. Bouton.
© 2013 John Wiley & Sons, Inc. Published 2013 by John Wiley & Sons, Inc.

## 268  Index

Big bluestem (*Andropogon gerardii*), 173, 182f
Biochemical conversion methods, 8–9
Bioenergy, 1
   breeding *Miscanthus* for, 67–81
   switchgrass, 2, 9
Bioenergy crop
   production systems, endophytes and, 256
   sorghum as, 90–93
   transgenic, 242
Bioenergy Science Center, DOE, 35, 45
Biofuel, 1
   U.S. on, 207
Biofuel Feedstock Genomics Resource, 35
Bioinformatics, *Miscanthus* incorporation of, 77–78
Biolistics transformation, 235–36
Biomass
   alfalfa, 208–11
   breeding approach, 3–4, 14–18, 14f, 67–81, 88, 90–93, 159–61
   cultivar development, 2–3, 8, 180, 185t, 188–90, 195–96, 196t, 214
   future outlook, 4, 23–24, 123–24, 180–81, 190–91, 196
   lignocellulosic feedstocks from, 1, 3, 7, 151–71, 208–9, 234–35
   quality for ethanol conversion, 24, 94
   sorghum, 93–94
   transgenics for, 233–48
   trials, of bluestems, 181
Biomass production, 137t, 155t–156t. *See also* Energy production
   of alfalfa, 221–22
   commercial level, 4
   of endophytes, 254–56
   of *Miscanthus*, 67–68
   quality, 7–8, 8f
   of sugarcane, 118–24
Biomass yield, 8, 23. *See also* Yield
   as alfalfa breeding target, 217–18
   canopy duration, 75, 75f
   of eastern gamagrass, 192, 193
   emergence rate, frost tolerance, and canopy development, 72
   flowering time, 58, 72, 74
   of *Miscanthus*, 238
   of *Miscanthus* × *giganteus*, 76
   of prairie cordgrass, 180
   QTL mapping on, 14
   research on, 2–3
   resource capture increase, 71, 71f
   senescence, 74–75, 74f

   of sugarcane, 118
   of switchgrass, 7, 13–14
   transgenic approaches for, 237–40
Biotechnology and Biological Sciences Research Council (BBSRC), 79
Biotic stress tolerance
   of alfalfa, 220
   cultivar development for, 8
   sorghum breeding for, 103
Birch, EG, 131
Bluestems
   abiotic stress tolerance of, 187
   big, 173, 182f
   biomass trials, 181
   cultivar development of, 185t, 188–90
   future goals, 190–91
   genetic variation and breeding methods of, 184–91
   germplasm resources for, 184–85, 185t
   importance of, 181–84
   little, 173, 183f
   molecular breeding of, 186
   pest resistance of, 187–88
   propagation of, 186–87
   sexual reproduction of, 185–86
Blummel, M, 97
BLUPs. *See* Best linear unbiased predictors
Boe, A, 184, 189
Bower, R, 131
Boye-Goni, SR, 102
Boyer, JS, 99
*Brachypodium distachyon*, 36, 60
Brazil
   ethanol use by, 2
   sugarcane production in, 118–20
Breeding approach. *See also* Transgenic approaches
   to feedstock development, 3–4
   of *Miscanthus*, 3, 67–81
   of sorghum, 88, 90–93
   of sugarcane, 3
   of switchgrass genetics, 14–18, 14f
   time and latitude effects on maize, 159–61
Breeding strategies
   of alfalfa, 213–17
   of *Miscanthus*, 68–69
Brummer, EC, 22
Buckler, ES, 22

Canada, *Miscanthus* breeding program in, 68
Canopy development, of *Miscanthus*, 72
Canopy duration, of *Miscanthus*, 75, 75f

Casler, MD, 13, 17
CCS. *See* Commercial cane sugar
Cellulose, 240–41. *See also* Lignocellulosic feedstocks
   sorghum conversion to ethanol, 94
   variation in *Miscanthus*, 59–60
Central America, switchgrass distribution in, 9
CGIAR. *See* Consultative Group on International Agriculture Research
Chen, SL, 61, 122, 190
Chou, CH, 53, 56, 122
Chromosome numbers and morphology
   of bluestems, 186
   of prairie cordgrass, 178
"Cimarron" switchgrass, 7, 8f
Clark, JW, 97
Clifton-Brown, JC, 59
Coal energy source, 106–7
Cobill, RM, 134–35
Codon substitution model, 37
Cold tolerance of *Miscanthus*, 58–59
Commercial cane sugar (CCS), 130
Composition
   of alfalfa, as breeding targets, 218–19
   biomass modification, transgenic approaches for, 240–42
   of energy canes for sugarcane improvement, 130–31, 130t
   in sorghum breeding, 93–95
   of sorghum for grain quality/starch, 96–97
Consultative Group on International Agriculture Research (CGIAR), 213
Conversion technologies, 2
   biochemical, 8–9
   enzymatic, 3
   for ethanol, 24, 94
   sorghum, 89–90, 89f
   thermochemical, 3, 9
Co-products of alfalfa, 223
Corn, 99
Corn feedstock, 1. *See also* Maize
   ethanol production from, 2
   hybrid production system, 3
Corredor, DY, 94
Cron, AB, 86
Crown libraries, 36
Cultivar development
   abiotic and biotic stress tolerance, 8
   of alfalfa, 214
   of bluestems, 185t, 188–90
   of eastern gamagrass, 195–96
   of prairie cordgrass, 180

research on biomass yield, 2–3
seed dormancy, 8
Cytogenetics of switchgrass, 12
Cytoplasmic genome in switchgrass, 42

Dale, J, 132
Daniels, J, 124
Department of Energy (DOE), U.S., 2
   Bioenergy Science Center, 35, 45
   Joint Genome Institute, 11, 34, 36, 45
   on petroleum consumption, 255
Deuter, M, 53
D'Hont, A, 131
Diseases
   of sorghum, 104, 106
   of sugarcane, 121, 129
Distribution
   of eastern gamagrass, 193
   of maize, 159–60, 161f
   of *Miscanthus floridulus*, 53
   of *Miscanthus sacchariflorus*, 54
   of *Miscanthus sinensis*, 51
   of *Miscanthus sinensis var. condensatus*, 53
   of sorghum, 84–86
   of switchgrass in Central America, 9
DOE. *See* Department of Energy
Drought, 99–100
Dwarfing genes (Dw), 86
Dwarfing of sorghum, 86, 89

E. *See* Environmental factors
Eastern gamagrass (*Tripsacum dactyloides*), 173, 191f
   abiotic stress tolerance of, 193–94
   biomass yield of, 192, 193
   cultivar development of, 195–96
   distribution of, 193
   future goals, 196
   genetic variation and breeding methods, 192–96
   importance of, 191–92
   pest resistance of, 194–95
   propagation of, 192–93
Economic weightings for energy canes, 136–38
Emergence rate of *Miscanthus*, 72
Endophytes
   bioenergy crop production systems and, 256
   in biomass crop production, 254–56
   consortia, 256–57
   of cool season grasses, 251
   defining, 249–50
   in genetic engineering of host plants, 258

Endophytes (*Continued*)
  microbial, 4
  novel compounds source, 257–58
  other fungal, 253–54
  of warm season grasses, 251–53
  of woody angiosperms, 253
Energy canes, 1, 117–41. *See also* Sugarcane
  herbicide resistance of, 132
  index selection, 136
  introduction, 117–18
  production systems, 118–24, 130t, 137t, 138–40
Energy canes
  breeding for energy production progress, 138–40
  economic weightings of, 136–38
  overall directions, 134–36
Energy canes improvement, 124–34
  breeding program features, 128–30
  composition for energy production, 130–31
  molecular genetics application, 131–34
  sugarcane breeding history, 127
Energy Independence and Security Act, 11, 207
Energy production. *See also* Biomass production
  endophytes system of, 256
  from maize stover, 154
Engineering and Physical Sciences Research Council (EPSRC), 79
Engler, D, 61
Environmental benefits of alfalfa, 223
Environmental factors (E), 15
  for plant adaptation, 240
Enzymatic conversion technology, 3
*Epichloë* endophytes, 251
EPSRC. *See* Engineering and Physical Sciences Research Council
*Erianthus*, 140
ESTs. *See* Expressed sequence tags
Ethanol
  biomass quality for, 24
  Brazil use of, 2
  corn feedstock production of, 2
  maize stover production of, 153
  sorghum cellulose conversion to, 94
  sugarcane production of, 2, 117
  U.S. production of, 2
*Eumiscanthus Miscanthus* genus
  *Miscanthus floridulus*, 52f, 53, 56, 69, 70f
  *Miscanthus sinensis*, 49–52, 55–56, 58, 61, 70f
  *Miscanthus sinensis var. condensatus*, 52–53, 56
Europe, *Miscanthus* breeding program in, 68

Evolution of sorghum, 84–86
Expressed sequence tags (ESTs), 21–22
  NAC domain-containing class, 39–40
  in switchgrass genomics, 11, 36–40
Ex situ phenotypic characterization, 69

Fertilization of switchgrass, 10–11
FISH. *See* Fluorescence *in situ* hybridization
Flowering
  of *Miscanthus sinensis*, 51–52, 58
  time, of *Miscanthus*, 58, 72, 74
Fluorescence *in situ* hybridization (FISH), 56
Forage quality and composition, as alfalfa breeding target, 218–19
Foxtail millet, 36
  linkage maps of, 41–42
Frost tolerance of *Miscanthus*, 72
Fuentes, RG, 13
*Fusarium verticillioides*, 252
Future outlook of biomass, 23–24, 123–24, 180–81, 190–91, 196

G. *See* Genes
GCA. *See* General combining ability
GEBVs. *See* Genomic estimates of breeding value
GEM. *See* Germplasm enhancement of maize
General combining ability (GCA), 88
Genes (G), 15
  Dw, 86
  maize single and transgenes, 165–67
Genetic diversity
  of *Miscanthus*, 61, 69
  of switchgrass, 12–13
Genetic improvement
  of *Miscanthus*, 57–58
  of plant species, 2
Genome
  resources, 3
  sequencing, of switchgrass, 34–36
  structure, of switchgrass, 12
Genome-enabled improvement, in switchgrass genomics, 42
  GEBVs and, 43, 44f
  phenotypic selection, 43–44, 44f
  QTLs in, 43
  RRLs sequencing, 43
Genome-Wide Association Studies (GWAS), 77–78
Genome-wide selection (GWS), 43, 44f
Genomic estimates of individual breeding value (GEBVs), 43, 44f

Genomic *in situ* hybridization (GISH), 56
Genotyping, 3
   MassARRAY methodology of, 19, 44
   SNP methodology, 19
   of sugarcane, 134–40
Germplasm collection and management of *Miscanthus*, 57
Germplasm enhancement of maize (GEM) program, 158
Germplasm Resource Information Network (GRIN), 88, 213
Germplasm resources
   for alfalfa, 213–14
   for bluestems, 184–85
   of *Miscanthus*, 49–66
   for prairie cordgrass, 176–77
Giamalva, MJ, 139
GISH. *See* Genomic *in situ* hybridization
Gonzalez-Hernandez, J, 177
Goud, JV, 95
Grain quality/starch composition of sorghum, 96–97
Grain sorghum, 90
   pest resistance of, 103
   QTLs of, 95–97, 98t
Grasses
   cool and warm season, 251–53
   cultivated, in U.S., 7
Grasses, underutilized, 173–96
   bluestems, 173, 181–91
   eastern gamagrass, 173, 191–96
   introduction, 173–74
   prairie cordgrass, 173–81
Greef, JM, 56
GRIN. *See* Germplasm Resource Information Network
Gustafson, DJ, 186
GWAS. *See* Genome-Wide Association Studies
GWS. *See* Genome-wide selection

Harvesting
   of maize, 151–52
   of sugarcane, 121–22
Haussmann, BIG, 100
Hazel, DL, 136
Heading, of switchgrass, 10
HECP. *See* Herbaceous Energy Crops Program
Heinz, DJ, 128
Herbaceous Energy Crops Program (HECP), 2
Herbicide resistance, of energy canes, 132

Heritability
   for biomass yield of switchgrass, 13–14
   trait, of *Miscanthus*, 78t
Heterosis, 3
   of alfalfa, 214–15
   of switchgrass genetics, 17–18
High-resolution melt (HRM), 19
Hirayoshi, I, 53
Hodkinson, TR, 56
Hodnett, GL, 106
Holland, JB, 86
Honda, M, 55
HRM. *See* High-resolution melt
Hunter, E, 91
Hybridization
   artificial of *Miscanthus*, 57
   FISH and GISH, 56
   intergeneric, 107
   interspecific, 106–7
   of *Miscanthus*, 68–69
   of sorghum, 87, 106–7
Hybrids
   corn system production of, 3
   of maize, 152–53, 159t, 160t
   of *Miscanthus*, 55, 61
   *Saccharum* × *Erianthus*, 140
   of sorghum, 86–87
   *Sorghum bicolor* x *S. halepense*, 106
   of switchgrass, 17f

ICARDA. *See* International Center for Agricultural Research in the Dry Areas
ICRISAT. *See* International Crops Research Institute for the Semi-Arid Tropics
Igartua, E, 102
indels. *See* Insertion and deletions
Index selection of energy canes, 136
Inheritance
   in sorghum breeding, 95–106
   in switchgrass genetics, 13–14
Insertion and deletions (indels), 19
   in switchgrass genomics, 42
Intergeneric hybridization, 107
International Center for Agricultural Research in the Dry Areas (ICARDA), 213
International Crops Research Institute for the Semi-Arid Tropics (ICRISAT), 88
Inter-simple sequence repeats (ISSR), 61
Interspecific hybridization, 106–7
Irvine, JE, 125
ISSR. *See* Inter-simple sequence repeats

Japan
  *Miscanthus* × *giganteus* from, 49–50
  *Miscanthus* hybrid plants in, 55
Jensen, E, 58
Joint Genome Institute, DOE, 34, 45
  on switchgrass ESTs, 11, 36

"Kanlow" switchgrass, 7, 11f, 18, 20, 40
*Kariyasu Miscanthus* genus
  *Miscanthus intermedius*, 55
  *Miscanthus oligostachyus*, 54, 56
  *Miscanthus tinctorius*, 54
Karyotype analysis, 55–56
Kennedy, AJ, 139
Kicherer, A, 59
Kim, HS, 61

Lafferty, J, 53
Land cover, 174t
Laser, M, 137
Laser capture microdissection (LCM), 38
LD. *See* Linkage disequilibrium
Leaf libraries, 36
Lee, DK, 177
Lelley, T, 53
Lewandowski, I, 59
Lignin, 9
  for plant growth, 241
  polymer, 3
  sorghum, 94
  variation in *Miscanthus*, 59–60
Lignocellulosic feedstocks, 3
  of alfalfa, 208–9
  of maize, 151–71
  from plant biomass, 1
  production of, 7
  transgenics for biomass of, 234–35
Linde-Laursen, I, 56
Linkage disequilibrium (LD) of sugarcane, 133
Linkage mapping
  development for *Miscanthus*, 60
  of foxtail millet, 41–42
  of rice, 40–41
  of sorghum, 40–41
  in switchgrass genomics, 40–42
Little bluestem (*Schizachyrium scoparium*), 173, 183f
Lorenz, AJ, 152
Lowland ecotypes of switchgrass, 8–10, 12, 20
  hybrid vigor of, 17f
Lowland-tetraploid, 18

Magalhaes, JV, 102
Maize, 151–67
  distribution of, 159–60, 161f
  hybrids, 152–53, 159t, 160t
  single genes and transgenes, 165–67
Maize attributes as biofuel crop
  grain and stover yield, 152–53, 157–59
  harvesting of, 151–52
  plant parts, 151–52
  soil health and, 152
Maize breeding for biofuels, 154–65
  methods, 164–65
  selection criteria, 154
  sustainability parameters, 163
  time and latitude effects, 159–61
Maize stover
  abiotic stress tolerance of, 157
  for energy production, 154
  for ethanol production, 153
  pest resistance, 157
  quality, 161–63
  yield, 152–53, 157–59
Makanda, I, 99
Marcarian, V, 102
Marker-assisted Selection (MAS), 77–78, 89
Markers
  mitochondrial, 42
  molecular, in switchgrass genetics, 18–19
  PCR types, 18
  RAPD of *Miscanthus*, 60
  RFLPs for *Miscanthus*, 60
  SSRs for *Miscanthus*, 60
Martinez-Reyna, JM, 18
Martin-Gardiner, M, 139
MAS. *See* Marker-assisted Selection
MassARRAY genotyping methodology, 19, 44
Matumura, M, 57
McNeilly, T, 102
*Medicago truncatula*, 39
Mendel's law, 86
Metabolom resources, 3
Microbial endophytes, 4
MicroRNAs (miRNAs), 39, 239
Milling of sugarcane, 122–23
Mineral content
  of *Miscanthus*, 59–60
  QTL analysis of *Miscanthus*, 60–61
  of switchgrass, 23
miRNAs. *See* MicroRNAs
MISCANMOD *Miscanthus* growth model, 71

*Miscanthidium*
  *M. ecklonii*, 57
  *M. junceus*, 57
*Miscanthus*, 1, 173
  biomass yield of, 71–72, 74–76, 238
  breeding program in U.S., 68
  European interest in, 68
  genetic improvement, 57–58
  high biomass of, 2
  Japan hybrid plants, 55
  propagation of, 77
  RFLP markers for, 60
  SSR markers for, 60
  trait heritability, 78t
*Miscanthus* breeding approach, 3, 67–79
  bioinformatics, MAS, and GWAS, 77–78
  biomass crop, 67–68
  breeding strategies, 68–69
  breeding targets, 70–77
  genetic diversity, 61, 69
  introduction, 67
  large-scale demonstration trials, 69
*Miscanthus floridulus*, 52f, 53, 56, 69, 70f
*Miscanthus* genus
  Eumiscanthus, 50–53, 56
    *Miscanthus floridulus*, 52f, 53, 56, 69, 70f
    *Miscanthus sinensis*, 49–52, 55–56, 58, 61, 70f
    *Miscanthus sinensis* var. *condensatus*, 52–53, 56
  Japan classification of, 51t
  Kariyasu, 54–56
    *Miscanthus intermedius*, 55
    *Miscanthus oligostachyus*, 54, 56
    *Miscanthus tinctorius*, 54
  Triarrhena, 53–54, 56, 69
    *Miscanthus sacchariflorus*, 49–50, 52f, 53–56, 70f, 73f
*Miscanthus* germplasm resources, 49–62
  agronomical traits, 58–60
  genetic improvement of, 57–58
  introduction, 49–50
  karyotype analysis, 55–56
  molecular resources, 60–61
  natural hybrids, 55
  phylogenetic relationships of, 56–57
  species of, genus of, 50–55
  transgenic, 61
*Miscanthus* × *giganteus*, 55–56, 68
  biomass yield of, 76
  cold tolerance, 58–59
  fertility through polyploidization, 50
  genetic modification of, 61
  from Japan, 49–50
  in U.S., 49–50
*Miscanthus intermedius*, 55
*Miscanthus ogiformis*, 55–56
*Miscanthus oligostachyus*, 54, 56
*Miscanthus sacchariflorus*, 49–50, 52f, 53–56, 70f, 73f
*Miscanthus sinensis*, 49–50, 52f, 56, 61, 70f
  distribution of, 51
  flowering of, 51–52, 58
  karyotype of, 55
  self-pollination rate of, 51
*Miscanthus sinensis* var. *condensatus*, 52–53, 56
*Miscanthus sinensis* var. *formosanus*, 56
*Miscanthus tinctorius*, 54
Missaoui, AM, 20, 34
Mitchell, KB, 18
Mitochondrial markers, 42
Molecular breeding
  of alfalfa, 215–16
  association mapping, 22
  of bluestems, 186
  of energy cane development, 131–34
  molecular mapping, 20, 21t, 22
  molecular markers, 18–19
  of prairie cordgrass, 178–79
  in switchgrass, 18–23
  transgenic approaches, 23
Molecular diversity studies of switchgrass, 12–13
Molecular mapping in switchgrass genetics, 20, 21t, 22
Molecular markers in switchgrass genetics, 18–19
Morphological traits of *Miscanthus*, 75–77
Muchow, RC, 126
Mukherjee, SK, 124
Murray, SC, 96–97
Mutations
  *Arabidopsis*, 238
  of maize, 165–67
  transgenic, of switchgrass, 9

NAC domain-containing class, 39–40
National Alcohol Program, Brazil, 2
National Center for Biotechnology Information (NCBI), 35
National Genetic Resources Program (NGRP), USDA, 12
National Plant Germplasm System, 22
Natoli, A, 97, 99

NCBI. *See* National Center for Biotechnology Information
NGRP. *See* National Genetic Resources Program
Nitrogen use efficiency (NUE), 102
Nutrient management of sugarcane, 121

Oakridge National Laboratory (ORNL), 2
O/I. *See* Output/input
Okada, M, 20
Oklahoma State University (OSU) switchgrass breeding program, 16
Olsen, A, 49
ORNL. *See* Oakridge National Laboratory
OSU. *See* Oklahoma State University
Output/input (O/I), energy, 118

Pairwise synonymous substitution rate of ESTs, 37–38
PCR. *See* Polymerase chain reaction
Perumal, R, 104
Pest resistance
  of bluestems, 187–88
  of eastern gamagrass, 194–95
  of maize stover, 157
  of prairie cordgrass, 179–80
  of sorghum, 103–4
Petroleum consumption, DOE on, 255
Petroleum production, in U.S., 207
Phenotypic selection and variability
  of switchgrass for biomass yield, 13
  in switchgrass genomics, 43–44
  within-family, 17
Phylogenetic relationships between *Miscanthus* species, 56–57
*Piriformospora indica*, 249
Plant biomass. *See* Biomass
Plant cell wall components of alfalfa, 209–11
Plant diversity trial, of *Miscanthus* at Aberystwyth, 71, 72f, 73f, 76t, 78
Plant growth, lignin for, 241
Polymerase chain reaction (PCR) -based marker types, 18
Polyploid, 3, 33
  switchgrass as, 9, 12
Polyploidization
  of *Miscanthus*, 58
  of *Miscanthus* × *giganteus*, 50, 58
Prairie cordgrass (*Spartina pectinata*), 173–81
  abiotic stress tolerance in, 179
  biomass yield of, 180
  cultivar development of, 180
  future goals, 180–81
  genetic variation and breeding methods, 176–80
  germplasm resources for, 176–77
  importance, 174–76
  molecular breeding of, 178–79
  pest resistance of, 179–80
  propagation of, 179
  sexual reproduction of, 177
  variations of, 176t
Propagation
  of bluestems, 186–87
  of eastern gamagrass, 192–93
  of *Miscanthus*, 77
  of prairie cordgrass, 179
Proteome resources, 3

QTLs. *See* Quantitative trait loci
Quality
  biomass, for ethanol conversion, 24, 94
  biomass production, 7–8
  forage of alfalfa, 218–19
  grain, of sorghum, 96–97
  of maize stover, 161–63
Quantitative trait loci (QTLs)
  of biomass yield, 14
  of grain sorghum, 95–97, 98t
  limits of, 133–34
  of *Miscanthus*, 60–61
  of sweet sorghum, 101t
  in switchgrass genomics, 43
Quinby, JR, 86

Ramdoyal, K, 139
Random amplified polymorphic DNA (RAPD) markers, of *Miscanthus*, 60
Recurrent selection for general combining ability (RSGCA), OSU procedure for, 16–17
Recycling of switchgrass, 23
Reduced-representation libraries (RRLs), in switchgrass genomics, 43
Rein, P, 122
Research
  on biomass yield, 2–3
  by ORNL, 2
Resource. *See also* Germplasm resources
  capture increase of *Miscanthus*, 71
  genome, 3
  metabolom, 3
  proteome, 3
  transcriptome, 3

Restriction fragment length polymorphisms
    (RFLPs), 18
  markers for *Miscanthus*, 60
RFLPs. *See* Restriction fragment length
    polymorphisms
Rhizomes
  of *Miscanthus*, 57
  of switchgrass, 10
Rice, linkage maps of, 40–41
Ritter, KB, 95
Roach, BT, 124
Robertson, MJ, 126
RRLs. *See* Reduced-representation libraries
RSGCA. *See* Recurrent selection for general
    combining ability

*Saccharum* × *Erianthus* hybrids, 140
*Saccharum officinarum*, 124–25, 127, 133
*Saccharum spontaneum*, 124–25, 131, 133,
    138–39
Sainz, MB, 132
Saline soils, 102
Samuel Roberts Noble Foundation, 22, 34, 35
Sand bluestem (*Andropogon hallii*), 173
SARDI. *See* South Australia Research and
    Development Institute
SCA. *See* Specific combining ability
*Schizachyrium scoparium*. *See* Little bluestem
*Sebacina vermifera*, 249, 253–54
Seed dormancy, 8
Self-pollination rate, of *Miscanthus sinensis*,
    51
Self-progeny of switchgrass, 20
Senescence, of *Miscanthus*, 74–75
Sequence-tagged sites (STSs), 18, 20–21
Sexual reproduction
  of bluestems, 185–86
  of prairie cordgrass, 177
"Shawnee" switchgrass, 11f
Shibata, F, 56
Shiotani, I, 47, 55, 56
Simple sequence repeats (SSRs), 18, 21–22
  markers for *Miscanthus*, 60
  for switchgrass, 42
Sinclair, TR, 126
SINGER. *See* System-wide Information Network
    on Genetic Resources
Single nucleotide polymorphism (SNP)
    genotyping methodologies, 19
Smith, HF, 136
SNP. *See* Single nucleotide polymorphism

Soil fertility/toxicity
  maize and, 152
  of maize stover, 157
  saline soils, 102
Sorghum, 1, 36, 88. *See also* Sweet sorghum
  adaptation, 89f
  conversion program, 89–90
  favorable characteristics of, 84t
  grain yield in U.S., 86–87
  linkage maps of, 40–41
*Sorghum bicolor x S. halepense* hybrids, 106
Sorghum breeding
  botanical description and evolution, 84–86
  breeding approaches, 88, 90–93
  composition in, 93–95
  genetic variation and inheritance, 95–106
  introduction, 83–84
  programs in U.S., 91
  traditional breeding and development,
    86–90
  wide hybridization, 106–7
Sorghum breeding for stress tolerance
  abiotic stress, 99
  biotic stress, 103
  drought, 99–100
  pests resistance, 103–4
  soil fertility/toxicity, 102
  sorghum diseases, 104, 106
  temperature, 100–101
*Sorghum halepense*, 106
*Sorghum propinquum*, 106
South Australia Research and Development
    Institute (SARDI), 213
*Spartina pectinata*. *See* Prairie cordgrass
Specific combining ability (SCA), 217
Srivastava, BL, 139
SSRs. *See* Simple sequence repeats
Stefaniak, TR, 94
Stephens, JC, 86
Stover. *See* Maize stover
Stress tolerance, 105t
  abiotic, 8, 99, 157, 179, 187, 193–94, 219–20,
    240
  biotic, 8, 103, 220
  sorghum breeding for, 99–106
STSs. *See* Sequence-tagged sites
Sugarcane. *See also* Energy canes
  breeding approach, 3
  ethanol production from, 2, 117
  nutrient management of, 121
  yield of, 125–26

Sugarcane production components
   growing, 120–21
   harvesting and transport, 121–22
   milling, 122–23
"Summer" switchgrass, 11f, 18, 20
Sweet sorghum, 90–92
   QTLs, 101t
   soluble carbohydrates, 97, 99
Switchgrass, 1, 153, 173
   "Alamo," 7, 20, 34, 40
   bioenergy issues of, 9
   biomass yield of, 7, 13–14
   breeding program at OSU, 16
   Central America distribution of, 9
   "Cimarron," 7, 8f
   ESTs, 11, 36–40
   fertilization, 10–11
   heading of, 10
   "Kanlow," 7, 11f, 18, 20, 40
   lowland ecotypes, 8–10, 12, 17f, 20
   mineral concentrations and recycling of, 23
   as model bioenergy species, 2
   molecular diversity studies, 12–13
   pheonotypic diversity analyses of, 12
   as polyploid, 9, 12
   rhizomes of, 10
   self-progeny of, 20
   "Shawnee," 11f
   "Summer," 11f, 18, 20
   transgenic mutations, 9
   upland ecotypes, 9–10, 13
Switchgrass Functional Genomics Server, 35
Switchgrass genetics, 7–24
   conventional breeding approaches, 14–18
   future directions of, 23–24
   genetic diversity, 12–13
   genome structure and cytogenetics, 12
   growth and development, 10–12
   inheritance, 13–14
   molecular breeding, 18–23
   origin and distribution, 9
   phenotypic variability, 13–14
Switchgrass Genome Project, 22
Switchgrass genomics, 33–45
   cytoplasmic genome, 42
   ESTs analysis, 11, 36–40
   genome-enabled improvement, 42–45
   genome sequencing, 34–36
   introduction, 33–34
   linkage mapping, 40–42
Synthetic cultivars of alfalfa, 214–15

Systematic recurrent selection, of switchgrass genetics, 15–17
System-wide Information Network on Genetic Resources (SINGER), 213

Takahashi, C, 56
Taliaferro, C, 13
Tanner, CB, 126
Taxonomy and crop physiology, 124–27
Temperature, 100–101
Tetraploid, lowland- and upland-, 18
Tew, TL, 134–35
TFs. *See* Transcription factors
Thermochemical conversion technology, 3
   high-lignin plants and, 9
Time
   effects on maize breeding, 159–61
   flowering, of *Miscanthus*, 58, 72, 74
Trait. *See also* Quantitative trait loci
   agronomical, of *Miscanthus*, 58–60
   heritability of *Miscanthus*, 78t
   integration of alfalfa, 216–17
   morphological, of *Miscanthus*, 75–77
Transcription factors (TFs), 237–38
Transcriptome resources, 3
Transgenic approaches, 9
   *Agrobacterium*-mediated transformation, 236–37
   in alfalfa, 218–19
   biolistics transformation, 235–36
   in *Miscanthus*, 61
   in switchgrass genetics, 23
Transgenic approaches for biomass improvement
   biomass composition modification, 240–42
   biomass yield improvement, 237–40
Transgenic mutations of switchgrass, 9
Transgenics for biomass, 233–42
   biofuels, 234
   biomass yield improvement, 237–40
   introduction, 233–35
   lignocellulosic biomass, 234–35
Transgenic technologies, 3
   in switchgrass genetics, 23
*Triarrhena Miscanthus* genus, 69
   *Miscanthus sacchariflorus*, 49–50, 52f, 53–56, 70f, 73f
*Tripsacum dactyloides*. *See* Eastern gamagrass

Ueng, JJ, 56
United States (U.S.)
   on biofuel, 207
   cultivated grasses in, 7

DOE, 2, 11, 34–36, 45, 207, 255
   ethanol production in, 2
   *Miscanthus* breeding program in, 68
   *Miscanthus* × *giganteus* in, 49–50
   petroleum production, 207
   sorghum breeding programs in, 91
   sorghum grain yield in, 86–87
Upland ecotypes of switchgrass, 9–10, 13
Upland-tetraploid, 18
U.S. *See* United States

Vinall, HN, 86
Vogel, KP, 18, 184, 188, 190

Wang, L-P, 139
Weimer, PJ, 192

Whole genome association (WGA) analysis, 22
Whole genome shotgun sequencing strategy, 34
Winter hardiness as alfalfa breeding target, 220–21
Within-family phenotypic selection, 17
Wu, X, 96, 131

Yield. *See also* Biomass yield
   maize grain and stover, 152–53, 157–59
   QTL analysis of *Miscanthus*, 60–61
   of sugarcane, 125–26
Yoshida, T, 102
Young, HA, 42
Yu, CY, 58
Yukimura, T, 57